Fundamentals of Electrical Machines

This accessible textbook offers a clear and concise introduction to the core principles of electric machines, focusing on the conversion of energy through mechanical power, electrical power, torque, magnetic fields, and rated voltage.

Covering the key processes behind electric machine theory, including mechanics and electromagnetism, *Fundamentals of Electrical Machines* enables readers to develop a systematic view of the technology. Including case studies with industry examples, the book is both theoretical and practical. Chapters cover topics such as basic concepts of electrical machines and transformers, electromagnetism, asynchronous machines, and synchronous machines. It also details DC machines and includes revision questions to help self-guided learning.

This book will be of interest to students in electrical engineering and engineers in all fields interested in electromechanical systems.

Fundamentals of Electrical Machines

Ants Kallaste, Toomas Vaimann,
Anton Rassõlkin, and Kuno Janson

CRC Press
Taylor & Francis Group
Boca Raton London New York

CRC Press is an imprint of the
Taylor & Francis Group, an **informa** business

Designed cover image: Ants Kallaste, Toomas Vaimann, Anton Rassõlkin, and Kuno Janson

First edition published 2026
by CRC Press
2385 NW Executive Center Drive, Suite 320, Boca Raton FL 33431

and by CRC Press
4 Park Square, Milton Park, Abingdon, Oxon, OX14 4RN

CRC Press is an imprint of Taylor & Francis Group, LLC

ISBN: 9781032518206 (hbk)
ISBN: 9781032518213 (pbk)
ISBN: 9781003404088 (ebk)

DOI: 10.1201/9781003404088

Typeset in Times
by Deanta Global Publishing Services, Chennai, India

Contents

Preface

The modern world is mechanised, automated, and increasingly robotic. In this world that is saturated with technology, the main motion-generating device is the electric motor. Every engineer should have at least a general idea of how an electric motor functions – what it can and cannot do.

The turn of the 20th and start of the 21st century was the peak period of development of information technologies. The peak period of development of electrical machines, meanwhile, occurred about a hundred years ago. The fundamentals of the science of mechanics were established earlier still. Electrical engineering is no less complicated than information technology. Mechanics, in turn, is no less complicated than electrical engineering. Ideally, every engineer should be versed in mechanics as well as electrical engineering and information technology. Unfortunately, the amount of knowledge contained in these fields is beginning to exceed the limits of human ability.

One hundred years ago, the author of a textbook on electrical machines would have written down everything they knew. They could do so because back then electrical engineering was only studied by those few who had the opportunity, interest, and innate ability. Old textbooks on electrical machines have not become outdated. The electrical machines in use today were also known a hundred years ago. However, today's students can no longer afford to devote themselves to studying electrical machines to the extent it was done in the past. A whole new and important field has arisen: information technology. And, of course, there is also electronics. The situation calls for a new, less voluminous textbook on electrical machines, which still manages to cover all of the fundamentals. The textbook you are holding is an attempt at just that. When you know the fundamentals, details can be found relatively easily online.

In their most concentrated and precise form, fundamentals are captured in mathematical equations. However, mathematical equations present information in an encoded format. To decode an equation, you need to know the meaning of the symbols used therein. You also need the ability to move from the general to the specific. This occurs when the symbols in an equation are replaced, whether in the mind or in writing, with numeric values. Without this, equations are useless. This ability can be learnt and honed by solving problems.

Each equation describes a relatively small part of the processes that take place in electrical machines. Engineers need to know all of the major processes, the connections between them, as well as the structural components that make up the machine. Knowing all of the major processes allows one to develop a systematic understanding as well as one's general technical acumen, giving rise to broader insight into the field of engineering. This acumen is rooted in the subconscious mind, which is very fast, but not the most accurate. Nevertheless, it is highly useful for solving problems, as it enables one to quickly discard what is irrelevant from consideration and focus on what matters.

The key terms when it comes to both electrical machines and transformers are energy, mechanical power, electrical power (including active power, reactive power, and apparent power), thermal power, torque, rotational speed, synchronous speed, rotating magnetic field, magnetic flux, magnetic flux density, magnetic field strength, magnetic circuit, magnetic resistance, parasitic inductance, induced electromotive force, induced mechanical force, magnetic permeability, magnetisation curve and its hysteresis, hysteresis loss and eddy current loss, no-load current, short-circuit current, rated current and rated voltage, switchover from motor function to generator function. Additional key terms concerning transformers include short-circuit voltage, vector group, voltage regulation, and equivalent circuit. The other side of the system is the set of different types, structures, and operating principles of the existing electrical machines and transformers.

In order to be able to describe the processes that take place in electrical machines, one also needs to be familiar with the laws of energy, mechanics, electrical engineering, and the branch of physics known as electromagnetism. For the sake of providing a proper understanding of the subject, brief descriptions are given at the beginning of the textbook of the areas of these disciplines which relate to electrical machines. The aim of the textbook is to enable learners to develop a systematic technological worldview. Knowledge that is unsystematic is worthless.

Authors' Biography

Ants Kallaste serves as a tenured associate professor at Tallinn University of Technology (TalTech), Estonia, where he has devoted his career to the research, development, and teaching of electrical machines and drive systems. As the head of the Electrical Machine Group (EMG) within the Department of Electrical Power Engineering and Mechatronics, his work focuses on advancing innovative and sustainable technologies in electrical engineering.

His academic and research efforts are centred around the design, diagnostics, and optimisation of electrical machines. Over the past decade, he has led projects ranging from the development of low- and medium-power radial flux machines to the creation of a three MVA ring generator for wind turbine applications. These research activities reflect his deep commitment to integrating sustainable energy solutions with high-performance engineering systems.

A significant portion of his research also explores diagnostic methodologies for electrical machines. He has investigated rotor and stator fault detection in induction machines, developed pre-diagnostic tools using mobile devices, and examined the application of inverse problem theory for early fault recognition. This work is aimed at improving machine reliability while reducing lifecycle costs and environmental impact.

An emerging area of his research is additive manufacturing in electrical machine design. By incorporating 3D printing into engineering workflows, Professor Kallaste aims to revolutionise how electrical machines are conceptualised and produced, enabling more sustainable and flexible manufacturing processes.

Alongside his research, he has made substantial contributions to academic publishing, authoring numerous scientific articles on robust design optimisation, energy-efficient drive systems, and environmental assessments of motor lifecycles. He actively participates in the international research community, serving as a co-chair or committee member at conferences and maintaining collaborations with both industry and academic partners.

Professor Kallaste remains passionate about educating the next generation of engineers. He believes that a solid foundation in electrical machine theory, coupled with practical experience and openness to innovation, is key to advancing the field and addressing global challenges related to energy efficiency and sustainability. His ongoing work continues to bridge rigorous engineering research with practical solutions that benefit both industry and society.

Anton Rassõlkin is a tenured associate professor at Tallinn University of Technology (TalTech), Estonia, where he leads research and teaching in the fields of electric drives, mechatronics, and autonomous systems. As the head of the Mechatronics and Autonomous Systems research group within the Department of Electrical Power Engineering and Mechatronics, he is committed to advancing innovative engineering solutions that bridge mechanical systems, electronics, and intelligent control.

Professor Rassõlkin's academic journey began at TalTech, where he earned his bachelor's, master's, and doctoral degrees in electric drives and power electronics. He further expanded his expertise internationally, obtaining an engineering diploma in automation from the University of Applied Sciences Giessen-Friedberg in Germany. His multidisciplinary background underpins his research focus, which includes the design, modelling, and control of electric drives, the development of electric and autonomous vehicles, and the creation of digital twins for software-defined mobility systems.

His work has contributed to advancing the understanding and application of mechatronic systems in sustainable transportation and industrial automation. By combining hardware innovation with advanced simulation and digital twin technologies, he aims to improve system performance, energy efficiency, and reliability in next-generation mobility solutions.

Beyond his research, Professor Rassõlkin plays an active role in the global engineering community. He is a senior member of the IEEE and serves as chair of the IEEE Estonia Vehicular Technology Society (VTS) section, fostering professional collaboration and promoting emerging technologies in transportation systems. He is also an active member of the Estonian Moritz Hermann Jacobi Society, supporting scientific exchange and the promotion of electrical engineering heritage.

Through his research, leadership, and professional engagement, Professor Rassõlkin continues to advance the frontiers of mechatronics and autonomous systems, contributing to technological progress and the education of future engineers equipped to address the challenges of an increasingly automated and connected world.

Toomas Vaimann is a research professor at Tallinn University of Technology (TalTech), Estonia, where he has dedicated his career to the advancement of electrical engineering, with a particular focus on electrical machines, diagnostics, and condition monitoring. His academic journey reflects a steadfast commitment to research excellence and engineering education.

At TalTech, he holds a leading role within the Electrical Machines Research Group, which is part of the Department of Electrical Power Engineering and Mechatronics. In this capacity, he has played a pivotal role in guiding research initiatives on the calculation, modelling, control, testing, and development of electrical machines, including motors, generators, and transformers. The group also provides expert consultations and fosters research collaborations that contribute to the progress of electrical engineering.

Professor Vaimann's research spans a wide spectrum, including diagnostics and condition monitoring of electrical machines, additive manufacturing applications in electrical engineering, and the development of digital twins for propulsion drives in autonomous electric vehicles. His scholarly output includes numerous peer-reviewed publications and has received significant citation recognition, attesting to the impact of his work.

In addition to his research, he has assumed leadership positions within the academic community. Notably, he served as Program Director for Energy Conversion and Control Systems at TalTech, where he was responsible for aligning academic programs with evolving industry demands and technological advancements. He has also contributed to the professional engineering community through his role as Secretary of the IEEE Estonia Section, supporting knowledge exchange and professional development.

His commitment to sharing knowledge extends beyond academia, as demonstrated by his keynote addresses at international conferences, where he presents innovative approaches to fault detection and other cutting-edge topics. Professor Vaimann remains focused on advancing the field of electrical engineering through research, collaboration, and mentoring the next generation of professionals to navigate a rapidly evolving technological landscape.

Kuno Janson is the associate professor emeritus and has dedicated his professional life to the advancement of electrical engineering education and research. His academic career has been closely associated with Tallinn University of Technology (TalTech), Estonia, where he served for many years as an associate professor in the School of Engineering before being honoured with the title of Associate Professor Emeritus.

Throughout his extensive teaching career, Professor Janson instructed a wide range of courses, including *Elektrimasinad* (Electrical Machines) at the bachelor's level, *Transpordivahendite elektromehaanika* (Electromechanics of Transport Vehicles) at the master's level, and advanced doctoral-level courses such as *Elektrotehnika erikursus II* (Special Course in Electrical Engineering II). His work also extended to distance learning and vocational education, consistently supporting accessible and impactful technical education.

Recognising the importance of flexible learning, he collaborated with colleagues to create e-learning versions of key courses. These resources, made available through Estonia's national e-learning platform Moodle, have expanded access to engineering education across the country and beyond.

His research has been primarily focused on electrical machines and power systems, with a special emphasis on sustainable energy solutions. Much of his work has addressed real-world challenges relevant to Estonia's unique geographical and infrastructural context.

In addition to teaching and research, Professor Janson has made significant contributions to educational literature. He has co-authored multiple textbooks and technical learning materials aimed at enhancing the quality of engineering education.

Throughout his career, he remained committed to deepening the understanding of electrical engineering among students and colleagues. His legacy includes not only his scholarly and instructional contributions but also his enduring impact on the development of Estonia's engineering education and the preparation of future generations of engineers.

1 Fundamentals of Electrical Machines and Transformers

In the 20th century, global annual electrical energy generation increased from approximately 10 terawatt-hours to 15,000 terawatt-hours, i.e. 1,500 times. This was the century of the introduction of electricity. The introduction of electricity has transformed our way of life.

If only a few hundred years ago everyone managed perfectly well without electricity, the way we conduct our daily affairs has now changed considerably and coping without electricity would be extremely challenging. We only fully realise how much we depend on electricity when the power unexpectedly goes out. On 14 August 2003, a major power outage occurred in the United States. In the span of an hour, more than 500 power plants were knocked offline, leaving around 50 million people without electricity. 350,000 people found themselves trapped in trains, elevators, or the subway. As everyone started reaching for their phones, mobile networks were overloaded and cellular service was interrupted. People were unable to withdraw cash from ATMs. Shops could no longer sell goods, because the cash registers didn't work. Water supply systems lost pressure and toilets stopped working. And, of course, all production companies were forced to stop their operations. Every single area of life was severely disrupted. The total damage was estimated at $6 billion.

1.1 THE ROLE OF ELECTRICAL MACHINES AND TRANSFORMERS IN THE GENERATION, USE, AND TRANSMISSION OF ELECTRICAL ENERGY

1.1.1 Why Do We Need Electrical Machines and Transformers?

Electrical machines and transformers play a rather important role in providing us with electricity on a daily basis. About 98% of electrical energy is generated by electric generators. More than 60% of the electrical energy we produce is consumed by electric motors. The majority of all mechanisms used today are powered by electric motors. Electric motors are clean, easy to start, and adjustable. Electric motors, unlike internal combustion engines, do not require a constant supply of air and fuel. And they are also suitable in applications where combustion-related emissions are undesirable. Thermal energy or mechanical energy can

DOI: 10.1201/9781003404088-1

be converted into electrical energy at a suitable location, then transmitted over long distances.

In electric motors, the rotational speed and torque can be adjusted within a much broader range than in thermal prime movers, and the energy conversion efficiency of the electric motor does not decrease nearly as much during adjustment. For example, a vehicle with a petrol engine needs a clutch, while the engine of an electric vehicle can also develop torque at zero speed and does not require one.

What electric motors do need, however, is power from an electrical power source, which is usually an electrical grid. An electrical grid is a power flow transmission system. Electrical grids are powered by generators located at power plants (Figure 1.1). The shaft of an electric generator is rotated at a thermal power plant either by a steam turbine, which is a type of thermal prime mover, or by another prime mover. Electric generators, electrical grids, and electric motors transfer the mechanical energy of a steam turbine to consumer appliances (loads). However, electrical grids also supply power to a variety of other appliances, such as luminaires, heaters, etc.

Power plants can supply power to a large number of electrical loads located hundreds of kilometres away from the plant. The electrical resistance in power lines that span such distances is quite high. Where there is a large number of loads, a large amount of power is drawn. The high current in high-resistance power lines results in high energy losses. This power loss increases in proportion to the current squared:

$$P_{\text{loss}} = I^2 R_l, \tag{1.1}$$

where

P_{loss} – power loss in the line;
I – line current;
R_l – line resistance.

Line losses can be reduced without reducing the transmitted power by increasing the voltage. For example, if you increase the voltage 100-fold, the current decreases 100-fold (while the power remains the same), and line losses decrease

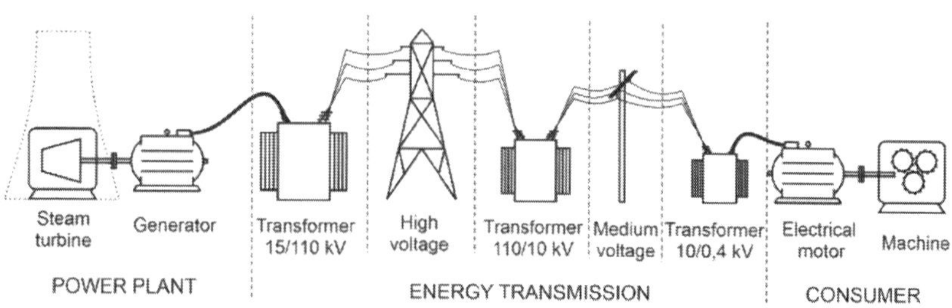

FIGURE 1.1 Generation and transmission of electrical energy to electrical loads.

10,000-fold. At high voltages, losses are greatly reduced. However, running 10,000-volt power lines into homes presents a safety hazard. To prevent this hazard, higher voltages are reserved for transmission, while points of consumption utilise lower voltages. The required shifts in voltage levels are achieved by means of transformers.

Power plant generators cannot be built for voltages higher than 10–20 kV. To overcome this limitation, transformer substations may be built at power plants to increase the voltages (Figure 1.1). Long-distance lines of transmission networks are operated at high voltages of 110 or 330 kilovolts. Meanwhile, shorter lines and branched distribution lines carry medium voltages of 10–36 kV. Electrical loads, however, are supplied with low voltages of 0.4/0.23 kV. Transformers that change voltage levels are vital for electrical grids. In both electrical machines and transformers, power flows are transmitted through air via a magnetic field. In addition to electrical circuits, they also contain magnetic circuits.

The power output of a power plant at a given moment is determined by the sum of the capacities of the loads, which changes arbitrarily in time as the loads are switched on and off. A power plant needs to be able to adjust its output according to demand from loads. It must always have a certain reserve capacity. If the sum of the load capacities exceeds the maximum output of the plant, the plant will no longer be able to operate and will automatically shut down, leaving all loads without power. This is what happened during the event described above.

Problem 1.1

A DC line is carrying over 80 kW of power at 400 V. The power loss in the line is 5% of the transmitted power. How many times over would the line voltage need to be increased to decrease the power loss in the line 10-fold?

Solution

1. The initial amount of power lost is

$$P_{\text{loss}1} = 0,05P = 0,05 \cdot 80 = 4\,\text{kW}.$$

2. DC power is calculated using the equation $P = U \cdot I$. Based on this, the initial current is

$$I_1 = P / U_1 = 80000 / 400 = 200\text{A}.$$

3. The resistance of the power line, according to equation (1.1), is

$$R_l = P_{\text{loss}1} / I_1^2 = 4000 / 200^2 = 0,1\,\text{W}.$$

4. In a line where the power loss is 10 times lower, the current can, again, be calculated using equation (1.1), with $P_{loss2} = 0,1P_{loss1}$:

$$I_2 = \sqrt{\frac{0,1P_{loss1}}{R_l}} = \sqrt{\frac{0,1 \cdot 4000}{0,1}} = 63,25\text{A}.$$

5. The voltage in the line at this new current would be

$$U_2 = P / I_2 = 80000 / 63,25 = 1265\text{V}.$$

6. Therefore, the voltage would need to be increased by

$$U_2 / U_1 = 1265 / 400 = 3,16 \text{ times}.$$

Revision questions

1. What would happen if the electrical grid of a city lost power for three days?
2. Where and with what machines is electrical energy produced?
3. Which machines consume more than half of all electrical energy produced?
4. How is electrical energy delivered to loads?
5. What determines the power output of a power plant at a given moment and how does that output change over time?
6. Why is electrical energy transmitted at a greater voltage than required by loads?
7. Which devices are used to change voltage levels?
8. What are the standard voltage levels for power plant generators, transmission lines, and distribution networks?

1.1.2 What Exactly Is the Energy and Power Supplied and Converted by Electrical Machines?

Before we can talk about electrical machines, it is necessary to have a basic understanding of energy and power. To understand what constitutes electrical power and electrical energy, mechanical power and mechanical energy. As well as thermal power and thermal energy, which in electrical machines is also called power dissipated and energy loss.

In the field of economics, energy is mainly viewed as a commodity that is produced, sold, and consumed. However, according to the laws of thermodynamics, energy can never actually be generated nor, therefore, produced. It can only be converted from one form to another. Thus, energy can also not be consumed. When energy is said to be produced and consumed, it is actually, strictly speaking, only converted.

In the vernacular, it is said that refrigerators, for example, consume electrical energy from the grid. The correct turn of phrase, however, would be that refrigerators receive electrical power via the grid, which is supplied with that power by a power plant. What is the difference? If a refrigerator were switched on and started to draw 100 W of electrical power from the grid, the power plant would increase its power output by 100 W plus losses. There is no reserve energy in a grid. Electrical energy is consumed at the moment in which it is produced. A grid without a power plant (or electric generator) cannot deliver power. A power plant, transmission grid, and loads form an energy system. Thus, refrigerators consume energy from an energy system and power from an electrical grid.

The easiest type of power to describe is DC power, which is given by

$$P = U \cdot I, \tag{1.2}$$

where P – electrical power in watts, W;
U – voltage in volts, V;
I – current in amperes, A.

Electrical energy is defined as the integral of electrical power over time. Assuming a power that is constant over time:

$$W_{el} = P \cdot \Delta t = U \cdot I \cdot \Delta t, \tag{1.3}$$

where W_{el} – electrical energy in watt-seconds, Ws;
Δt – time between the end and start of consumption in seconds, s.

Equation (1.3) shows the amount of energy in watt-seconds or joules that has passed through an electrical circuit over a given period of time, i.e. which has already left the electrical circuit and now exists in some other place and form (heat, mechanical energy, or some other type of energy). This amount of energy can be measured easily with an electricity meter. In metres, the measured electrical energy is usually expressed in kilowatt-hours (kWh); 1 kWh = $3.6 \cdot 10^6$ Ws.

While the weight of ordinary goods can be determined instantly by weighing, and the goods will still be there after it, electricity meters measure the amount of electrical energy throughout the period of consumption of the electrical power. When the measurement is completed, the measured electrical energy is no longer available. It has been consumed (or, more precisely, converted to another form).

Moreover, even during measurement, it would not be possible to indicate the location of any amount of electrical energy. Electrical energy is unique and is difficult to compare to ordinary material objects. The voltage, current, and time in equation (1.3), too, are not material objects.

Electrical energy means energy supplied to loads from an energy source by means of an electrical grid or circuit, measured either by an electricity meter or indirectly. Electrical energy is energy that has passed through the electrical grid, not energy inside the electrical grid.

What flows through a grid is electrical power. At a power of 10 kW, for example, 10,000 watt-seconds, or joules, of energy pass through the grid every second. But, paradoxically, this amount of electrical energy is also not in the electrical grid at any time. Power is a measure of the intensity of an energy flow. For illustrative purposes, it can be compared to the intensity of a water flow in a pipe. An electrical grid is like a bucket without a bottom. Everything you pour into it will flow out through the bottom. Energy, meanwhile, is the figurative water in the bucket, delivered via a pipe. Note, however, that while there is a certain amount of water in a pipe at any given moment, there is never any electrical energy in an electrical grid.

Revision questions

1. How can DC power be calculated?
2. How would DC power change if the current were doubled and the voltage remained the same?
3. How would DC power change if the current were halved and the voltage were doubled?
4. How is DC energy defined?
5. Which type of device is used to measure electrical energy?
6. Which units of measurement are used for electrical energy?
7. 2 kWh of electrical energy passed through an electricity meter over one hour, after which consumption stopped. Where could those 2 kWh of electrical energy now be located?
8. What makes up an electrical grid and what makes up an energy system?
9. What is the measure of energy flows transmitted via an electrical grid?
10. A consumer switched on a 100 W electric light bulb. How did this affect the power output of the power plant?
11. How much energy can be stored inside an electrical grid?
12. Why can energy not be compared to material objects?
13. What would happen if an electrical grid were powered only by wind turbines and the wind stopped blowing?

Let us now consider energy and power in electrical machines in a specific instance. Electrical machines are either motors or generators. The purpose of an electric motor is to impart motion to something. An appliance (such as the flour mill in Figure 1.2) has been connected to the shaft of an electric motor, performing work and consuming mechanical power at a rate of 10 kNm/s (equivalent electrical power = 10 kW). The electric motor can run the mill because the motor itself is receiving 11 kW of electrical power through a set of wires. 1 kW of this power is lost in the motor as the motor heats up during operation. The motor converts the supplied electrical power into the mechanical power required by the mill. The motor is a power converter.

The electrical power is supplied by another electrical machine: a generator. The generator receives 12 kNm/s of mechanical power from a prime mover (steam turbine in Figure 1.2) and converts it into 11 kW of electrical power and 1 kJ/s of heat losses. The generator, like the motor, is a power converter. However, the generator performs the opposite conversion of power to the motor, converting mechanical power into electrical power. In power conversion processes in electric machines, some power is always lost (heat losses occur in the machine).

Stored energy and power

Mechanical power, too, needs to come from somewhere. It can be obtained by converting chemical energy stored in fuel (burning the fuel), as shown in Figure 1.2. Burning oil shale produces heat. Let us explore this example in more detail.

In this example, 72 kg of oil shale with a calorific value of 8.9 MJ/kg is burned in the furnace of a steam boiler over 4 hours, releasing 640 MJ of thermal energy. The majority of the heat (576 MJ) is transferred from the flue gases in the boiler to water, causing it to evaporate. A small amount of the heat escapes through the smokes stack with the smoke. The steam is fed to a steam turbine, which converts 30% of the steam's thermal energy into mechanical work (173 MJ over four hours). The turbine rotates, producing 12 kNm/s of mechanical power, which is

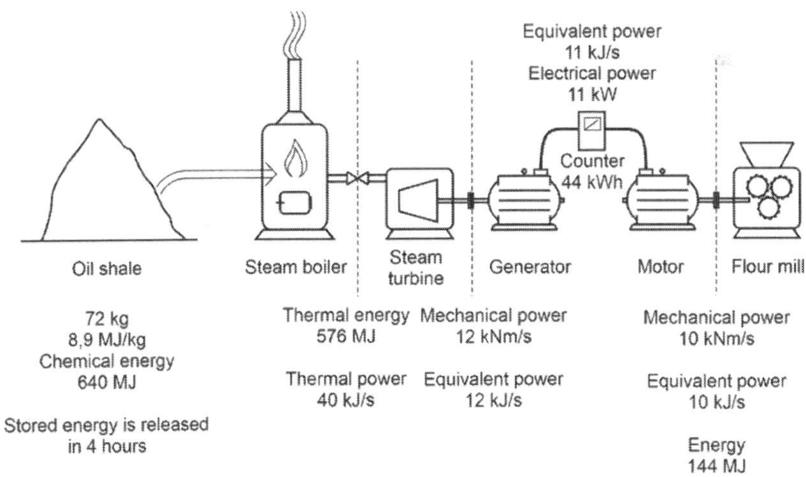

FIGURE 1.2 Energy conversion and transmission chain through which chemical energy stored in oil shale is delivered to a flour mill.

delivered to the generator via a shaft. Next, the generator converts the mechanical power into electrical power, as we learned above. Initially, the energy inside the oil shale was in a steady state, and had been so for a long time. In other words, the energy was stored in the oil shale. Oil shale is known as an energy carrier. In this example, part of the energy stored in the oil shale was delivered to the flour mill through multiple conversion stages. Most of it, however, was lost. Losses are inevitable. Losses are measured by energy conversion efficiency. In the instance above, the useful output was 10 kN m/s of mechanical power over four hours. Multiplying the power by the time (four hours equals 14,400 seconds), we get 144,000 kN m of mechanical energy, i.e. work that is equivalent to 144 MJ of energy. Energy conversion efficiency is the ratio between spent energy (640 MJ) and useful energy (144 MJ), in this case 22.5%.

The key takeaways from this example are that each type of energy has its own units of measurement, that conversion intensity is described by power, that heat losses occur in every conversion process, and that there must always be an energy source from which the converted energy is obtained as well as a final load receiving the energy flow. Meanwhile, the amount of energy transmitted is determined by the multiple of the power and the time.

Oil shale is not the best energy carrier. Petrol is better, as it does not produce ash during combustion. In the internal combustion engine of a car, the chemical energy stored in petrol is converted into mechanical power, which makes the car move. However, cars can also be made to move by an electric motor powered by a battery. A battery is an energy storage device, which is a special kind of energy carrier. Energy storage devices enable energy to be loaded into and extracted from them as desired. From a battery, electrical power can be transmitted through wires to an electric motor, which converts it into mechanical power.

Idling and loading of electrical machines

An electric generator can only deliver power and energy if there is a load to consume that electrical power (i.e. to convert the electrical energy into another type of energy). Figure 1.3a depicts one such situation. The mechanical energy of the wind causes a rotor connected to the shaft of an electric generator to rotate. The generator converts the mechanical power into electrical power and delivers it to the grid via wires. If, however, the generator were disconnected from the grid using a switch (Figure 1.3b), it could no longer convert the power contained in the wind. Instead, the generator would idle. Wind energy would continue to be supplied, but there would be no load to consume the electrical power. The conversion process could not occur. In conversion processes, one type of energy can only be transformed into another in equivalent amounts. Zero power on one side of the converter would result in zero power on the other side as well, regardless of the direction of the conversion process.

An electric motor will only output as much mechanical power as is consumed by the appliance it is used to run. This is illustrated in Figure 1.4a. An appliance is connected to the shaft of an electrical machine. The mechanical torque in the shaft of the electrical machine is exactly as high as the resistive torque of

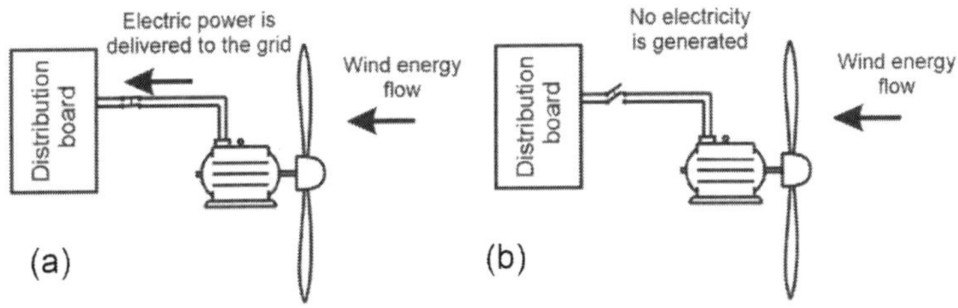

FIGURE 1.3 (a) A wind turbine (generator) is receiving wind energy. The rotor converts it into rotating mechanical power. The generator converts this mechanical power into electrical power, which is delivered to the electrical grid via wires. (b) The wind makes the rotor and the shaft of the generator rotate, but a switch is open in the electrical circuit and thus no electrical power can be generated. The generator is idling.

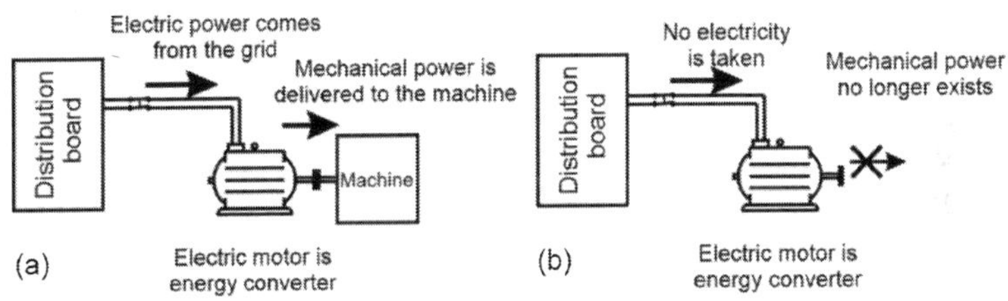

FIGURE 1.4 (a) An electric motor is receiving electrical power from the grid via wiring. The motor converts it into mechanical power, which is supplied to an appliance via a shaft. (b) The electric motor cannot convert the electrical power, as the mechanical power cannot exit via the shaft. The electric motor is idling.

the appliance. If the load (appliance) were disconnected from the shaft of the motor (Figure 1.4b), the mechanical power could no longer exit via the shaft. The motor could continue to rotate, but the electrical energy would not be converted to mechanical energy. In such cases, the motor is said to be idling. The fact that it says 'Power: 10 kW' on the data plate of the motor would be irrelevant. A power of 10 kW only means that the appliance should not place a higher load than that on the motor.

What are the similarities and differences between different types of energy?
The presence of stored energy may not always be evident. However, when energy is converted from one type to another, it is accompanied by various observable phenomena (motion, temperature change, electric current, etc.). Energy can be measured through the phenomena that occur in such conversion processes and through their duration. That is also why different types of energy have different units of measurement. Electrical energy is measured in watt-seconds (Ws) or kilowatt-hours (kWh), mechanical energy is measured in newton-metres (N m), thermal energy used to be measured in calories (cal) and, to an extent, still is today. But there is also one common unit of measurement, the joule (J), that can be used for all types of energy. The following equation holds:

$$1\,\text{Ws} = 1\,\text{Nm} = 0{,}24\,\text{cal} = 1\ \text{J}. \tag{1.4}$$

This equation should be read as saying that converting 1 Ws of electrical energy yields 1 Nm of mechanical energy, which in turn, when converted, yields 0.24 cal of thermal energy, and all three of the above amounts of energy are equivalent to 1 J of energy. Equation (1.4) holds only if there are no conversion losses. Electrical energy, mechanical energy, and thermal energy are quite different in nature, but in equation (1.4) they are equated because the given amounts of energy are equivalent in conversion processes. For example, 1 Ws of electrical energy would yield 1 N m of mechanical energy when converted.

This is similar to how water can occur as a liquid, as ice, or as water vapour, which are also quite different in character. However, while we know what these three states of water share in common – the water molecule (H_2O) – it is not known what electrical energy, mechanical energy, and thermal energy share. 'Energy molecules' have yet to be discovered.

The essential nature of energy that is fed to an electric motor via wires and extracted via a rotating shaft – and which can neither be created nor destroyed – is still a mystery. It is beyond the understanding of mankind. Energy is expected to be composed of some kinds of elementary particles that are much smaller than all elementary particles currently known to us. Energy conversion processes also take place at the quantum level. Electrons orbiting a nucleus can change their orbit, emitting the energy of a quantum of light in the process. The energy contained in a quantum of light is minuscule compared to the amounts of energy that pass through electrical machines.

Problem 1.2

An electric vehicle needs to deliver a set of goods to 75 km away and then return. The vehicle is travelling at a velocity of v = 90 km/h, at which its electric motor must output a power of P = 40 kW. The energy conversion efficiency of the electric motor is η_M = 0.92 and that of the converter between the motor and the battery is η_{con} = 0.94. How much energy will be drawn from the battery for the journey?

Solution

1. The total travel distance (s) equals

$$s = 75 \cdot 2 = 150\,\text{km}.$$

2. The travel time (t) equals

$$t = s\,/\,v = 150\,/\,90 = 1{,}67\text{h}.$$

3. The amount of energy (W_M) consumed by the electric motor during the journey equals

$$W_\mathrm{M} = P \cdot t = 40 \cdot 1{,}67 = 66{,}67\mathrm{kWh}.$$

4. The amount of energy drawn from the batteries (W_bat) equals

$$W_\mathrm{bat} = W_\mathrm{M} / \left(\eta_M \cdot \eta_{con}\right) = 66{,}67 / \left(0{,}92 \cdot 0{,}94\right) = 77{,}09\mathrm{kWh}.$$

Revision questions

1. What are the three types of power that can be converted by electrical machines?
2. Which type of electrical machine can be used to convert electrical power into mechanical power?
3. Which type of electrical machine can be used to convert mechanical power into electrical power?
4. Why are power and energy also converted to heat in electrical machines, and is it desirable or not?
5. What is meant by the energy conversion efficiency of an electrical machine?
6. How are mechanical power and energy obtained at a thermal power plant?
7. In which form is energy delivered to a thermal power plant?
8. Where can stored energy be found?
9. Where does the power put out by an electric motor go?
10. What determines the amount of power put out by an electric motor?
11. How much power does an electric motor put out if no appliance is connected to its shaft, but the electric motor is rotating?
12. How much power does a generator put out if it is rotating, but there are no loads connected to the generator?
13. What amount of mechanical energy is equivalent to 1 Ws of electrical energy?
14. What amount of thermal energy is equivalent to 1 Ws of electrical energy?
15. What do mechanical energy, electrical energy, and thermal energy share in common?

1.2 MECHANICAL AND ELECTRICAL COMPONENTS OF ELECTRICAL MACHINES

As discussed above, electrical machines are tasked with converting mechanical power into electrical power and, conversely, to convert electrical power into mechanical power. Now let us take a look at which components electrical machines should contain in general and how their design enables them to perform these tasks.

The most common types of electrical machines are **DC machines**, **synchronous machines**, and **induction machines**. All of these electrical machines can be designed to produce either **rotary motion** or **linear motion** (the latter are known as **linear electrical machines**). Most electrical machines, nevertheless, are rotary.

Figure 1.5 shows a cutaway view of a medium-power induction motor. The housing, stator, and rotor of the motor are shown in cross-section for a view of their internal structure. The terminal box contains terminals which are connected to 3-phase conductors and a protective earth leading to the electrical grid. When the phase conductors are energised, the rotor of the electric motor starts to rotate along with the shaft. The shaft rests on ball bearings positioned behind end plates. The protruding end of the shaft is intended to be connected to an appliance. At the other end of the shaft is a cooling fan. You can also partially see the motor windings in the figure, with the end connections cut through. Note, however, that Figure 1.5 does not provide a complete overview of the design of induction motors, as electric motors are quite complex.

Figure 1.6 shows another induction motor, which is partially disassembled. Figure 1.6b shows the main components of the electrical machine – the stator and the rotor – separately. The stator is a stationary hollow cylinder positioned inside

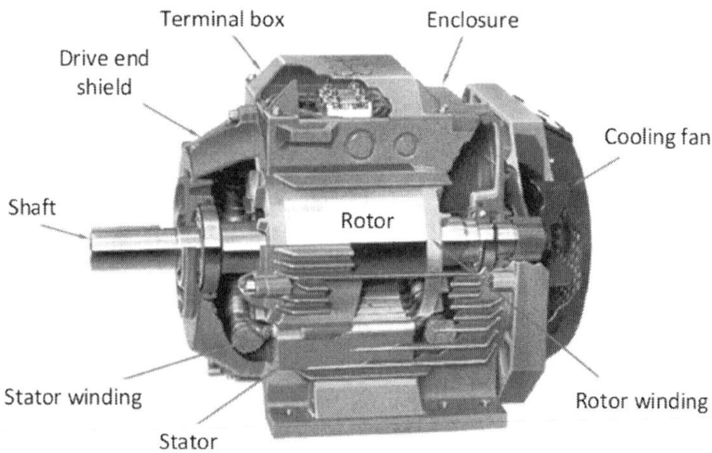

FIGURE 1.5 Cutaway drawing of an induction motor.

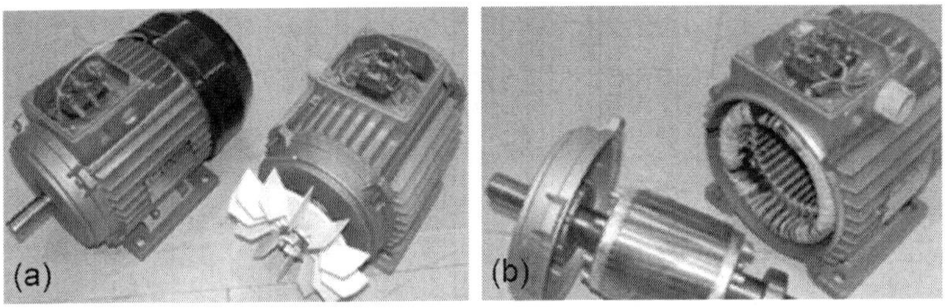

FIGURE 1.6 (a) Induction motor with and without fan cover; (b) rotor of an induction motor with a shaft, one end plate, and bearings, and stator with housing.

the housing and only visible from the end. The rotor is a rotating cylinder that is positioned inside the stator.

DC machines and synchronous machines are different from induction machines, but the main types of electrical machines are structurally similar enough that their main components can be discussed together.

A rotor (1) is placed inside a stator (2) in a way that leaves an **air gap** (3) between the stator and the rotor (Figure 1.7a, b). A **shaft** (4) resting on **bearings** (5 and 6) passes through the rotor (1). The stator (2), which takes the form of a hollow cylinder, is pressed into another hollow cylinder (7), called the **housing** of the electrical machine. The outer surface of the housing (7) does not always have to be cylindrical. This surface may, for example, have ribs for better heat dissipation, as in the case of the induction motor in Figure 1.5. The outer surface of the housing may also be rectangular. The bearings are positioned inside **end plates** (8 and 9), which are fixed to the housing (7). Resting on the bearings, the shaft is able to rotate along with the rotor.

The inner surface of the **stator (2) holds the stator winding** (10). The outer surface of the rotor (1) holds the **rotor winding** (11). In some types of machines, **permanent magnets** may be used in place of either the stator winding (10) or the rotor winding (11). The current flowing through the windings or the placement of the permanent magnets causes **pairs of magnetic poles** to form on the outer surface of the rotor (1) and the inner surface of the stator (2). Each pole pair consists of one north pole and one south pole, such as the **north pole** 12 and the **south pole** 13 in the figure. In Figure 1.7, both the rotor and the stator have two pairs of poles. The smallest possible number of pole pairs is one, and the number of pole pairs on the rotor of an electrical machine is usually equal to the number of pole pairs on its stator. There is no maximum limit to the number of pole pairs.

Magnets produce a pulling force between a north pole and a south pole. Since the rotor (1) is able to rotate relative to the stator (2), the rotor moves into a position where each north pole on the stator is facing a south pole on the rotor and each south pole on the stator is facing a north pole on the rotor.

Windings consist of **winding wire**, which is usually a copper wire with a thin insulating coating. This wire is used to wind **coils.** The windings used in electrical machines are designed and installed on stators and rotors (Figure 1.8) in

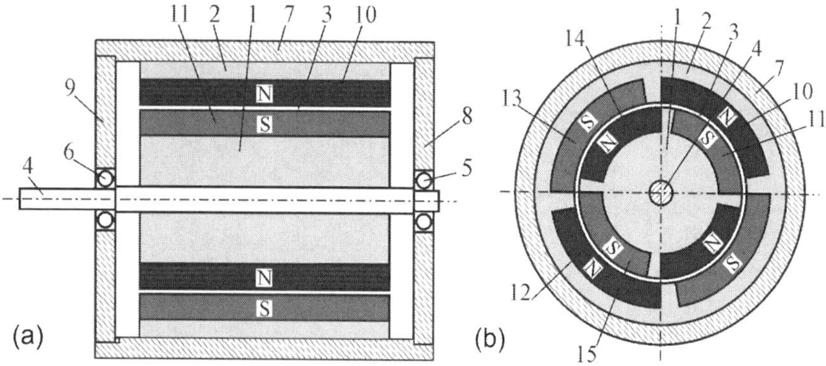

FIGURE 1.7 Cross-sectional view of a standard cylindrical electrical machine along the axis of rotation (a) and perpendicular to the axis of rotation (b).

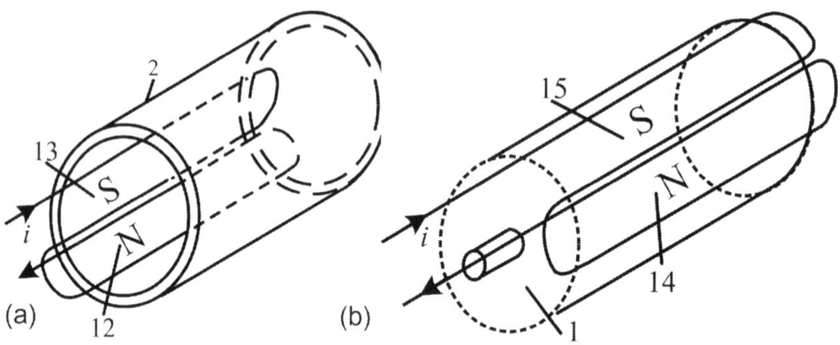

FIGURE 1.8 Creation of pole pairs on the inner surface of a stator (a) and on the outer surface of a rotor (b) by running a current through winding wires.

such a way that north poles (12) and south poles (13) form alternately on the inner surface of the stator (2) as current passes through the winding (Figure 1.8a). Similarly, north poles (14) and south poles (15) form on the outer surface of the rotor (1) (Figure 1.8b).

Magnetic poles are needed in electric motors to cause the motor to rotate and develop torque. The construction of magnetic poles, the magnetic fields generated by the supply of current to pole windings, and the forces generated by magnetic fields are discussed below. Next, we will explore the parameters used to describe the rotation of electrical machines, their torque, and how to determine the power of an electrical machine.

Revision questions

1. Which parts of an electrical machine rotate?
2. Which parts of an electrical machine are stationary?

3. How are pairs of magnetic poles created in an electrical machine?
4. Which position will the magnetic poles on a rotor assume in relation to the magnetic poles on the stator, if the rotor can rotate freely?

1.3 MECHANICS OF ROTATIONAL MOTION

Every electrical machine contains an electrical component and a mechanical component. A discussion of the electrical component requires a knowledge of electrical engineering, while a discussion of the mechanical component requires a knowledge of mechanics. As most electrical machines in use are rotating machines, let us now explore the parameters relating to rotational motion. These are angle of rotation, rotational speed, angular velocity, angular acceleration, torque, moment of inertia, work, and mechanical power.

In electrical machines, rotation occurs around a fixed axis. The direction of rotation of the machine is determined by looking at the shaft from the side of the appliance. The direction of rotation can be **clockwise** (CW) or **counterclockwise** (CCW). The direction of rotation is said to be positive in the case of counterclockwise rotation and negative in the case of clockwise rotation.

Angle of rotation θ

In the case of translational motion, the displacement of a body from an initial position to a final position is measured by distance s, expressed in metres. Rotational motion, however, involves a body rotating around its axis and moving from an initial angular position to a final angular position. This kind of movement is measured by angle of rotation θ . The natural unit of rotational motion is the revolution. One revolution equals 360° or 2π radians. The radian is defined as follows: where the length of the arc of a circle is equal to the radius of that circle, the angle between the radii at the ends of that arc is equal to one radian, 1 rad $\approx$ 57.3°.

Angular velocity ω

Angular velocity is a measure of the change in the angle of rotation of a rotating body per unit time. Angular velocity is considered to be positive in the case of counterclockwise rotation. Angular velocity is analogous to the linear velocity in translational motion. Linear velocity is given by the equation

$$v = \mathrm{d}s / \mathrm{d}t[\mathrm{m/s}] \tag{1.5}$$

and angular velocity is given by the equation

$$\omega = \mathrm{d}\theta / \mathrm{d}t[\mathrm{rad/s}] \tag{1.6}$$

As angle of rotation is measured in radians, the unit of angular velocity is radians per second (rad/s).

The shaft speed of electrical machines is also described using other units in addition to rad/s. Often, the speed is expressed in revolutions per second or revolutions per minute. Conventionally, different symbols are used for different units. Herein, we will be using the following symbols for angular velocity: ω_{mech} – radians per second (rad/s); f_{mech} – revolutions per second (rps); n_{mech} – revolutions per minute (rpm). The following relationships apply between these shaft speeds:

$$n_{\mathrm{mech}} = 60 f_{\mathrm{mech}}; f_{\mathrm{mech}} = \omega_{\mathrm{mech}} / 2\pi; \omega_{\mathrm{mech}} = 2\pi f_{\mathrm{mech}}. \quad (1.7)$$

The index ‘mech’ in the symbols above indicates the mechanical nature of the quantity. If there is no risk of confusion between mechanical and electrical quantities, the index is often omitted.

Angular acceleration α

Angular acceleration is the rate of change of angular velocity over a period of time. Angular acceleration is considered to be positive if it increases in algebraic terms. Angular acceleration is analogous to the linear acceleration in translational motion. Linear acceleration is given by the equation

$$a = \mathrm{d}v / \mathrm{d}t \left[\mathrm{m/s^2} \right] \quad (1.8)$$

Angular acceleration is given by the equation

$$\alpha = \mathrm{d}\omega / \mathrm{d}t \left[\mathrm{rad/s^2} \right]. \quad (1.9)$$

As the unit of angular velocity ω is rad/s, the unit of angular acceleration α is radians per second squared (rad/s^2).

Torque T

In the case of linear motion, a force applied to a moving object (body) causes a change in velocity. When there is no force, the velocity of the body is constant. For a rotating object, angular velocity is constant until it is affected by torque. The higher the torque acting on the object, the faster the rate of change of the object’s angular velocity.

What is torque? It may be described as the force of rotation of an object. To better understand torque, let us imagine a cylinder that can rotate around its axis. If a force were applied to this cylinder so that the line of action of the force passed through the axis of the cylinder (Figure 1.9a), the cylinder would not rotate. If, however, the force were applied so that it acted on the cylinder to the right of the axis (Figure 1.9b), the cylinder would begin to rotate counterclockwise. The torque or the rotational effect on the cylinder depends firstly on the amount of force applied and secondly on the distance between the axis of rotation and the line of action of the force.

In the situation depicted in Figure 1.9a, the object cannot start moving translationally in the direction of the force F, because the stationary axis of rotation A produces a counteracting force Q. The force F acting on the body at point B

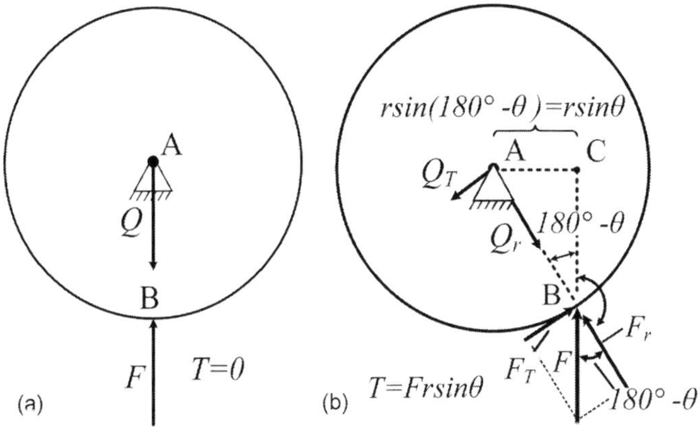

FIGURE 1.9 (a) Force is applied to the cylinder with the line of action passing through the axis of rotation, $T = 0$. (b) Force is applied to the cylinder with the line of action not passing through the axis of rotation and the torque T acting counterclockwise.

(Figure 1.9b) can be separated into two components, one acting along the radius r (F_r) and the other along the tangent line to the circle (F_T). In Figure 1.9a, the tangential component of the force is equal to zero, and the body will not begin to rotate. Rotation is caused only by the tangential component F_T, because the axial counteracting force Q_T is not acting directly opposite to this force component, but parallel to it and at a distance equal to the radius r.

Compared to translational motion, the generation of rotational motion is also unique due to the relevance of the distance between the line of action of the force and the axis of rotation (line segment AC in Figure 1.9b). This distance is called the **lever arm**. The product of a force and the corresponding lever arm is called **torque *T***. Torque fully describes the rotation-generating effect. The greater the lever arm, the easier it is to make the body rotate. A nut is easier to loosen if the arm of the wrench is longer.

In the case of the geometric structure shown in Figure 1.9b, the lever arm can be determined via the angle θ. The angle θ is located between the radial line passing through the point of force application and the line of action of the force F. The lever arm can be given as the product $r\sin(180° - \theta)$ or, simplified, $r\sin\theta$. Therefore, torque

$$T = Fr\sin\theta. \tag{1.10}$$

The tangential component F_T of the force can also be given by $\sin(180° - \theta)$:

$$F_T = F\sin\theta. \tag{1.11}$$

Consequently, torque can be given also by the equation:

$$T = F_T r. \tag{1.12}$$

A higher force F applied near the axis of rotation has the same rotation-generating effect as a smaller force F_T applied farther away (the line segment AC is shorter than AB).

The unit of torque in the International System of Units (SI) is the newton-metre (N m).

Newton's second law for rotational motion. Newton's law for an object moving in a straight line describes the relationship between the force applied to the object and the resulting acceleration. This relationship is given by the equation

$$a = F / m, \tag{1.13}$$

where F is the force applied to the object, m is the mass of the object and a is the resulting acceleration. In the SI system, the unit of force is the newton (N), the unit of mass is the kilogram (kg), and the unit of acceleration is the metre per second squared (m/s^2).

An analogous equation describes the relationship between the torque applied to an object and the resulting angular acceleration. This relationship is called Newton's law for rotation, which is given by the equation

$$\alpha = T / J, \tag{1.14}$$

where T is the applied torque in newton-metres and α is the resulting angular acceleration in rad/s^2. The factor J carries the same meaning as the mass of the object in the case of linear motion. It is called the moment of inertia of the object, and its unit is the kg·m^2.

In mechanics, the moment of inertia is defined by the equation

$$J = mr_g^2, \tag{1.15}$$

where m – mass of the rotating body, kg;
r_g – radius of gyration of the body, m.

The radius of gyration measures the average distance of a rotating mass from the axis of rotation. In flywheels, it is desirable for the moment of inertia to be as large as possible. To achieve this, the main mass is placed far out in the rim of the wheel, near the outer edge, where only spokes connect the mass to the hub and axle. The determination of the radius of gyration is not discussed here.

Linear velocity v

In addition to angular velocity, the different points of a rotating body also have a linear velocity. Linear velocity depends on the distance of the observed point from the axis of rotation. This distance is the radius r of the path travelled by the observed point (point B in Figure 1.9b). In one revolution of the shaft, point B covers a distance of s = $s = 2\pi\ r$. Based on equation (1.6), the time it takes to complete one revolution is $t = \theta / \omega = 2\pi / \omega$. This leads us to the relationship between angular velocity and linear velocity

$$v = \frac{s}{t} = \frac{2\pi r}{2\pi / \omega} = \omega r. \tag{1.16}$$

While the linear velocities of different points on a rotating body are different, the angular velocity is the same for all points. Linear velocity is highest in points located on the outer surface of a wheel, while on the axis, linear velocity always equals zero.

Work or consumed energy W

For linear motion, work is defined as the integral of the force over the displacement (distance). It is given by the equation

$$W = \int F \mathrm{d}s, \tag{1.17}$$

where the force is assumed to be collinear (identical in direction) to the direction of motion. In special cases where the force applied collinearly to the direction of motion is constant, work can be given by the equation

$$W = Fs. \tag{1.18}$$

In the SI system, the unit of work and energy is the joule (J).

In the case of rotational motion, work is the integral of torque over the angle of rotation, and it is given by the equation

$$W = \int T \mathrm{d}\theta \tag{1.19}$$

or in the case of a constant torque

$$W = T\theta\ . \tag{1.20}$$

Power P

Power is the rate at which work is done, i.e. the work done in a unit of time. The equation for power is

$$P = \mathrm{d}W / \mathrm{d}t. \tag{1.21}$$

Power is usually measured in joules per second (J/s) or watts (W), but may also be expressed in horsepower (hp). Conversions between the units above can be performed based on the relationships 1 J/s = 1 W and 1 hp = 735.5 W.

Given this definition and assuming a force that is constant and collinear to the direction of motion, we get

$$P = \mathrm{d}W / \mathrm{d}t = \mathrm{d}(Fs) / \mathrm{d}t = F(\mathrm{d}s / \mathrm{d}t) = Fv. \tag{1.22}$$

Similarly, assuming a torque that is constant, we can derive the equation for power in the case of rotational motion:

$$P = \mathrm{d}W / \mathrm{d}t = \mathrm{d}(T\theta) / \mathrm{d}t = T(\mathrm{d}\theta / \mathrm{d}t) = T\omega. \tag{1.23}$$

The equation $P = T\omega$ is essential in the study of electrical machines, as it allows us to calculate the mechanical power on the shaft of a motor or generator. In this equation, the units of measurement are the watt (W), the newton-metre (N m), and the radian per second (rad/s); if different units are used, factors must be added to the equation to convert the units.

Problem 1.3

The grinding wheel of an electric grinder is attached directly to the shaft of an electric motor. The motor's rotational speed n is 2,850 rpm and the energy conversion efficiency is $\eta_M = 0.85$. The diameter of the grinding wheel is $D = 15$ cm, and during grinding, a resistive force of $F = 40$ N acts along the tangent line to the wheel.

Solution

1. The lever arm r equals half the diameter of the grinding wheel:

$$r = D/2 = 15/2 = 7{,}5\text{cm} = 0{,}075\text{m}.$$

2. The resistive torque T is

$$T = F \cdot r = 40 \cdot 0{,}075 = 3\text{Nm}.$$

3. The angular velocity ω of the motor is

$$\omega = n \cdot 2\pi / 60 = 2850 \cdot 2\pi / 60 = 298{,}5\text{rad} / \text{s}.$$

4. The mechanical power P_{mech} of the motor is

$$P_{mech} = T \cdot \omega = 3 \cdot 298{,}5 = 895{,}4\text{W}.$$

5. The power drawn from the grid is

$$P_{el} = P_{mech} / \eta_M = 895{,}4 / 0{,}85 = 1053\text{W}.$$

Revision questions

1. Which units are used to measure the angle of rotation of the rotor of an electrical machine?
2. Which units are used to measure rotational speed and angular velocity in electrical machines?
3. What determines the torque of an electrical machine and what does torque describe?
4. Why is it easier to loosen a nut with a wrench that has a longer arm?
5. How is Newton's second law expressed for rotational motion?
6. How can one calculate the linear velocity of a point on a rotating body?
7. How can one calculate the work done in rotational motion?
8. How can one calculate the power for rotational motion?

1.4 FIELDS AND SPACE: PERMANENT MAGNETS AND ELECTROMAGNETS

In electrical machines and transformers, energy is always transmitted through magnetic fields in the air. The power drawn from a magnetic field in an air gap makes the shaft of an electric motor rotate, and the magnetic field transmits the energy flow, or power, from one transformer winding to another. Understanding magnetic fields is essential.

Magnetic fields are a component of light. Light is an electromagnetic field that moves at extremely fast speeds in both vacuum and air. An electromagnetic field is a flow of energy. Solar panels on the roof of a house convert the flow of energy of sunlight into electrical power. Electromagnetic fields have two components. In addition to a magnetic field, they include an electric field. Electric fields and magnetic fields can also be stationary in space. In such cases, they are called static fields.

Generating an electric or magnetic field in an area of space requires putting energy into it. This input energy then becomes electric field energy or magnetic field energy. Unlike light, magnetic fields penetrate all materials to an extent.

Static magnetic fields are usually generated using either permanent magnets or electromagnets, but natural sources of magnetic fields also exist. Every current-carrying wire also has a magnetic field around it.

According to a Greek legend, magnetism was discovered by a shepherd wandering the mountains of western Turkey. Upon stepping onto an outcrop of magnetite, his feet suddenly became stuck to the ground, as the magnetite pulled on the nails in his sandals. The site of the legend is a magnetite-rich region near the city of Magnesia, which is also the source of the name of this mysterious and useful force inherent in certain metals (iron, cobalt, nickel), which can attract or repel these metals.

A magnetic field is always generated around every current-carrying wire (Figure 1.10a). Magnetic fields are invisible. In drawings, the direction and strength of magnetic fields are indicated using **lines of force**. The greater the density of the lines of force, the stronger the field. Between two magnetic fields, a mechanical force is generated. The directions of the lines of force indicate the direction of the generated force. The magnetic field of a current can be amplified by winding the current-carrying wire into a coil (Figure 1.10b).

Magnetic fields can also be generated using various permanent magnets (Figure 1.10c, d, e), without the use of any current-carrying wires. Permanent magnets have two **poles**. The lines of force leave the north pole, which is denoted as **N**. They enter the south pole, which is denoted as **S**. The magnetic field of a permanent magnet is caused by the unidirectional orientation of the atoms of the material in space. The electrons orbiting the atomic nucleus constitute an electric current, which also has a magnetic field. Unidirectionally oriented atoms yield unidirectionally oriented magnetic fields. The sum of the magnetic fields of the orbiting electrons is expressed in the form of a permanent magnetic field.

Magnetic fields are considered to be composed of two parts. These are the ***H*-field** and the ***B*-field**. *The* H-field is also simply called the magnetic field. This field is the 'invisible part' of an electric current in the space near the wire. The H-field generates something like mechanical pressure or electrical voltage in the space. This results in a **magnetic flux**. The unit of magnetic flux is the weber (Wb). Magnetic flux can be compared to a flow of water or electric current, except no one has managed to detect magnetic flux particles. The field of magnetic flux density is called the B-field, and it is measured in Wb/m^2. Magnetic flux density is also known as **magnetic induction**, which is measured in teslas (T); 1 T = 1 Wb/m^2. The existence of two different names for the same physical object is due to the fact that as the science of physics has progressed, our understanding of empty space, or vacuum, has changed. The concept of magnetic flux density is in line with the modern understanding of empty space.

The B-field results from the H-field and its direction is identical to that of the latter. However, the strength of the B-field also depends on the properties of the medium.

$$B = \mu H,$$

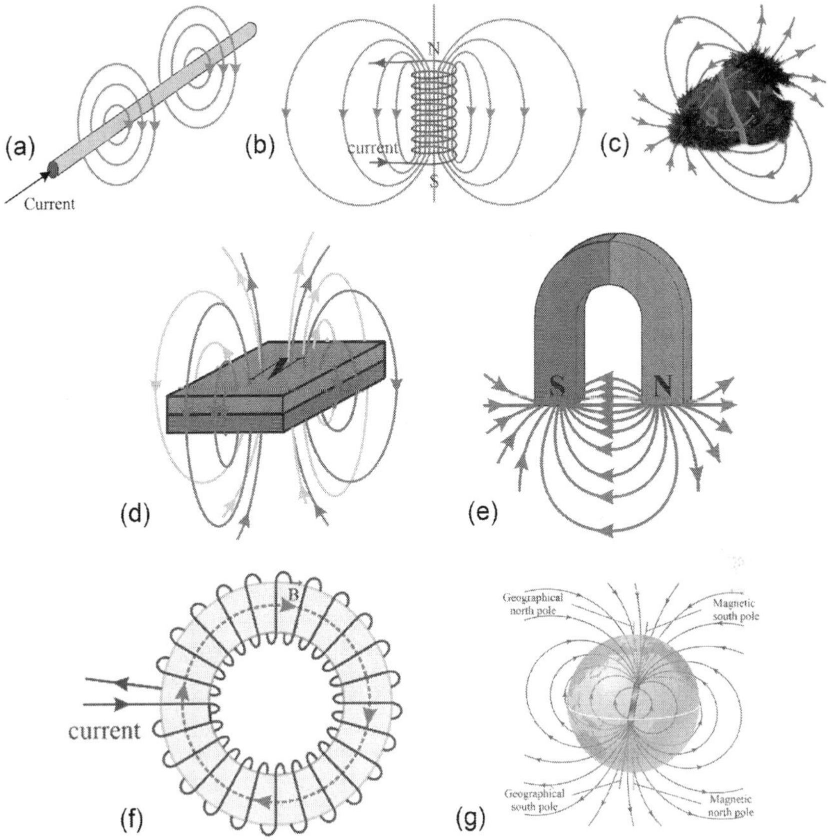

FIGURE 1.10 Magnetic field sources: (a) magnetic field of a current-carrying electrical wire; (b) current-carrying coil; (c) piece of natural magnetite; (d) neodymium magnet for wind turbines; (e) horseshoe-shaped permanent magnet made of steel; (f) magnetic field in a toroidal steel core fitted with a current-carrying winding; (g) Earth's magnetic field.

where

B – magnetic flux density, Wb/m^2;

μ – magnetic permeability of the medium, H/m;

H – magnetic field strength, A/m.

The magnetic permeability μ of iron and other **ferromagnetic substances** is several hundred to a few thousand times greater than that of air and vacuum. That is also the reason why iron is used for B-field amplification.

The high magnetic permeability of ferromagnetic substances arises from their internal structure, which is discussed in more detail below.

Figure 1.10f shows a toroidal steel core inside a current-carrying coil. The magnetic field of the coil exerts a mechanical force on the magnetic fields of the atoms of the core. This causes its iron atoms to adopt a unidirectional orientation in space, as a result of which the magnetic field of the atoms combines with the magnetic field of the coil. This yields a field with a greater total strength. The field is amplified several hundred or even several thousand-fold. The lines of force of the toroidal core's magnetic field close inside the core.

If one were to place a cylindrical steel core in the cylindrical coil shown in Figure 1.10b, this magnetic field would also be amplified, but the lines of force of the magnetic field would emerge into the air from the ends (poles) of the core. Two magnetic poles, as in a permanent magnet, would be generated. Out in the air, the field would look roughly the same as in the case of a coil without a core, except it would be much stronger. Coils with a steel core where the magnetic field of the core intrudes into the air are called **electromagnets**.

If one were to interrupt the current in a coil with a core, thermal motion would cause the orientation of the iron atoms in space to become chaotic, resulting in a smaller total field. Switching the current in a coil on and off enables us to switch a magnetic field of the required strength on and off. This is how electromagnets work.

Figure 1.10g shows the Earth's magnetic field. Near the geographical north pole, there is a magnetic south pole inside the earth. The north pole end of the small permanent magnet used in a compass always points towards the Earth's magnetic south pole (which is near the north pole). Compasses work because the opposite poles of two magnets always attract.

In steel cores with a certain composition, the magnetic field does not disappear after the current in the coil is switched off. After magnetisation by a current, such cores become **permanent magnets**. The materials of permanent magnets are unique in that thermal motion is unable to significantly change the orientation of their atoms in space, as a result of which the magnetic field of the core remains in place even after the current disappears from the surrounding coil.

Objects can be placed in a space occupied by a magnetic field in the same way as in a space where there is no magnetic field, with the exception of ferromagnetic substances. When a body composed of ferromagnetic substances is inserted into a magnetic field, another field, extending outward from the body, is generated inside the ferromagnetic substance. A mechanical force arises between these two fields. The inserted body will be pulled towards the source of the magnetic field. In order to study the arising forces, as well as other magnetic field-related phenomena, it is necessary to be able to describe magnetic fields.

Mathematical description of fields

Describing magnetic fields and other types of fields requires indicating the strength and direction of the field for each point in space. Describing non-directional quantities, such as temperature, only requires indicating the strength of the field. Here is a simple example.

Say we need to investigate the temperature of the exterior wall of a room. We can use a contact thermometer to measure the temperature at different points of the wall and record the results of our measurements in a figure of the wall drawn on paper. It is likely that the wall will be warmer under the ceiling than near the floor, and any windowpanes will be cool. Doing this would give us an overview of the distribution of the temperature on the wall. These kinds of temperature analyses are known as two-dimensional temperature fields (2D field). The relevant dimensions are the width and height of the wall.

If we were interested in the air temperature at all points in the room and if we were to measure it, we would get a three-dimensional temperature field (3D field). Temperature has no direction. Non-directional fields are called **scalar fields**.

If we were to open the door and a window in our room, there would be a draught. We could then measure the velocity of the air at all points in the room. This would give us a three-dimensional field of the velocity of the air. Here, the direction of motion of the air is relevant. Velocity is a directional quantity, i.e. a vector quantity. Thus, we would have a **3D vector field** of the movement of the air. The magnetic fields in electrical machines are also 3D vector fields. To better understand changing magnetic fields, we can imagine them as analogous to wind whose direction and magnitude are changing in a space.

It follows from the foregoing that fields generally describe the distribution of a physical phenomenon (such as temperature) in space. In this respect, a field is a geometric and mathematical concept. However, fields are also used as a physical concept when talking about a magnetic field or an electric field. For example, a field of a magnetically induced force is measured in Tesla (unit symbol T). It might be said, for example, that the magnetic induction is 0.5 T or 1 T.

By definition, 1 T is a magnetic induction in which a force of 1 N per 1 m of length acts on a wire positioned perpendicularly to the direction of the field when an electric current of 1 A passes through the wire.

From this, it follows that a magnetic field is defined via a measurable mechanical force. And a magnetic field itself is a physical phenomenon that exists in reality, but can only be measured indirectly.

The strength of a magnetic field can be 0.5 T at one point in space and 0.6 T at another. If we wish to give a name to the distribution of a magnetic field in space, it should be 'magnetic field'. However, it is usually not called that. In common parlance, the term 'magnetic field' refers to two different things: first, a physical phenomenon and, second, the distribution of this phenomenon in space. The fact that we use language in this way is unfortunate (terminology should be unambiguous), but needs to be kept in mind.

In electrical engineering, a magnetic field is considered to be a mathematical vector field. This approach allows us to both design electrical machines as well as to study the processes that take place in electrical machines. Nevertheless, the physical side of magnetic fields is still of interest to us here.

Physical nature of magnetic fields

What does a magnetic field consist of? There are probably two components: space and energy put into that space. Energy was already discussed above. At the end of the 20th century, physicists reached new positions on empty space or vacuum (*Brian Greene. The Fabric of the Cosmos: Space, Time, and Texture of Reality. Publisher Alfred A. Knopf, 2004*).

Today, empty space is no longer considered to be absolutely empty. Instead, it is known to be composed of tiny particles called **superstrings** and **branes**. These particles are 10^{14} times smaller than an atomic nucleus. Compared to a superstring, an atomic nucleus is about as large as the Earth is compared to a speck of dust. Empty space also has a weight – albeit minuscule. However, because the universe contains an enormously greater amount of empty space than it does matter, empty space still weighs more in total than matter.

Superstrings are thread-like in structure. They have only a single dimension: length. They have no diameter. The ends of superstrings stick to branes, forming something like a three-dimensional web. These superstrings are the universal building blocks of the world. These tiny threads can oscillate in various ways like the strings of a musical instrument. When superstrings are supplied with energy, their oscillation changes, which can result in their assemblage turning into an electron, a quark, or another elementary particle with mass and energy. Therefore, superstrings are also energy storage mechanisms. The oscillation of superstrings can also become such that we would see empty space, or vacuum, at the macro level. By adding energy in a certain way, we could also get a space with a magnetic field.

In principle, vacuum and energy can form all particles of matter and the whole visible world. Superstrings are so small that we are not yet able to verify their existence directly today. So instead, we are attempting to do so indirectly. Space that we have previously believed to be a void consists of elements with finite dimensions.

The generation of a magnetic field in an empty space surrounding a current-carrying wire can also be described in the light of the superstring theory.

The magnetic field of a current increases as the current increases. The rate of increase of a current in an electrical circuit is given by the equation

$$\mathrm{d}i \,/\, \mathrm{d}t = U \,/\, L. \tag{1.24}$$

Here, U is the component of the voltage applied to the circuit that drops on the inductance (there exist also voltage components that drop on resistance and capacitance). L is the inductance of the circuit. Inductance describes the location of the electrical circuit in space. Inductance will be discussed in more detail later on.

Applying voltage to an electrical circuit creates an electric field in the circuit and in the space around the circuit. The electric field causes both the current and the magnetic field surrounding the current to increase. Energy is introduced into the electrical circuit and the magnetic field.

The energy enters the electrical circuit through wires. However, some of this energy leaves the circuit wire via superstrings into space and becomes stored in the space around the wire. The oscillation energy of superstrings in the space near the wire manifests as a magnetic field at the macro level. The processes occurring in an electrical circuit are always accompanied by the emergence of fields and changes in the structure of the space around the electrical circuit.

1.5 CONSTRUCTIONAL MAGNETIC POLE AND FLUX POLE

The term 'pole' is used to talk about two different things in electrical machines. A pole can be a component of a magnetic circuit with a winding coil. Such poles (Figure 1.11a) are visible mechanical constructions which emit an invisible magnetic flux. These mechanical constructions may be referred to as **constructional magnetic poles**. However, a pole can also mean the invisible magnetic flux together with the surface on the mechanical construction whence the magnetic flux exits (in the case of a north pole) or into which the magnetic flux enters (in the case of a south pole). This entry or exit point of the invisible magnetic flux is the magnetic flux pole or simply **flux pole**. A flux pole is the surface of a constructional pole (surface of the pole shoe in Figure 1.11) whence the magnetic flux exits or into which it enters. Constructional poles along with their flux pole (Figure 1.11b) move in unison (e.g. with the rotor of a synchronous machine), but the surface of a flux pole remains stationary with respect to the pole construction.

Structurally, constructional poles can be either electromagnets, as in Figure 1.11, wherein the winding coils of a magnet are supplied with direct current, or **permanent magnets**. The advantage of permanent magnets is that they do not need to be supplied with electricity, but their downside is that their magnetic flux is not adjustable.

Two different designs exist for the electromagnetic poles of electrical machines. They are called **salient poles** and **non-salient poles**. The constructional poles shown in Figures 1.11 and 1.12a are salient poles.

In the case of non-salient poles (Figure 1.12b), the outer surface of the rotor has grooves along the axis of the machine, called **slots**. Winding coils are fitted in the slots in a particular arrangement. Because the rotor rotates, current is supplied to the winding coils via slip rings on the shaft. Stationary brushes are pressed against the rotating slip rings. In Figure 1.12b, part of the winding housed in the

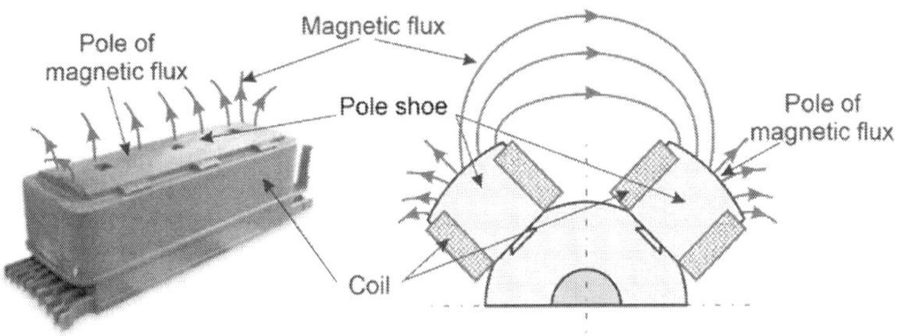

FIGURE 1.11 (a) Constructional magnetic pole; (b) cross-sectional view of two constructional magnetic poles.

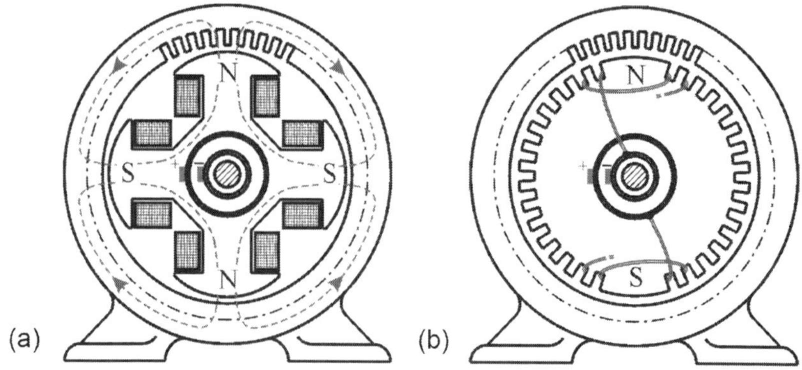

FIGURE 1.12 Salient poles (a) and non-salient poles (b) on the rotor of a synchronous machine.

slots is not shown. Two flux poles are formed on the surface of the rotor (N and S in Figure 1.12b). These flux poles rotate along with the rotor.

1.6 DESIGN AND OPERATING PRINCIPLE OF STEPPER MOTORS

Figure 1.13 shows a cross-sectional view of a stepper motor with a permanent magnet-type rotor. The rotor is a permanent magnet positioned on the axis of the machine. The inner surface of the stator holds four constructional magnetic poles fitted with winding coils, installed at 90° intervals. The magnetic flux of these poles can be switched on and off and the direction of the magnetic flux can also be changed.

The coils of stator poles 1 and 3 are connected in series, the current passing through them is indicated as i_a, and they form one pole pair. By changing the direction of the current, you can change the direction of the magnetic flux. When the direction of the current is positive (as in Figure 1.13), pole 1 is the south pole and pole 3 is the north pole. When the current i_a flows in the opposite (negative) direction, pole 1 is the north pole and pole 3 is the south pole. Similarly, a second pole pair is formed by poles 2 and 4, the current in the windings of which is indicated as i_b.

When current i_a is switched on and flows in the positive direction and current i_b equals zero, the attractive forces present between the pole pairs of the stator and the rotor are such that the north pole of the rotor is held facing the south pole of the stator and the south pole of the rotor is held facing the north pole of the stator (as in Figure 1.13).

The rotor maintains this position as long as the currents in the stator do not change. The rotor can be made to move in steps by changing the currents in the stator. This is illustrated in Figure 1.14. In the motor in question, this motion can either occur in 90° steps, which is known as full-step drive, or in 45° steps, which is called half-step drive. For full-step drive, current is supplied to a single pole pair at a time, causing the rotor to assume a position where the opposite poles of

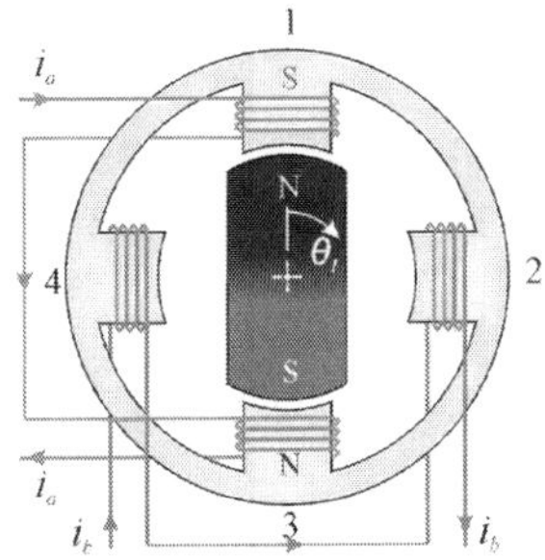

FIGURE 1.13 Schematic cross-sectional view of a stepper motor with a two-pole permanent magnet rotor.

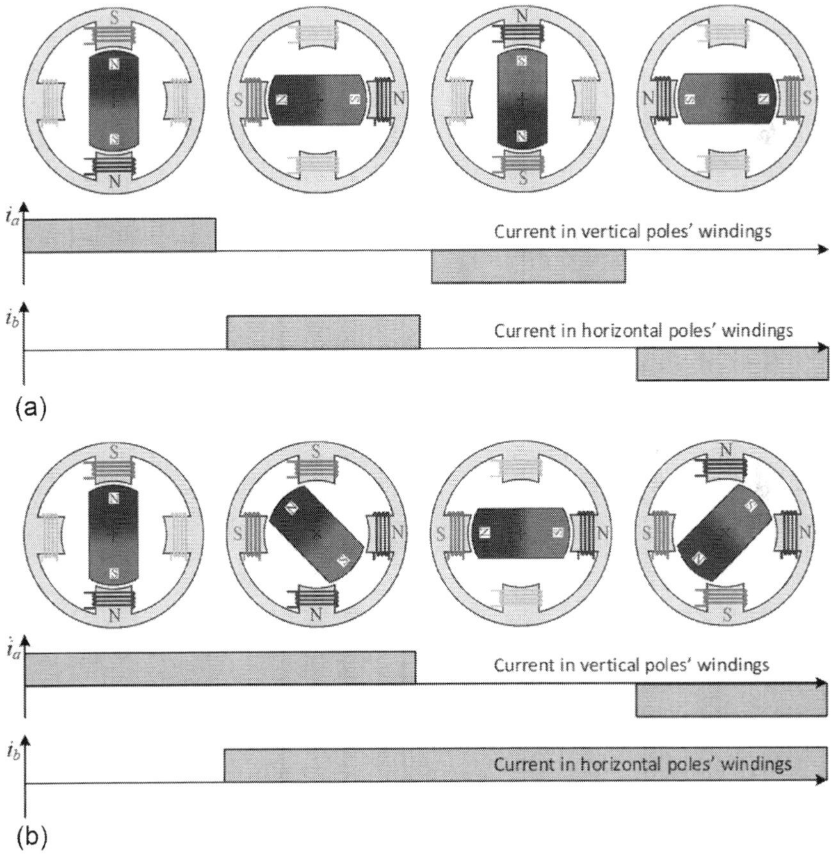

FIGURE 1.14 Stepwise movement of the rotor of a stepper motor according to the switching of the stator poles; (a) full-step drive; (b) half-step drive.

the rotor and the stator are facing each other (N and S opposite each other, separated by an air gap).

Figure 1.14 shows the switching on and off and changing of the direction of stator winding currents i_a and i_b. In addition, it shows how the changing of the currents in the stator windings causes the flux poles to shift on the stator, thereby making the rotor move.

In Figure 1.14a, the south pole of the stator is initially located at the top, then on the left, then at the bottom, and finally on the right. In three steps, a movement of 270° counterclockwise occurs. The changes in the position of the magnetic poles of the stator cause the rotor to rotate to an equal extent and in the same direction (counterclockwise). A stepper motor is a special type of synchronous motor in which the stator's magnetic field is made to rotate by changing the current in the windings of the constructional poles by means of switches. Semiconductor switches are used as the switches.

If we wished for the next step from the starting position to occur clockwise instead of counterclockwise, we would need to make current i_b flow in the negative direction instead of the positive direction. This would cause the south pole to shift not to the left, but to the right. The direction of rotation can be changed after any step.

In half-step drive (Figure 1.14b), unlike in full-step drive, the current i_a is not changed in the second step, while current i_b is also switched on in the positive direction. This results in two south poles existing next to each other on the stator: one at the top and another on the left. In response, the rotor then rotates 45°, stopping between the two south poles on the stator. Table 1.1 lists the changes in the currents in the stator windings as the stepper motor rotates in 45° increments over one revolution.

Stepper motors that are in use also include ones where the rotor is not a permanent magnet, but is made of magnetically soft iron. Instead of featuring a cylindrical design, such rotors have two opposite sides cut off (Figure 1.15). The stator, meanwhile, has three pairs of constructional poles (six poles in total). In each pair of poles, the windings are connected in series. The current is switched on and off in these windings, but – unlike in the permanent magnet-type rotor above – the direction of the current is not changed.

When current i_a is switched on in poles 1 and 4, the rotor assumes the position shown in Figure 1.15. Then the magnetic field leaving pole 4 passes through the rotor and the two air gaps and enters pole 1. In this position of the rotor, the

TABLE 1.1

Stator currents during the movement of the rotor in a stepper motor in 45° increments over one revolution

θ_1	0°	45°	90°	135°	180°	225°	270°	315°	360°
i_a	+	+	0	–	–	–	0	+	+
i_b	0	+	+	+	0	–	–	–	0

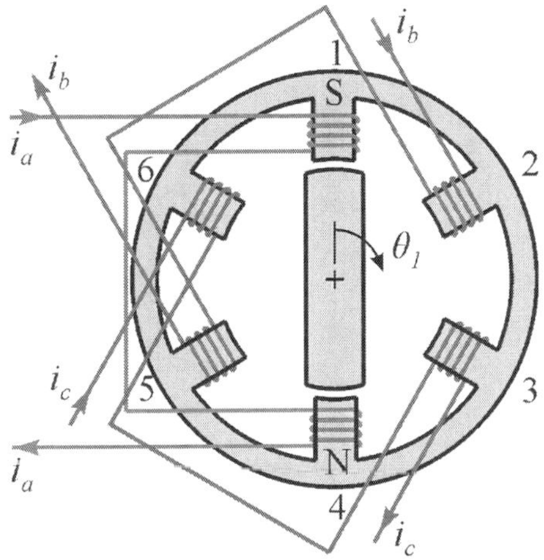

FIGURE 1.15 Schematic cross-sectional view of a stepper motor with a reluctance rotor.

resistance, or **reluctance**, of the magnetic circuit between poles 1 and 4 is the smallest, as the path of the magnetic lines of force through the air is the shortest there. Air is about 1,000 times harder to pass through for magnetic fields than iron. If, with the rotor in the position depicted in the figure, current were supplied to poles 2 and 5, the magnetic lines of force travelling from pole 5 to pole 2 would have to take a much longer path through the air than the lines of force going from pole 4 to pole 1. The reluctance of the magnetic circuit would then also be much greater between poles 2 and 5 than between poles 1 and 4. The reluctance between the poles depends on the position of the rotor, which is why these types of motors are called **reluctance motors**. The operation of reluctance motors is based on the fact that the resistance presented to the magnetic flux by the section of the magnetic circuits that comprises the air gap depends on the position of the rotor.

A magnetic field always exerts on the parts of the magnetic circuit a mechanical force that attempts to reduce the length of the path of the lines of force. Where the path of lines of force passing through the air is shorter, the magnetic resistance, or reluctance, of the magnetic circuit is lower. If the rotor of the reluctance stepper motor were initially in the position shown in Figure 1.15 and we were to turn off the magnetic flux of poles 1 and 4 and then turn on the magnetic flux of poles 2 and 5 (by switching on the current i_b), then the distances between pole 5 and the lower end of the rotor and pole 2 and the upper end of the rotor would at first be relatively long. These distances would then be traversed by the magnetic flux, which would generate mechanical forces that would attempt to reduce the distances between the ends of the rotor and the flux-producing poles 2 and 5. The forces would turn the rotor clockwise until the distances between the ends of the rotor and poles 2 and 5 are minimised. This would result in the rotor rotating clockwise by one step of 60°. If instead of poles 2 and 5 we were to supply current to poles 6 and 3, the rotor would rotate counterclockwise by one step.

If a magnetic flux were generated simultaneously in two pole pairs, the rotor would stop between the two magnetic flux-producing poles. For example, if we were to simultaneously switch on currents i_a and i_b, the rotor would stop between poles 1 and 2. This would involve the rotor shifting 30° away from pole 1, i.e. by half a step.

Table 1.2 lists the changes in the currents in the stator windings as the reluctance stepper motor rotates in 30° increments over half a revolution.

Figure 1.16 shows a variant of the stator and rotor of a reluctance stepper motor with four pole pairs.

Stepper motors are designed with steps ranging from 90° to 0.72°. Small steps require numerous poles, and a single shaft can even have several pole systems. The number of steps can range from 100 to 10,000 per second. If the number of steps exceeds the upper limit for the given type of motor, so-called step losses will occur. This means that while the corresponding electrical impulse is still transmitted to the winding, it is so short that the rotor is unable to complete the required step within its duration.

TABLE 1.2
Stator currents during the movement of the rotor in a reluctance stepper motor in 30° increments over half a revolution

θ_1	**0°**	**30°**	**60°**	**90°**	**120°**	**150°**	**180°**
i_a	+	+	0	0	0	+	+
i_b	0	+	+	+	0	0	0
i_c	0	0	0	+	+	+	0

FIGURE 1.16 Stator and rotor of a reluctance stepper motor.

Stepper motors with a permanent magnet-type rotor usually utilise steps of up to 5°. These motors have a higher inertia than reluctance stepper motors but develop braking and holding torque when power is removed from the motor. In reluctance stepper motors, on the other hand, if you were to remove the power, the motor would go into idle mode.

Reluctance stepper motors are preferred at lower steps and higher speeds.

Multi-pole stepper motors have a relatively low energy conversion efficiency and are normally utilised in low-power automation systems. They are easy to control and monitor via a computer.

Reluctance motors with a low number of pole pairs are also used in medium-power variable-speed drives in place of induction motors with a frequency converter.

1.7 MAGNETIC FLUX POLES MOVING IN A STATIONARY MAGNETIC CIRCUIT

In order to make a stepper motor rotate, it is necessary to change the currents in the windings of the stator's poles so that the position of the flux poles begins to shift along the circumference of the stator. In other words, the flux poles begin to rotate step by step, while the constructional poles remain stationary. To do this, the stepper motor must incorporate an electronic switch for switching the current in the windings on and off.

Alternatively, rotating flux poles can be attained without a switch by using three-phase alternating current and a three-phase winding installed in the slots on the inner surface of the stator in a particular arrangement (Figure 1.17). The flux poles would then be moving within a stationary magnetic circuit. **Changing the intensity and direction of the current** in the coils positioned in the slots **changes the position of the flux poles on the inner surface of the stator along the circumference.**

The changing of the instantaneous values of the winding currents of the three phases generates, on the inner surface of the stator, a total magnetic field with rotating poles (Figure 1.17 shows the position of the poles at a moment when the instantaneous value of the current is maximal in phase A).

The rotating magnetic field is obtained by means of three windings. These windings are supplied with voltage from the different phases of a three-phase alternating current system (Figure 1.17). Each winding consists of two parts positioned in separate slots. The winding fed from phase A may be positioned, for example, in slots 1 and 4 (Figure 1.18). These connections between winding parts positioned in different slots are called **end connections**. For a better understanding of the arrangement of the windings, see Figure 1.18b, which shows a stator split by the plane m-m and opened out.

The starting points of the windings are marked with the letters A, B, and C and are supplied with voltage. The ending points of the windings, marked X, Y, and Z, are joined together (see also Figure 1.17). The slots are positioned along the

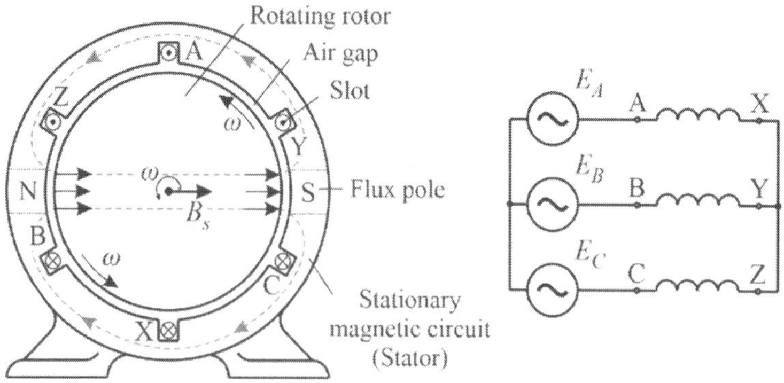

FIGURE 1.17 Cross-sectional view of and power supply diagram for the stator of an alternating-current machine; the direction of the magnetic field corresponds to the moment of maximum current in phase A.

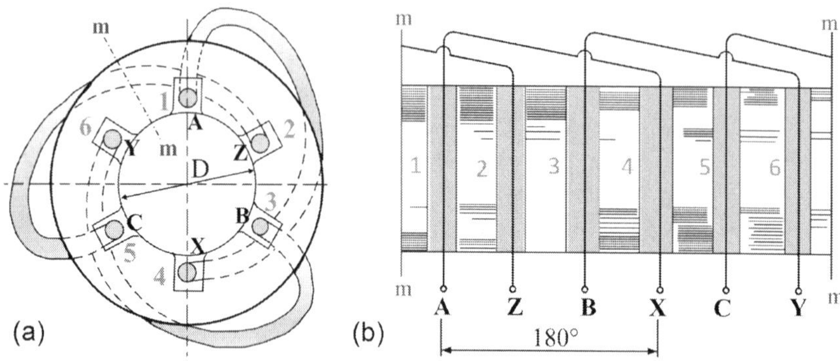

FIGURE 1.18 (a) Spatial distribution of end connections between windings going from one slot to another; (b) same end connections with windings in a stator shown as split and opened out.

circumference in the sequence A-Z-B-X-C-Y, with the starting and ending points alternating. Only such a sequence ensures a rotating magnetic field. This is the simplest form of a three-phase winding. There is only one winding coil in each phase and each coil only has one turn of wire. Electrical machines usually have more slots, more than one coil in each phase, and multiple turns of wire in each coil. However, even the simplest three-phase winding will generate a rotating electric field in the same way as more complex windings.

When current flows through a winding coil (e.g. A-X), the current generates a magnetic field that is perpendicular to the plane of the coil. When current flows through all three coils, three magnetic fields with different orientations are generated. As the coils are spaced 120° apart, so are the magnetic fields. The joining of the three fields of the three currents results in a single total field. This is illustrated in Figure 1.19.

At the instant $t = 0$, the current in phase A is equal to zero, the current in phase B is negative and equal in intensity to the current in phase C, which is positive (Figure 1.19a). These current directions are also shown in the cross-sectional view of the stator, where a dot indicates that the current is flowing towards the reader and a cross indicates that the current is flowing towards the plane of the figure (Figure 1.19b). In this case, a positive current is introduced from the starting point of the winding and a negative current from the ending point of the winding. Thus, the current in the first side of winding C is flowing towards the plane of the figure, and the current in the second side of the winding is leaving the plane of the figure and flowing towards the ending point Z of the winding. As the current in winding B is negative, it enters from the ending point Y of the winding and leaves from the starting point B. Due to having the winding coils arranged in this way, at the instant $t = 0$, in the two slots on the left, the current is flowing towards the plane of the figure, while in the two slots on the right, it is leaving the plane of the figure. The figure shows the resulting lines of force of the magnetic field around these current-carrying wires. The lines of force leave the inner surface of the stator at the top of the figure, which means that this is the location of the north pole (N) of the flux. The same lines of force enter the inner surface of the stator at the bottom of the figure – the location of the south flux pole (S). The three-phase winding in question generates a single pole pair.

At the instant $t = 30°$, the currents in the three-phase current system have obtained the following new values (Figure 1.19a): phase A has a positive current equal to half the peak value, the negative current of phase A has increased to the peak value, and the positive current of phase C has dropped to half the peak value. The currents are now entering slots C, Y, and A and leaving slots Z, B, and X. As

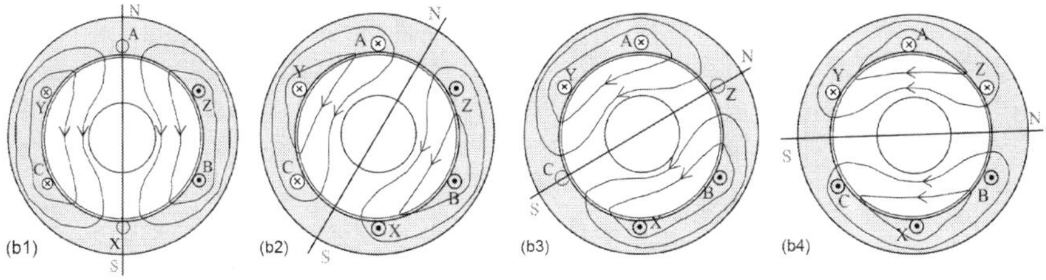

ωt = 0; iA = 0;
iB = −86.6; iC = 86.6

ωt = 30°; iA = 50;
iB = −100; iC = 50

ωt = 60°; iA = 86.6;
iB = −86.6; iC = 0

ωt = 90°; iA = 100;
iB = −50; iC = −50

FIGURE 1.19 Change in the instantaneous values and direction of the currents of a three-phase current system over time (a) and the corresponding change in the direction of the currents in the stator winding and the change in the direction of the total magnetic field of the currents at four time instants (b).

a result, the axis of the total magnetic flux has shifted clockwise by 30°, compared to the instant $t = 0$.

By the instant $t = 60°$, the currents of the three-phase current system have once again obtained new values, and the axis of the total magnetic flux of the currents has shifted by 60°, compared to the instant $t = 0$. The increase in the angle of the change of direction of the magnetic flux is equal to the progression of time in electrical degrees. When one full alternating current cycle has elapsed, the change in the direction of the magnetic flux will be 360°, or one complete revolution. Therefore, **the rotational speed of the field is proportional to the frequency of the alternating current**. At a frequency of 50 Hz, the magnetic field will be rotating at 50 rps, or 3,000 rpm.

A more detailed study of how the total magnetic flux of the coils of the three phases changes over time would reveal that the resulting magnetic flux density is a constant, measuring 1.5 times the flux density produced by a single coil at the instant of maximum (peak) current. In a stationary stator, the three-phase winding described above generates two flux poles which, as they rotate, are constantly shifting in the iron in the stator and extend into the air gap. This is one way to generate so-called travelling flux poles in iron.

The rotational speed of the field is what determines the rotational speed of the synchronous machine, as we will see in the following sections. A rotational speed of 3,000 rpm is often too high. The rotational speed can be reduced two-fold by using the following method. Instead of distributing the set of three-phase stator windings across the entire circumference of the stator (i.e. 360°), it is distributed across only half the circumference of the stator (i.e. 180°), as shown in Figure 1.20a. With the windings arranged in this way, the magnetic field will only travel 180° over one alternating-current cycle, which means that the travel speed of the field will be two times lower. As a result, the rotational speed of the machine will also be two times lower. As speed decreases, however, so does power. To prevent this from happening, a second three-phase set of winding coils (Figure 1.20b) may be fitted on the stator, which will increase the torque by two times. As a result, the power will remain the same, the rotational speed will be twice as low, and the torque will be twice as high (compared to the case shown in Figure 1.17).

Placing two sets of three-phase windings on a stator results in four rotating magnetic poles (Figure 1.20c) with the sequence N–S–N–S. Coils of the same phase in different coil sets are usually connected in series (A1–X1–A2–X2), as shown in the diagram of windings opened out onto a plane in Figure 1.20d. The diagram has been obtained from a cross-sectional view of the stator between coil sides A1 and Y2. The current passes from one slot to another via end connections outside the slots. Stators may be fitted with more than two three-phase coil sets to achieve an even greater decrease in the rotational speed of the field.

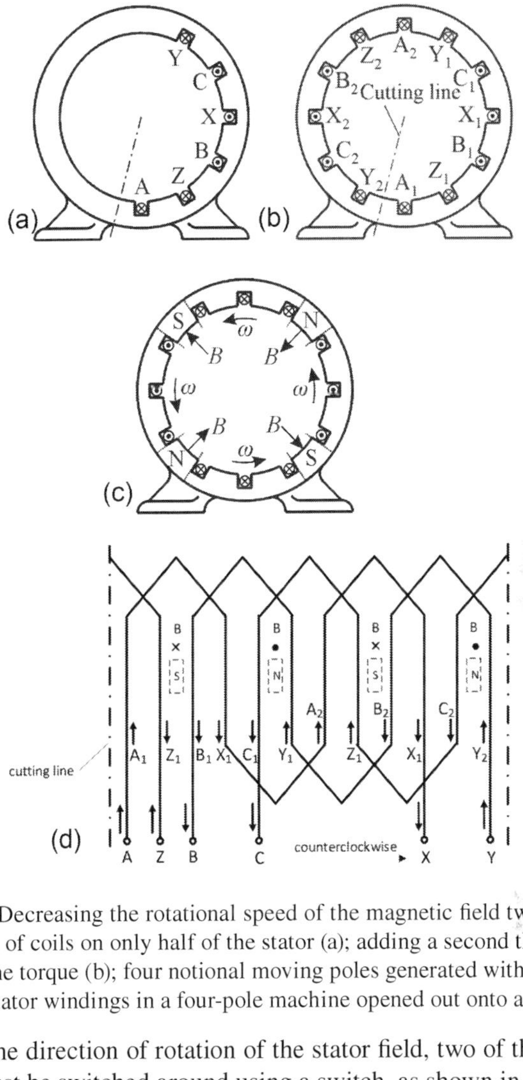

FIGURE 1.20 Decreasing the rotational speed of the magnetic field twofold by placing a three-phase set of coils on only half of the stator (a); adding a second three-phase set of coils to double the torque (b); four notional moving poles generated with two sets of coils (c); diagram of stator windings in a four-pole machine opened out onto a plane (d).

To change the direction of rotation of the stator field, two of the three power-phase wires must be switched around using a switch, as shown in Figure 1.21.

When the switch is in position 1, the power supply of electromotive force E_B is connected to the start of the machine's phase B winding and the power supply of electromotive force E_C to the start of the machine's phase C winding (Figure 1.21a). This causes the field to rotate counterclockwise. When the switch is set to position 2, the power supply of electromotive force E_B is connected to the winding that was originally the phase C winding (original phases are indicated in brackets in the figure). In essence, the phase B and C windings swap places on the stator. The former C and Z are now B and Y, and the former B and Y are now C and Z. Since the field on the stator rotates in the direction A–Z–B–X–C–Y–A (as seen

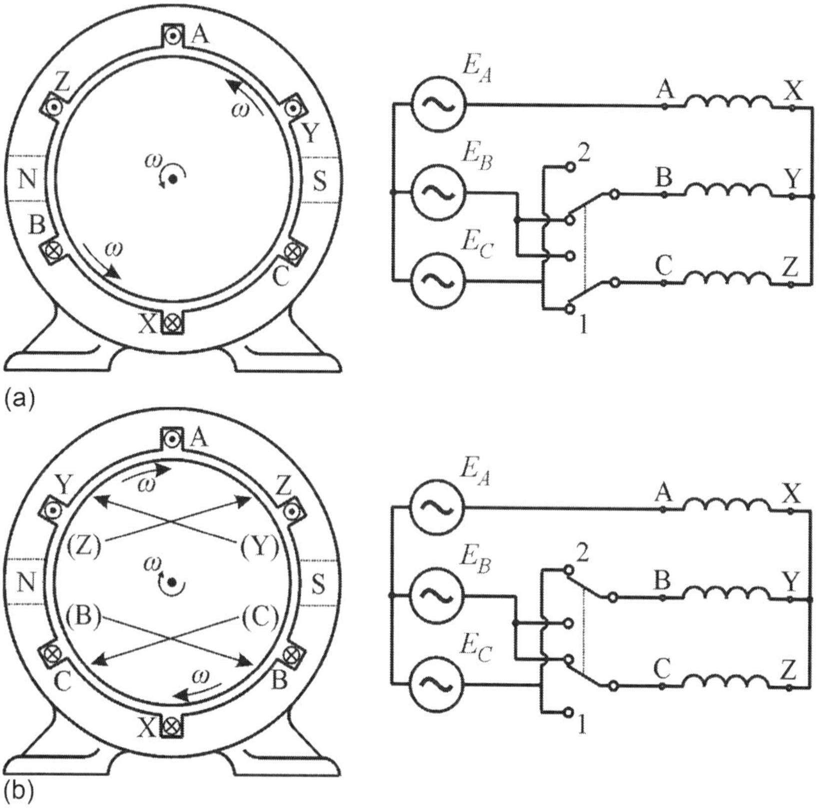

FIGURE 1.21 Swapping over from counterclockwise rotation (a) to clockwise rotation (b) by switching two power phases around.

above), moving in this order in Figure 1.21b corresponds to a clockwise direction of rotation, which is opposite to the initial direction.

1.8 DESIGN AND OPERATING PRINCIPLE OF SYNCHRONOUS MACHINES

In synchronous machines, the stator is fitted with a three-phase AC winding, which generates a rotating magnetic field extending from the inner surface of the stator into the air gap. This magnetic field consists of one or more pairs of flux poles.

Figure 1.22 shows one possible design for a synchronous machine with two pole pairs (i.e. four poles). The rotor has four constructional poles. These poles are shown in the figure without windings. Such a rotor with the windings and the magnetic fields generated by the windings is shown in Figure 1.12a. The magnetic field leaving the north poles passes through the air gap, enters the stator, and then exits the stator above the south pole to return to the rotor via the air gap there.

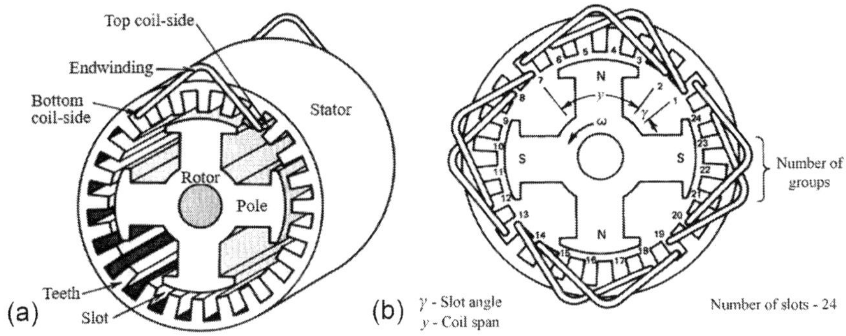

FIGURE 1.22 Design of a synchronous machine with two salient pole pairs (a) and arrangement of one phase of a two-layer winding on the stator (b).

The stator has 24 slots divided between three phases to provide 8 slots per phase winding (Figure 1.22b). While in the simplest winding shown in Figure 1.14 there was only one coil side in each slot, here there are two coil sides in each slot. Such windings are called double-layer windings.

Double-layer distributed winding

In double-layer windings, each slot houses two sides belonging to different coils (see Figure 1.22b, slot 1). Moreover, the coils usually consist not of one turn, but of a number of them (several dozen turns). The coils are arranged in such a way that one side of each coil is positioned at the bottom of a slot, making it the lower coil side, while the other side of the same coil is positioned in another slot, above the lower side of another coil, making it the upper coil side.

Another difference compared to the simplest possible three-phase winding (see Figure 1.18) is that, instead of one coil, there are several coils (2, 3, 4, etc.) in each phase. Such multi-coil windings are known as **distributed windings**. The stator of the machine shown in Figure 1.22b has 24 / 4 = 6 slots per pole (the machine has four poles). To obtain a rotating field, the currents of all three phases must be used to generate each pole field (as shown above). Thus, in this case, we have 6 / 3 = 2 slots per phase per pole. Figure 1.22b shows the winding of only a single phase. One phase of each pole here is made up of a group of two coils. In a double-layer winding, the number of slots and coils is equal: each coil has two sides and each slot also holds two coil sides. Increasing the number of coils and slots allows to reduce the depth of the slots and to cover the entire inner surface of the stator with slots, which is advantageous in terms of reducing losses as well as the weight and dimensions of the machine.

However, with a large number of slots, positioning and connecting the winding coils is more difficult. To facilitate the process, winding diagrams depicting the inner surface of the stator together with the windings are used. In these diagrams, the stator is shown as split and opened out onto a plane. Figure 1.23a presents a diagram of the winding of one phase of the stator depicted in Figure 1.22, with the stator shown as split between slots 1 and 24. The slots have not been drawn, but their numbers are given in the diagram. Lower coil sides are shown with dashed lines. Figure 1.23 shows all four phase A coil groups connected in series. In principle, parallel and mixed connections are also possible.

When preparing a coil connection diagram (Figure 1.23a), care should be taken to ensure that the directions of the currents of the coil sides are the same as for the simplest possible winding, the arrangement of which on a stator with 24 slots is shown in Figure 1.23b. There, the winding of a single phase occupies 4 slots, with the windings of three phases occupying 12. 12 slots remain empty, because the winding is not distributed (in reality, this variant is of course not very practical, but it is useful for explaining the distribution of windings). By comparing the distributed and concentrated windings shown in Figure 1.23, we see that the four coil sides A_1, X_1, A_2, and X_2 of the concentrated winding are mirrored by four so-called **phase belts** A_1, X_1, A_2, and X_2 in the distributed winding. For example, the phase belt X_1 is made up of the winding coil sides in slots 18, 19, and 20, and this phase belt is equivalent to the single coil side of the concentrated

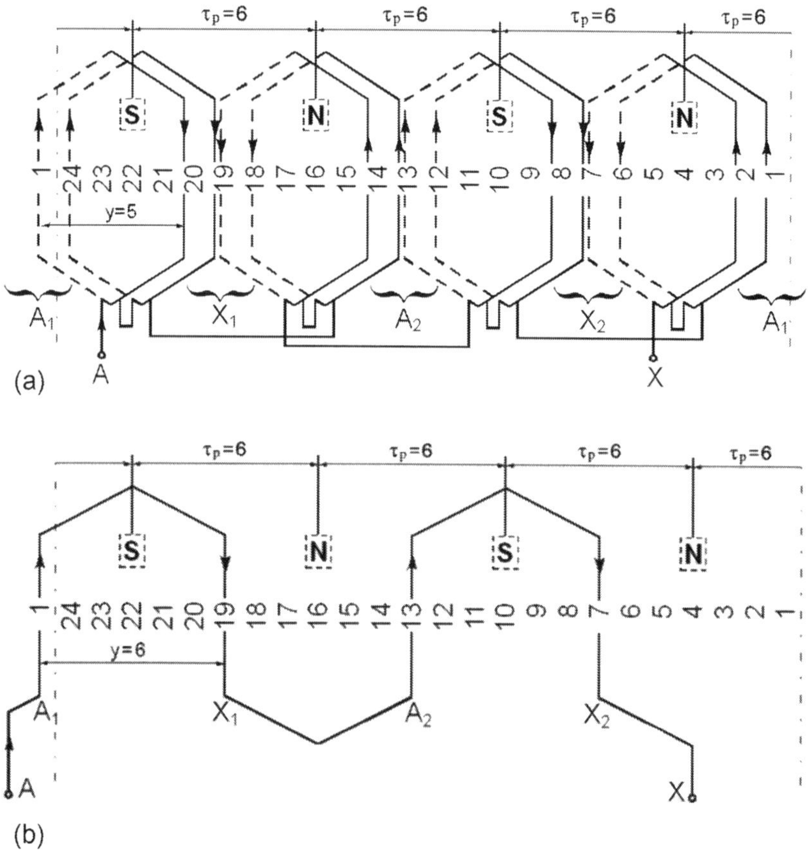

FIGURE 1.23 Phase A winding in a four-pole machine: (a) double-layer distributed winding; (b) simplest (concentrated) single-layer winding.

winding in the 19th slot. In the case of the distributed winding, three slots house four coil sides. In each of these coil sides, the direction of the current is identical to that of the current in the single coil side of the concentrated winding. This identical direction of current is ensured by the appropriate selection of connections between the starting and ending points of the winding coils. The currents in the distributed winding produce the same distribution of magnetic poles on the inner surface of the rotor as the currents in the concentrated winding.

The distance between magnetic poles along the inner surface of the stator is called the **pole pitch** and is denoted by τ_p. Pole pitch is measured both in metres and the number of slots. Pole pitch in units of length is given by

$$\tau_p = \frac{\pi D}{2p}, \tag{1.25}$$

where D – inner diameter of the stator;

p – number of pole pairs.

Pole pitch in slots is given by

$$\tau_p = \frac{Z}{2p}, \tag{1.26}$$

where Z is the number of slots. In Figure 1.23, $\tau_p = 6$ slots.

Stator windings are also characterised by the number of slots q per pole and phase:

$$q = \frac{\tau_p}{m} = \frac{Z}{2pm}, \tag{1.27}$$

where m is the number of power phases (in most cases, $m = 3$, but $m = 2$ is not uncommon either).

In the winding shown in Figure 1.23, $q = 2$, but the winding of each phase belt occupies three slots. Looking at the phase belt X_1, we see that slot 18 holds one coil side, slot 19 holds two, and slot 20 again holds one. Once the windings of all three phases have been placed on the stator, a coil side of phase belt C_1 will be positioned above the phase A coil side in slot 18 and a coil side of phase belt B_1 will be positioned below the phase A coil side in slot 20. Thus, slots 18 and 20 will include coil sides from the two adjacent belts, while in slot 19 both coil sides will belong to the same phase belt (see also the winding arrangement in Figure 1.18b).

The distance between the two sides of a coil is called the **winding pitch** y, which can also be measured both in units of length and the number of slots. In the simplest possible winding shown in Figure 1.23b, the winding pitch y is equal to the pole pitch τ_p. Such a pitch is called a **full pitch**. In distributed windings, the winding pitch may differ slightly from the pole pitch. Usually, the winding pitch is made to be a little shorter than the pole pitch. In Figure 1.23a, the pole pitch τ_p

= 6, while the winding pitch $y = 5$. Such a pitch is called a **short pitch**. Another relevant term here is **relative winding pitch** β :

$$\beta = \frac{y}{\tau_{\mathrm{p}}}. \tag{1.28}$$

The relative winding pitch is the ratio between the winding pitch and the pole pitch. The relative winding pitch of the winding shown in Figure 1.23a is $\beta = 5 / 6 = 0.833$.

Operating principle of synchronous motors

The windings of all three phases placed on the stator are connected in star. The machine functions as a motor when a three-phase voltage is applied to the windings (Figure 1.24b).

The winding of the poles on the rotor of a synchronous machine is called the **excitation winding**. The winding on the rotating rotor is supplied with power via two **slip rings**, which rotate along with the shaft, and stationary **brushes** that are in contact with the rings (Figure 1.24a). A more complicated brushless excitation winding supply system is also used, where the excitation current is obtained by rectifying the current of a small alternator positioned on the rotor shaft.

When current is present in both the stator winding and the excitation winding, both currents generate their own magnetic field, which combine. When the fields combine, forces are generated which pull each south pole on the rotor to face one of the north poles (blue) produced on the stator and each north pole on the rotor to face one of the south poles (red) produced on the stator, as shown in Figure 1.25.

Figure 1.25 shows that four closed loops of magnetic lines of force arise in the machine. The total current of each of the loops of magnetic lines of force is composed of the currents of both the stator winding and the excitation winding. What is more, the directions of the currents in all of the wires in a loop are identical at all times. The magnetic field leaving the north pole of the rotor passes through the air gap to enter the south pole of the stator, branches into two, and travels along both branches down the magnetic circuit of the stator all the way to the north pole of the stator, where it passes through the air gap again and enters the south poles of the rotor. In the rotor, the loops of the magnetic lines of force close.

When the magnetic field of the stator starts to rotate (e.g. as the frequency is raised above zero by means of a frequency converter), the rotor, too, starts to rotate. Due to the pulling forces between the north and south poles, the constructional poles of the rotor 'stick' to the flux poles of the stator and begin to follow the movement of the flux poles. The rotor begins to rotate synchronously with the magnetic field of the stator.

If an appliance that produces a load torque is connected to the shaft of the machine, the rotor poles will become slightly offset with respect to the stator poles, but the rotor will continue to rotate synchronously with the stator field. Such a situation is illustrated in Figure 1.26.

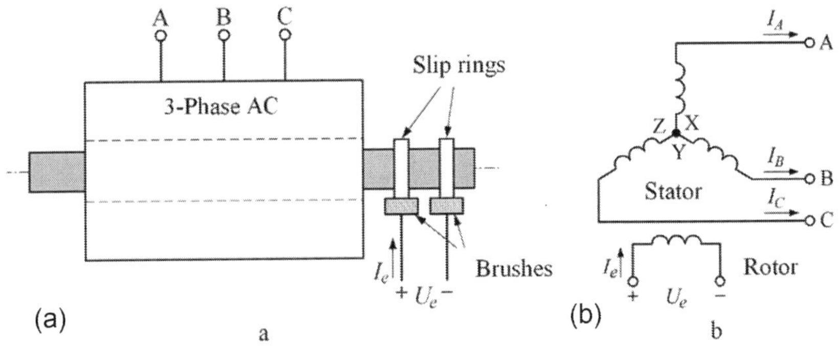

FIGURE 1.24 Transmission of excitation current to a synchronous machine (a) and a diagram of the stator winding and the rotor winding (b).

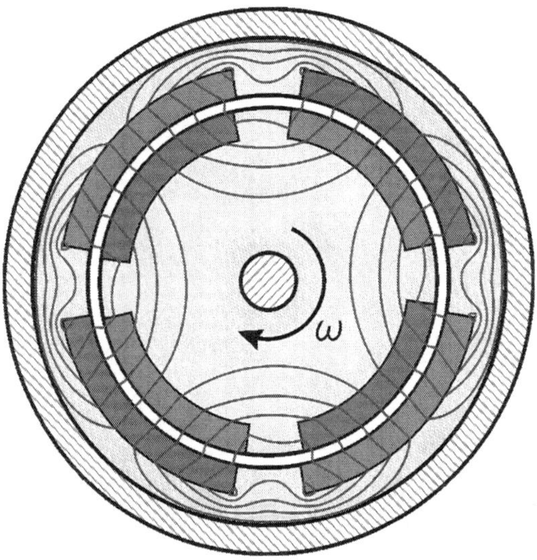

FIGURE 1.25 Mutual positions of stator and rotor poles in a synchronous machine with two pole pairs at idle.

Increasing the load torque will increase the offset between the poles. The corresponding angle of offset is called the **load angle**.

When the shaft of a synchronous machine is connected to a diesel engine or steam turbine that tries to make the synchronous machine rotate faster than the stator field, the synchronous machine will begin to operate as a generator and develop a resistive torque. In such a case, the rotor poles will shift in the opposite direction, compared to when operated as a motor, with respect to the stator poles, but the synchronous machine will not allow itself to rotate faster than the stator field determined by the mains frequency. Even in generator mode, a synchronous machine will rotate synchronously with the grid. When operated as a generator, a synchronous machine will supply power to the electrical grid.

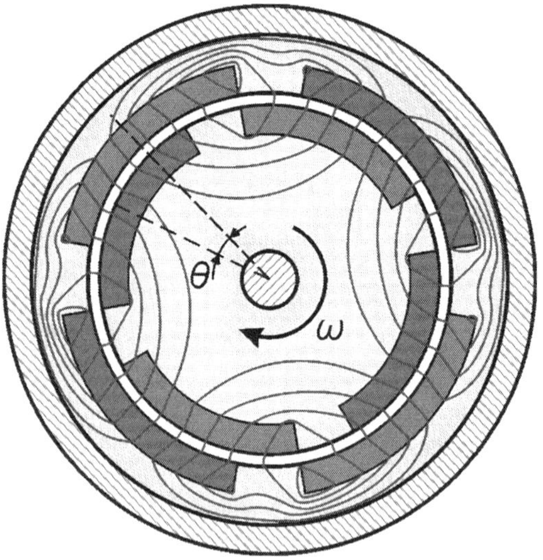

FIGURE 1.26 Shifting of the rotor poles relative to the stator poles due to load torque.

If we were to switch off the excitation current in a conventional synchronous motor with salient poles, such as that described above (Figure 1.12a), causing the rotor to stop producing a magnetic field, then this synchronous motor would continue to operate as a reluctance motor. The rotor would continue to rotate synchronously with the magnetic field produced by the three-phase winding, and the motor could even develop a certain torque. This torque would, however, be smaller than the one the motor could develop if the rotor were excited. A synchronous motor with non-salient poles (Figure 1.12b), on the other hand, cannot be operated without excitation current. This is because, in the case of a rotor with non-salient poles, the magnetic resistance, or reluctance, of the air gap is not dependent on the angular position of the rotor. The reluctance is constant because the air gap between the stator and rotor of the machine is the same across the entire circumference.

1.9 BASIC CONCEPTS OF DIRECT-CURRENT CIRCUITS

1.9.1 Interpreting Electrical Processes through Mechanical Equivalents

Electric currents are only perceptible to us when they pass through our body. The experience is generally unpleasant. A current of 0.1 A is lethal to humans, unless it is very short in duration. Current, voltage, and resistance, as well as magnetic and electric fields, can generally only be identified with the appropriate measuring instruments. Electrical concepts can be easier to understand through figurative mechanical equivalents. Figure 1.27 shows an interpretation Ohm's law, which is mathematically expressed as follows:

$$I = U / R, \tag{1.29}$$

where I – current in amperes, A;

U – voltage in volts, V;

R – Resistance in ohms, Ω.

Here, voltage is compared to water pressure, electric current to water flow, and variable resistance in an electrical circuit to a valve (water tap) that can be opened or closed to various degrees.

The greater the pressure and the more the tap is opened – i.e. the lower the outflow resistance – the more water comes out of the pipe. The parallel drawn with electrical circuits is as follows: **in an electrical circuit, the greater the voltage and the lower the resistance, the greater the intensity of the current.**

1.9.2 Ideal and Real-World Voltage Sources

Mathematically, Ohm's law is quite simple, but it is useful to have an in-depth understanding of the manifestation of this law in commonly used electrical circuits (i.e. the typical processes in those circuits).

Consider a direct-current circuit consisting of a voltage source U = 12 V, a load R (variable resistor), connection wires, and a switch K (Figure 1.28). When we close the switch, we get a closed circuit in which the intensity of the generated current is determined by Ohm's law. The figure shows how the current in

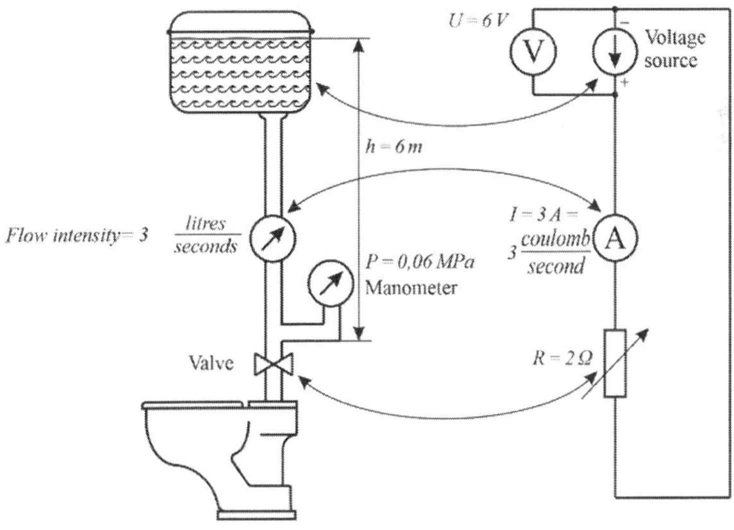

FIGURE 1.27 Mechanical equivalents for a figurative characterisation of electrical quantities.

the circuit changes when we adjust the resistor R within the range of 8–1 Ω. The current increases from 1.5 amps to 12 amps. The voltage of the voltage source is independent of the current and remains equal to 12 volts. Such a voltage source is called an **ideal voltage source**. According to equation (1.18), if we brought the resistance R down to zero, we should get an infinitely large current.

In reality, infinitely large currents are impossible. There are two reasons for this. First of all, the resistance of a circuit can never be brought down all the way to zero. The connection wires, too, have a resistance. For example, the resistance of a one-metre length of copper wire with a cross-section of 1.5 mm^2 is about 0.0115 Ω. Second, in **real-world voltage sources**, the voltage begins to drop as the load current increases. In general, voltage sources contain electrically conductive parts and they have a resistance. But voltage drops also occur for other reasons.

To take into account the internal voltage drop of a voltage source, the decrease in voltage accompanying an increase in load current is imagined to be due to a hypothetical resistance R_0, called the **internal resistance**. This internal resistance is also imagined to be connected in series to an ideal voltage source. The voltage of this ideal voltage source is also referred to as the **electromotive force** (emf), denoted E.

An equivalent circuit of the real-world voltage source is shown in Figure 1.29 in the section marked by the dashed line. The equivalent circuit consists of the internal resistance R_0 and the emf source E. The terminal voltage of the real-world voltage source is given by

$$U = E - IR_0. \tag{1.30}$$

Figure 1.29 also shows how the voltage of the real-world source would change, given an internal resistance of $R_0 = 0.1$ Ω. In this case, if we were to bring the resistance R of the resistor down to zero, the intensity of the generated current would be $I = E / R_0 = 12 / 0.1 = 120$ A. This is roughly comparable to a battery. Short-circuiting a flashlight battery would only produce a current of a few amps because of the higher internal resistance of the battery.

The circuit in question is also subject to energy conversion processes. When current passes through resistor R, the resistor heats up. Electrical energy is converted into thermal energy. The amount of energy converted per unit time (e.g. a second) is called **power**

$$P = W / t, \tag{1.31}$$

where P – power in watts, W;
W – work, or energy, in joules, J;
t – time in seconds, s.

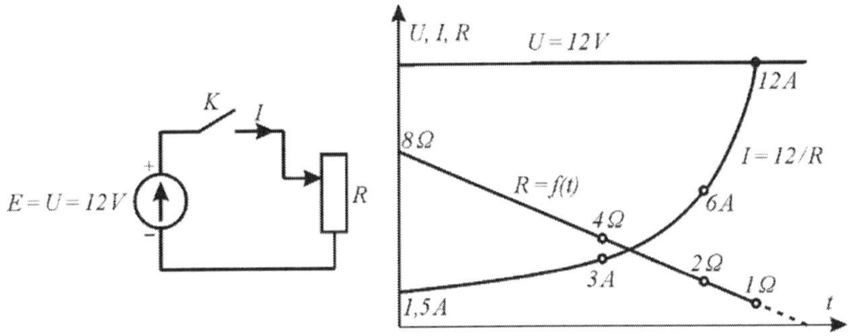

FIGURE 1.28 Loading of an ideal voltage source.

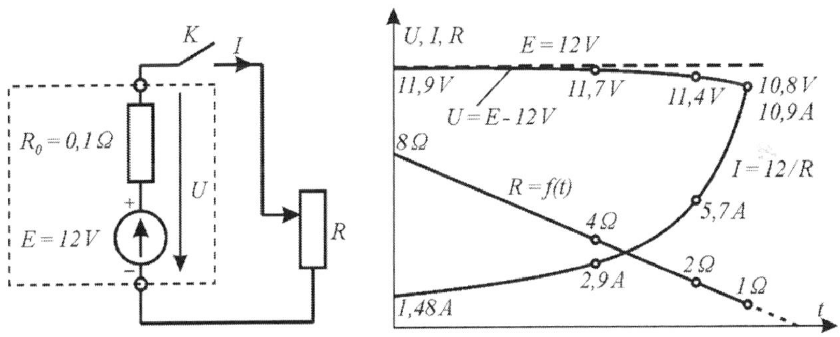

FIGURE 1.29 Loading of a real-world voltage source.

The **thermal power** P emitted by a **resistor** can be found on the basis of the Joule–Lenz law:

$$P = RI^2. \tag{1.32}$$

Using Ohm's law, the thermal power of a resistor can be expressed in two additional ways:

$$P = UI \tag{1.33}$$

and

$$P = U^2 / R. \tag{1.34}$$

The electrical power required to obtain the thermal power is drawn from a voltage source; more precisely, from the source of electromotive force inside the voltage source. The mathematical expression for this electrical power is analogous to the expression (1.22):

$$P = EI. \quad (1.35)$$

The thermal power I^2R_0 emitted at the internal resistance of the voltage source is useless and is called the **power dissipated** by the source.

The operation of the electrical circuit shown in Figure 1.29 is typical for most use cases of electricity. This kind of operation can be characterised as follows.

- The resistance of the circuit changes due to some external factor, often unpredictably. In general, voltage and current do not affect the magnitude of the resistance.
- The voltage is usually selected based on electrical safety or other criteria. The magnitude of the supply voltage is not greatly dependent on current or resistance.
- Current is the most frequently changing parameter, depending on both resistance and voltage. Keeping the current within certain limits can often be problematic.

There also exists a considerable number of other possible modes of operation (continuously regulated voltage sources, constant current sources, etc.).

1.9.3 Active and Passive Modes of Operation of a Voltage Source

In the electrical circuit in Figure 1.30, in the current outside of the voltage source U_1 is directed away from the positive terminal and towards the negative terminal, while the current inside the voltage source is directed away from the negative terminal and towards the positive terminal. This direction of current flow corresponds to the direction of flow of positive charges. By convention, charge carriers are considered to carry a positive charge. The fact that electrons are known to actually carry a negative charge is disregarded for historical reasons.

If the current inside a voltage source flows from the negative terminal to the positive terminal, the voltage source is in **active mode. In active mode, energy is drawn from the voltage source and transmitted to an external circuit.**

Voltage sources can also be operated in **passive mode**. This is depicted in Figure 1.30a. The circuit contains two voltage sources U_1 and U_2, which are connected in series, but the voltage of the voltage source U_2 is reversed compared to that of the voltage source U_1. The total voltage is equal to the difference between the voltages of the two voltage sources. In the specific case shown in the figure, the total voltage $U = U_1 - U_2 = 12 - 6 = 6$ V, and the polarity of the total voltage corresponds to the polarity of U_1. The polarity of the total voltage determines the direction of the resulting current. The current passes through the source U_1 by flowing from the negative terminal to the positive terminal, which corresponds to an active mode of operation. In the source U_2, however, the current flows from the positive terminal to the negative terminal. In the resistor, R, too, the current travels from the positive terminal to the negative terminal. Both the voltage source U_2 and the resistor R are supplied with energy. The energy supplied to the resistor is

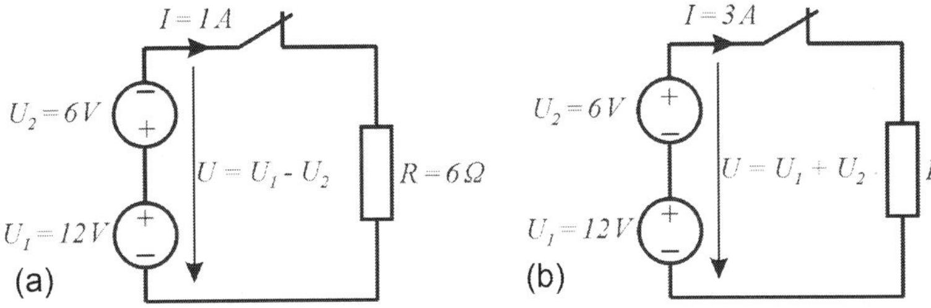

FIGURE 1.30 Voltage source U_2 in passive mode (a) and in active mode (b).

converted into heat. If the voltage source U_2 is a battery, energy will be stored in this battery, i.e. the battery will be charged. **The source U_2 draws energy from the external circuit and is in passive mode.** The resistor will always be in passive mode, regardless of the direction of the current. Resistors are considered **passive elements**. If we were to reverse the polarity of the voltage source U2 (Figure 1.30b), both voltage sources would be in active mode and the total voltage would be equal to the sum of the voltages of the voltage sources U_1 and U_2.

Capacitors, like batteries, are also energy-storing elements. During charging, a capacitor draws energy from the external circuit and is in passive mode (Figure 1.31a).

During discharging, a capacitor is in active mode (Figure 1.31b). Capacitors differ from voltage sources in that a capacitor does not contain a 'pump' that pumps in new charges to replace outgoing ones. Due to this, when a capacitor is discharged, the charge stored in the capacitor is depleted relatively quickly. In a battery, new charges are transmitted to the terminals by chemical forces, while in a generator this function is fulfilled by electromagnetic forces.

1.10 DESCRIPTION OF PROCESSES IN ELECTRICAL MACHINES

1.10.1 Generation of an Electromotive Force and a Mechanical Force in a Magnetic Field

Electric motors are also called prime movers. The motive force referred to in the name is generated by a magnetic field. And magnetic fields are also responsible for the voltage generated in an electric generator. Let us take a closer look at these processes.

The movement of a magnetic flux in space or its change over time generates a voltage in an electrical conductor. The conductor thereby becomes a voltage source. Second, a current-carrying conductor placed in a magnetic field becomes affected by a mechanical force. **The generated voltage and force are called the fundamental phenomena of magnetic fields.** Since these fundamental phenomena are measured not via the *H*-field, but the *B*-field, it might be more accurate to call them the fundamental phenomena of a magnetic flux density field.

This generation of a voltage and force is important, firstly, because it makes it possible to detect the presence of a magnetic field. Secondly, it is important

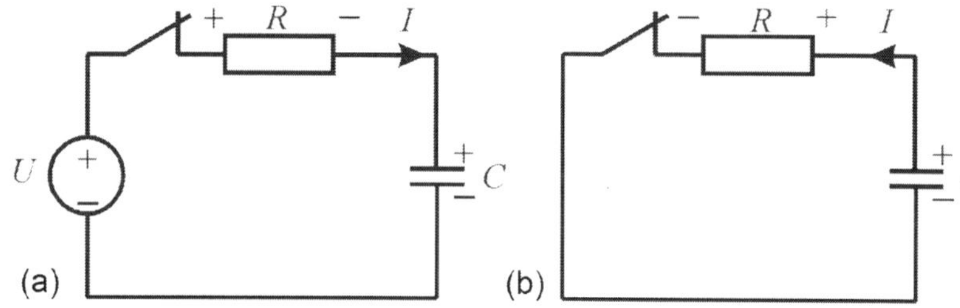

FIGURE 1.31 Capacitor charging is a passive mode of operation (a); capacitor discharging is an active mode of operation (b).

because the generation of the voltage can be utilised for generating electrical energy with generators, while the generation of the force can be utilised for obtaining mechanical energy with electric motors. This generation of voltage and force is shown in Figure 1.32.

The voltage and force generated by a magnetic field are also called **induced voltage** and **induced force**. When a voltage is induced in a closed circuit, it is accompanied by a current. In an open circuit, no current is generated and the voltage is maximal. The open-circuit voltage is called the **electromotive force** (abbreviated emf) and is denoted by E.

As far as the induced voltage and emf are concerned, it does not matter whether the magnetic field is moving relative to a stationary conductor or vice versa – whether the conductor is moving in a stationary magnetic field.

Figure 1.32a depicts a situation where a conductor with a length of l is moving with a velocity of v in a homogeneous magnetic flux density field B, perpendicular to the lines of force. The magnitude of the electromotive force induced in the conductor can be calculated using the equation

$$E = vBl. \tag{1.36}$$

The direction of the induced electromotive force can be determined by **the right-hand rule: when the magnetic flux density vector is pointing towards the palm of your right hand and the conductor is moving in the direction indicated by your outstretched thumb, the direction of the electromotive force induced in the conductor is indicated by your outstretched fingers.**

In general, if the velocity vector $\boldsymbol{v}$ and the flux density vector $\boldsymbol{B}$ are not perpendicular to each other and the direction of movement of the conductor in space does not coincide with the direction of the cross product $\boldsymbol{v} \times \boldsymbol{B}$ as in Figure 1.32a, the induced electromotive force can be found as follows:

$$\boldsymbol{E} = \left(\boldsymbol{v} \times \boldsymbol{B}\right)\boldsymbol{l}. \tag{1.37}$$

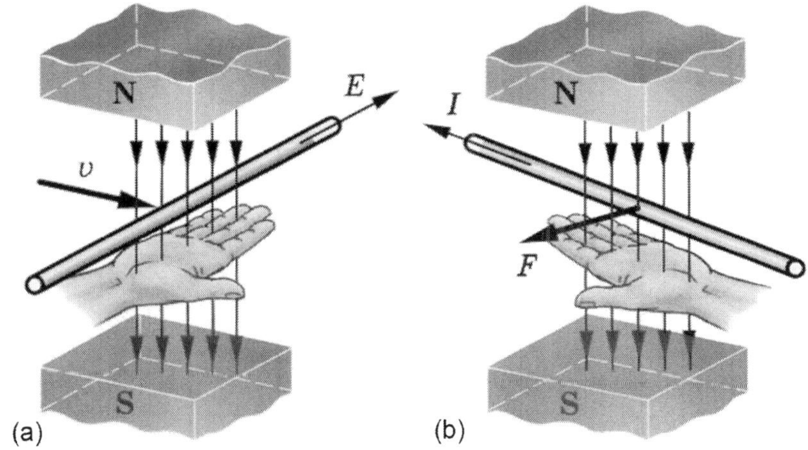

FIGURE 1.32 Electromotive force E induced in a conductor moving in a magnetic field (a) and force F induced in a current-carrying conductor held in a magnetic field (b).

The length vector $\boldsymbol{l}$ of the conductor is directed along the length of the conductor and towards the end where the angle made with the vector $\boldsymbol{v} \times \boldsymbol{B}$ is smallest. The positive end of the conductor is the one where the vector $\boldsymbol{v} \times \boldsymbol{B}$ exits the conductor.

Figure 1.32b depicts a situation where a stationary conductor with a length of l is carrying a current of i while positioned in a magnetic field B, perpendicular to the lines of force. The resulting force can be found as follows:

$$F = iBl. \tag{1.38}$$

The direction of the induced force can be determined by **the left-hand rule: when the magnetic flux density vector is pointing towards the palm of your left hand and the direction of the current in the conductor coincides with the direction of your outstretched fingers, the direction of the induced force is indicated by your outstretched thumb.**

The induced force is generally expressed as

$$\boldsymbol{F} = i\left(\boldsymbol{l} \times \boldsymbol{B}\right), \tag{1.39}$$

where the direction of the length vector $\boldsymbol{l}$ of the conductor coincides with the direction of the current.

The fundamental phenomena of magnetic fields discussed above are utilised in all electrical machines. Below is a description of a **linear direct-current machine** (abbr. 'linear DC machine'). This is a simple and easily comprehensible electrical machine, the principle of operation and characteristics of which are identical to those of rotating generators and motors. Linear machines produce translational motion, while conventional machines produce rotational motion.

The linear DC machine is supplied with power from a battery via the resistor R (Figure 1.33). At points A and B, a voltage is transmitted to a smooth, frictionless pair of rails. Along this 'railroad track', there is a constant uniform magnetic field, directed from top to bottom, perpendicular to the plane of the figure. Lying across the rails, there is an electrically conductive metal bar. In real-world linear DC machines, the metal bar is replaced by a winding coil, which is supplied with current via flexible conductors and which is also moving in the uniform magnetic field.

The movement of this bar is based on four fundamental equations for the machine:

1. Equation for force acting on a conductor in a magnetic field:

$$\boldsymbol{F} = i\left(\boldsymbol{l} \times \boldsymbol{B}\right),$$

where $\boldsymbol{F}$ is the force acting on the conductor, i is the intensity of the current in the conductor, $\boldsymbol{l}$ is the length of the conductor, the direction of which is determined by the direction of the current, and $\boldsymbol{B}$ is the magnetic flux density vector.

2. Equation for electromotive force induced in a conductor moving in a magnetic field:

$$e_{\text{ind}} = \left(\boldsymbol{v} \times \boldsymbol{B}\right)\boldsymbol{l},$$

where e_{ind} is the emf induced in the conductor, $\boldsymbol{v}$ is the velocity of the conductor, $\boldsymbol{B}$ is the magnetic flux density vector, and $\boldsymbol{l}$ is the length of the conductor in the magnetic field.

3. Kirchhoff's voltage law for the machine in question. In the figure, we can see that

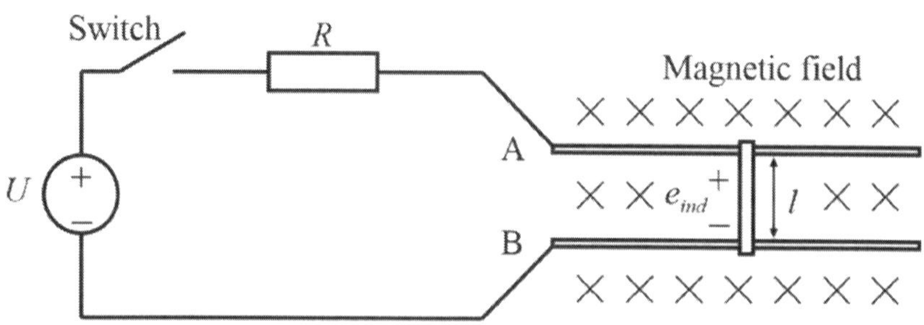

FIGURE 1.33 A linear DC machine.

$$U - iR - e_{ind} = 0,$$

where U is the voltage of the battery and R is the resistance of the resistor. This yields

$$U = e_{ind} + iR.$$

4. Newton's law, regarding the acceleration of the bar, says

$$F = ma,$$

where F is the force acting on the bar, m is the mass of the bar, and a is the acceleration of the bar.

Now let us explore the main characteristics of our simple DC machine, using these four fundamental equations.

1.10.2 Starting the Linear DC Machine as a Motor

To start the machine, you need to switch it on. This will cause a current to be produced in the bar, which can be calculated using Kirchhoff's voltage law:

$$i = \left(U - e_{ind}\right) / R.$$

Since the bar is not moving at first, $e_{ind} = 0$ and $i = U/R$. In the bar, current flows from the positive end to the negative end. If there is a current present in a conductor which is positioned in a magnetic field, then, according to equation (1.38), the magnetic field will induce a force in the conductor. The magnitude of this force is $F_{ind} = iBl$, and in Figure 1.34 it is directed to the right.

Therefore, the bar starts to accelerate to the right (according to Newton's law $a = F/m$). As the bar begins moving, an electromotive force $e_{ind} = vBl$ is induced in the bar (the upper end of the bar is the positive end).

As the electromotive force e_{ind} increases, the current i decreases: $i\downarrow = (U - e_{ind}\uparrow)/R$ (see 'Start-up' in Figure 1.35). As a result, the bar will eventually reach a steady speed of v_0 where the force acting on the bar is zero. When this happens, the induced electromotive force e_{ind} will be equal to the voltage U, and the speed of the bar at that time will be derivable from the equation $U = e_{ind} = v_0 Bl$ as $v_0 = U/Bl$.

The bar will continue to travel at this no-load speed forever, unless it is affected by an external force. At the end of the motor start-up process, it will have a speed of v_0, an induced emf of e_{ind}, a current of $i = 0$, and an induced force of $F_{ind} = 0$, as shown in Figure 1.35. This is a so-called ideal no-load scenario, where friction losses are not taken into account.

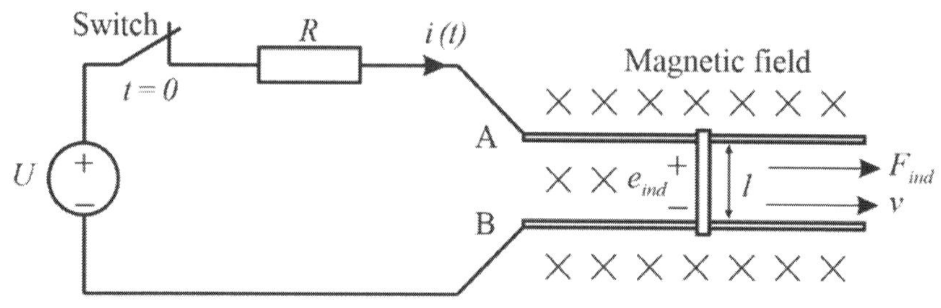

FIGURE 1.34 Starting of the linear DC machine.

1.10.3 The Linear DC Machine as a Motor

Let us assume that our linear DC machine is initially operating in the steady no-load conditions described above. Now, an external load is applied to the machine in the form of a force F_{load} acting on the bar (opposite to the direction of motion, see Figure 1.36). The net force acting on the bar in the direction opposite the direction of motion will then be $F = F_{\text{load}} - F_{\text{ind}}$, which will have a slowing effect on the bar. However, as the speed decreases, so does the induced emf, as $e_{\text{ind}}\downarrow = v\downarrow Bl$. As the induced emf decreases, the current in the bar rises:

$$i\text{-} = \left(U - e_{\text{ind}}^{-}\right)/R.$$

This in turn increases the induced force: $F_{\text{ind}} = i\uparrow Bl$. The end result of all this is that the induced force will rise until it is equal to the load force. When this happens, the bar will be moving at a steady speed that is slightly lower than the no-load speed. When a load force is applied to the bar, it will change its velocity v, induced emf e_{ind}, current i, and induced force F_{ind} as shown for the period t_1–$t3$ in Figure 1.35.

There is now an induced force acting on the bar in its direction of motion, and power is being converted from electrical form to mechanical form to keep the bar in motion. The power being converted is

$$P = e_{\text{ind}} i = F_{\text{ind}} v.$$

An amount of electric power equal to $e_{\text{ind}}\, i$ is drawn from the bar and is replaced by mechanical power in the amount of $F_{\text{ind}}\, v$. Since power is converted from electrical to mechanical form, the bar is operating as a motor. The above can be summarised as follows:

1. A force F_{load} is applied opposite to the direction of motion, which constitutes the entire force F acting on the bar.
2. The resulting acceleration $a = F/m$ is negative, causing the bar to slow down ($v\downarrow$).

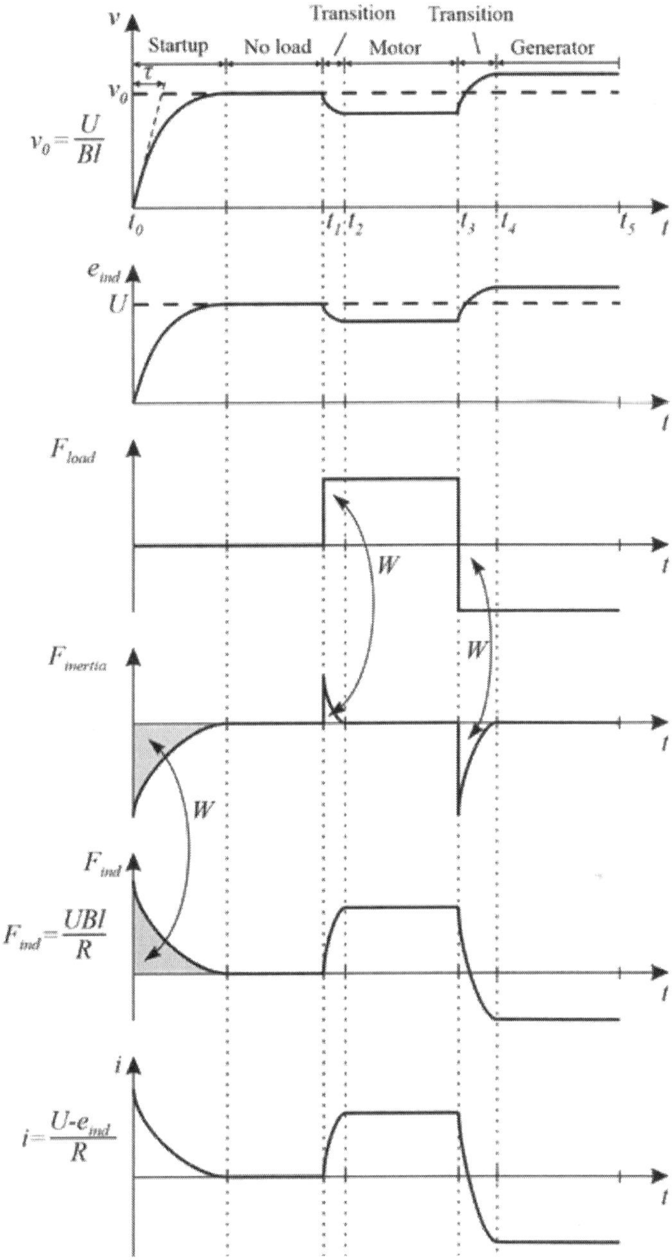

FIGURE 1.35 Operating processes of the linear electrical machine.

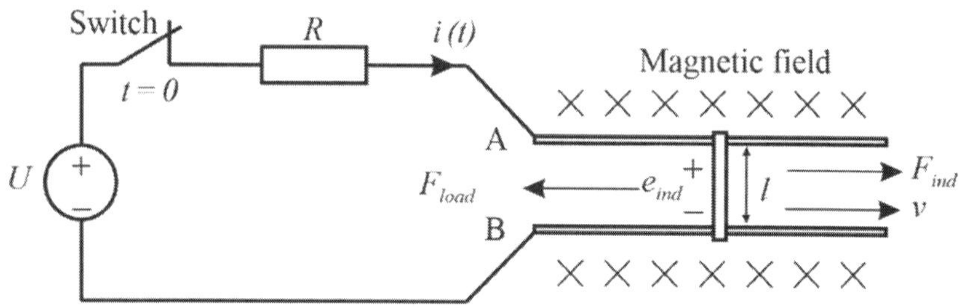

FIGURE 1.36 The linear DC machine as a motor.

3. The decrease of the electromotive force $e_{\text{ind}} = v\downarrow Bl$ causes the current $i = (U - e_{\text{ind}}\downarrow)/R$ to increase.
4. The induced force $F_{\text{ind}} = i\uparrow Bl$ increases until it becomes equal to the force acting on the bar $|F_{\text{ind}}| = |F_{\text{load}}|$ at a slightly lower velocity v.
5. An amount of electrical power equal to $e_{\text{ind}}\, i$ is being converted into mechanical power equal to $F_{\text{ind}}\, v$ and the machine is acting as a motor.

A rotating DC motor acts in an identical manner when it is loaded: as a load is added to its shaft, the speed begins to decrease, causing the internal emf to decrease and the current to increase. As the current increases, the induced torque increases until it becomes equal to the load torque at a new, lower speed.

Recall how in a linear motor, the power converted from electrical form to mechanical form can be calculated using the equation $P = F_{\text{ind}}\, v$. In a rotating motor, the converted power is given by the equation $P = T_{\text{ind}}\omega$, where the induced torque T_{ind} is the rotational equivalent of the induced force F_{ind} and the angular velocity ω is the rotational equivalent of the linear velocity v.

1.10.4 The Linear DC Machine as Generator

This time, let us apply a force F_{load} to the bar in the direction of motion. Now the bar will start to accelerate. As the speed increases, the emf ($e_{\text{ind}} = v\uparrow Bl$), too, will increase and will surpass the battery voltage. When $e_{\text{ind}} > U$, the current will reverse direction and will then be given by the equation

$$i = \left(e_{\text{ind}} - U\right) / R.$$

With the current now flowing from bottom to top in the bar (Figure 1.37), an induced force $F_{\text{ind}} = iBl$ will be acting on the bar, directed to the left. The direction of the induced force can be determined by the right-hand rule. The induced force acts in resistance to the force applied to the bar.

Eventually, the induced force will become equal and opposite to the applied force and the bar will be moving faster than before (time period t_3–t_5 in Figure 1.35). When this happens, the battery will be charging. This means that our linear machine is now acting as a generator, converting mechanical power $F_{\text{ind}} v$ to electrical power $e_{\text{ind}} i$.

The above process can be summarised as follows:

1. A force F_{load} is applied to the bar in the direction of motion, constituting the entirety of the force F acting on the bar.
2. Acceleration $a = F/m$ is positive, so the bar speeds up (v↑).
3. The electromotive force $e_{\text{ind}} = v\uparrow Bl$ increases and thus so does the current $i = (U - e_{\text{ind}}\uparrow)/R$.
4. The induced force $F_{\text{ind}} = i\uparrow Bl$ increases until $|F_{\text{ind}}| = |F_{\text{load}}|$ at a higher velocity v.
5. Mechanical power in the amount of $F_{\text{ind}} v$ is now being converted to electric power in the amount of $e_{\text{ind}} i$, and the machine is acting as a generator.

Again, a real DC generator functions in the same way: when a torque is applied to its shaft in the direction of motion, the speed of the shaft increases, the internal emf increases, and a current flows from the generator to the load.

The mechanical power which is converted to electrical power in a rotating generator can be calculated using the equation

$$P = T_{\text{ind}}\omega.$$

It is an interesting fact that the same machine can act **as both a motor and a generator**. The only difference is that in one case the external force is applied in the direction of motion (when operated as a generator), while in the other it is opposite to the direction of motion (when operated as a motor). Electrically, the machine acts as a generator when $e_{\text{ind}} > U$ and as a motor when $e_{\text{ind}} < U$. In both cases, both an induced force (motor effect) and an induced electromotive force (generator effect) are generated. The same is also true of other types of machines: both effects are always present, and whether the motor behaves as a motor or generator is determined solely by the direction of the total external force relative to the direction of motion.

Another interesting fact here is that the speed of our machine was higher when operating as a generator and lower when operating as a motor, but the direction

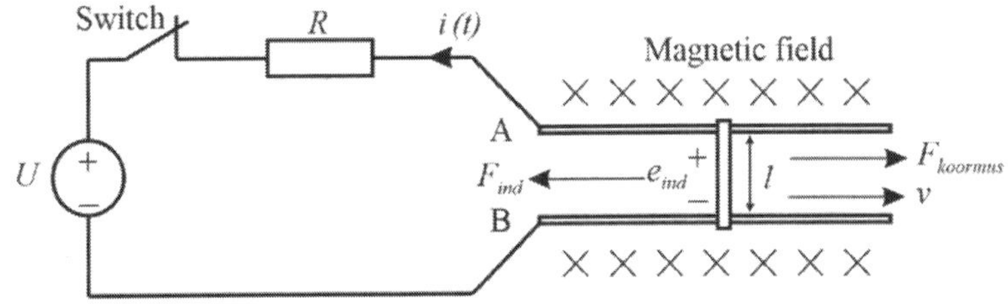

FIGURE 1.37 The linear DC machine as generator.

of rotation was the same in both cases. If anyone thinks that motors turn one way and generators the other way, they are mistaken. The only difference between the two modes of operation is a slight change in speed.

Based on the example of our simple linear machine, we can also conclude the following:

- Speed never changes instantly. The reason is the mechanical inertia of the machine. As the speed of the machine changes, the kinetic energy stored in its moving parts also changes. When the machine is started, this energy is drawn from the power supply (grey areas in Figure 1.35). As speed decreases, kinetic energy is returned either to the load (generally an appliance) or back to the electrical circuit. A situation in which the speed does not change is called **steady-state operation**. Steady-state operation can be operation with no load, operation as a motor, operation as a generator, or another mode of operation. A shift from one mode of operation to another is called the **changeover process** or simply **changeover**; the way the system behaves during this process is known as the **transient response**. The start-up process shown in Figure 1.35 is a changeover from standstill to no-load operation.
- The oscillograms for induced electromotive force e_{ind} and velocity v are similar, given a constant magnetic flux density B, because $e_{\text{ind}} = vBl$. Velocity and electromotive force always change in unison and in the same direction. The direction of motion or rotation can be changed by changing the polarity of the supply voltage.
- The oscillograms for induced force F_{ind} and current i are also similar, because $F = iBl$. Force (or torque in the case of rotational motion) and current always change in unison and in the same direction.
- The current is positive when the machine is operated as a motor, but becomes negative when it is operated as a generator. A situation where the current is zero and the machine is moving is an ideal no-load scenario. No-load operation is neither part of operation as a motor nor operation as a generator.
- Operation as a generator and operation as a motor to are possible in both directions of rotation of the machine (clockwise or counterclockwise). Therefore, there are a total of four different modes of operation: clockwise motor mode, clockwise generator mode, counterclockwise motor mode, and counterclockwise generator mode.

1.10.5 Necessity of Limiting the Starting Current of a DC Machine

Consider a machine which is supplied with a DC voltage of 250 V and which has an internal resistance of 0.10 Ω (these values are characteristic of medium-power DC machines).

At start-up, the speed of the machine is zero and the electromotive force is $e_{\text{ind}} = 0$. There will be a current of

$$i_{\text{start}} = \frac{U}{R} = \frac{250\,\text{V}}{0{,}1\,\text{W}} = 2500\,\text{A}.$$

This is an extremely high current – more than ten times over the rated current of the machine. Such a high current could damage the machine.

The easiest way to limit the current is to insert an additional resistance R_{start} into the circuit during starting. As the speed increases, e_{ind} will also increase, after which the additional resistance can be removed.

1.10.6 Overview of Energy Conversion Processes in an Electrical Machine

If we were to supply power to a machine from an electrical grid, the machine would begin to rotate. Electrical energy would be converted into mechanical energy. If, however, the same electrical machine were made to rotate by means of an external device, such as a steam turbine, it would instead supply power. In that case, mechanical energy would be converted into electrical energy. An electrical machine is a **reversible energy converter**.

Electrical machines can be said to consist of three basic components: an electrical component, a mechanical component, and a magnetic component connecting the two (Figure 1.38a).

The relationship between these basic components is illustrated in more detail in Figure 1.38b. In reality, there are two electrical components. One of them is stationary. This component is located on the stator of the machine. The second electrical component is able to rotate and is located on the rotor. The magnetic component is linked to the stationary electrical component (the current I_1 produces its own magnetic field), the rotating electrical component (the current I_2 produces its own magnetic field), and the mechanical component (the flux density B and the current I_2 produce a torque T). The magnetic field travels from the stationary component to the rotating one via an air gap. The air gap is part of the magnetic component.

Energy conversion is mediated by a special environment: **the magnetic flux density field**, or the B-field (the B-field and the H-field are components of the magnetic field). The B-field acts simultaneously on both the **electrical** and the **mechanical** system (see Figure 1.38c).

The electrical system usually consists of a winding in the electrical machine, connecting wires, **a converter** for voltage and current conversions, and an electrical grid for supplying power. In simpler cases, there is no electrical converter (e.g. if there is no need to adjust the speed of the motor) and the electrical machine is connected directly to the grid. The state of the electrical system is characterised by voltage, current, and other electrical quantities.

The mechanical system usually consists of an electrical machine rotor, **a clutch** connecting the electrical machine to the appliance, and an appliance. The state of the mechanical system is characterised by rotational speed, torque, and other mechanical quantities.

If the machine is operated as a motor, electrical power P_{el} is transformed into mechanical power P_{mech} via the B-field. The power (or energy flow) of a magnetic field is called **electromagnetic power** P_{em}.

If the machine is operated as a generator, the power flow is opposite in direction: mechanical power is transformed into electrical power via electromagnetic power.

During the conversion process, the proportion of energy that is released as heat into the surrounding environment is lost (see Figure 1.38c). In the electrical system, losses occur due to the electrical resistance of the winding. As current passes through a winding, the winding heats up. The resulting losses are also called **copper losses** $\boldsymbol{P_{Cu}}$, because windings are often made of copper. The magnetic fields in electrical machines largely travel along a magnetic circuit made of electrical steel. A changing magnetic field gives rise to hysteresis and eddy currents. This causes the magnetic circuit to heat up. The resulting losses are called **iron losses** $\boldsymbol{P_{Fe}}$. Losses also occur due to friction in the mechanical system. A distinction is made between losses in bearings and in the ventilation system. They are jointly called **mechanical losses** $\boldsymbol{P_{mech\ loss}}$.

Losses are characterised by **energy conversion efficiency** η . For motor mode, energy conversion efficiency is given by

$$\eta_M = \frac{P_{mech}}{P_{el}}. \tag{1.40}$$

For generator mode, energy conversion efficiency is given by

$$\eta_G = \frac{P_{el}}{P_{mech}}. \tag{1.41}$$

The energy conversion efficiency of electrical machines at rated operation (i.e. at the rated torque and rated speed) mostly falls between 0.7 and 0.9. Energy conversion efficiency depends on the type and size of the machine. In very large machines, energy conversion efficiency can be as high as over 0.98. The energy conversion efficiency of smaller machines with a power of less than 10 W is less than 0.5.

Electromechanical converters include **electromagnets**. These are widely used in **relays** and **contactors**, which are electrically controlled switches. Electromechanical converters are also used in **sensors**, or **measuring transducers**, which output an electrical signal in the presence of mechanical motion.

Describing electromechanical converters requires using a whole host of electrical, mechanical, and magnetic parameters.

Although the direction of power flow in electrical machines is variable, generators and motors are still constructed slightly differently. The net power produced by a generator is electrical power P_{el}, which in the case of a DC generator can be calculated via the voltage U and current I on the generator's terminals:

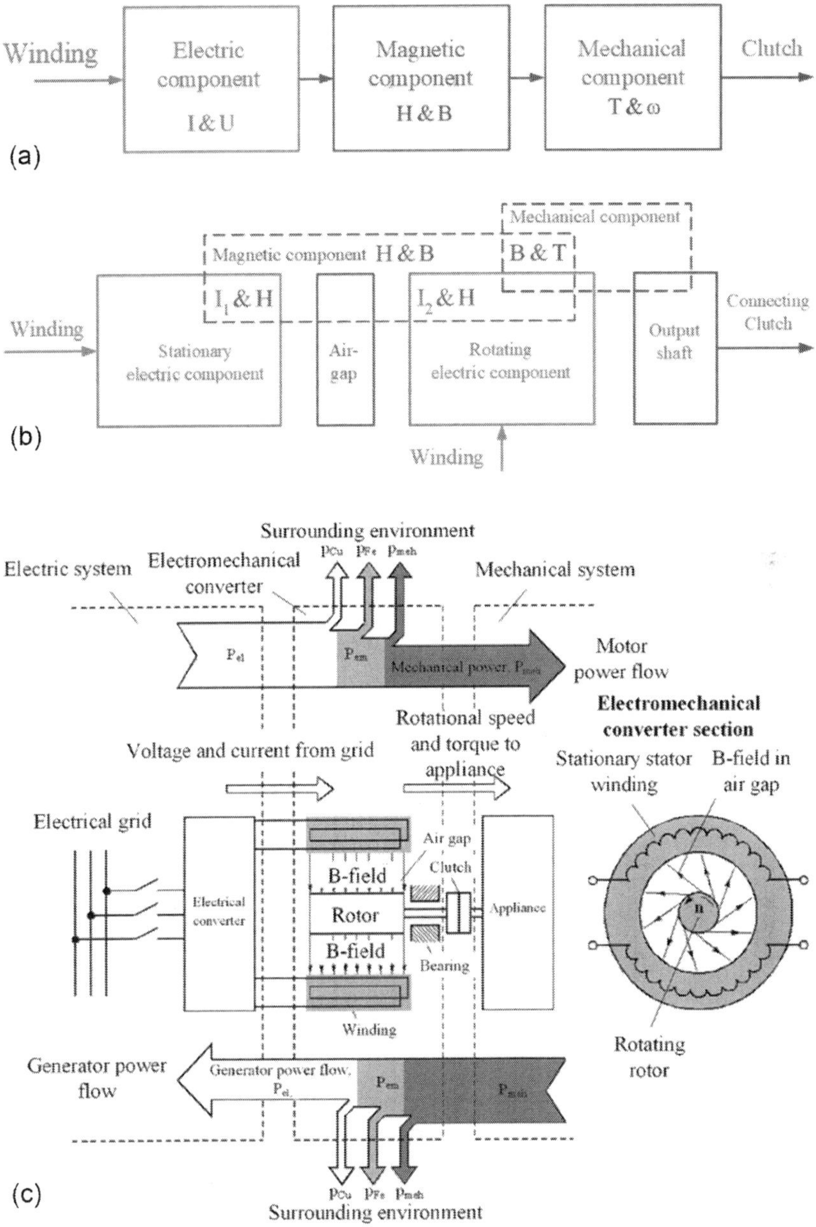

FIGURE 1.38 Basic components of an electrical machine (a), the links between them (b), and exchange of energy between the electrical system and mechanical system via the magnetic field in the converter (c).

$$P_{el} = UI. \tag{1.42}$$

AC generators are mostly three-phase systems, and their power can be expressed as follows:

$$P_{el} = \sqrt{3}U_l I_l \cos\varphi. \tag{1.43}$$

Here, U_l and I_l are, respectively, the line voltage and line current on the generator's terminals, while $\cos\varphi$ is the power factor.

The net power produced by a motor is the mechanical power on the motor's shaft, which can be expressed via angular velocity and torque:

$$P_{mech} = T\omega. \tag{1.44}$$

Electric motors are used to run appliances. The power of a motor is selected based on the torque needed to start the appliance and the required speed of the appliance.

1.11 DESIGN AND OPERATING PRINCIPLE OF TRANSFORMERS. CONVERSION OF ELECTRICAL ENERGY IN AN IDEAL TRANSFORMER

A transformer is a device that shifts AC electrical energy from one voltage level to another by means of a magnetic field. For example, to step voltages down from 10 kV and 50 Hz to 230 V and 50 Hz. Transformers do not adjust frequency. Neither do they work with direct current. The design of a transformer is simple. A transformer consists of two or more windings wrapped around a common ferromagnetic **core** (Figure 1.39).

In the simplest transformers, there is one **primary winding** with w_1 turns of wire, into which electrical energy is fed, and one **secondary winding** with w_2 turns of wire, from which energy is transmitted via a switch (S_2) to a load (R). At frequencies of $f < 150$ Hz, the core is made from 0.35 to 0.50-mm thick sheets of electrical steel.

When an AC voltage is applied to the primary winding, an alternating current is generated in this winding. Transformers bear some similarity to iron-core coils, which we discussed in Section 1.4. What makes them different is that the steel-sheet core of a transformer forms a closed loop with no start or end, and thus no magnetic lines of force generally extend out into air from the core. Another distinctive feature of transformers is that transformer windings are supplied not with direct current, but with alternating current. The alternating current in the primary winding generates in the core of the transformer a magnetic field with a changing direction and magnitude, which, like the current, also follows the law of sines.

This magnetic field, which changes according to the law of sines, passes through all of the turns of wire in both windings of the transformer. When the changing magnetic field passes through a turn of wire, it induces an electromotive force in that turn. The electromotive force in a turn of wire is determined by Faraday's law

$$e = -\mathrm{d}F\,/\,\mathrm{d}t,$$

where e – electromotive force in volts;

Φ – magnetic flux in webers (the nature of a magnetic flux is explored in more detail in the following chapter);

t – time in seconds.

Note that if a magnetic flux Φ is present, but is not changing in magnitude, as is the case with direct currents, then $\mathrm{d}\Phi/\mathrm{d}t = 0$ and $e = 0$. Thus, when a magnetic flux that is constant in magnitude passes through a turn of wire in a winding, it does not induce an electromotive force in the turn.

Since the turns of wire in a winding are connected in series, a greater number of twist results in a greater electromotive force in the winding. Given w_1 turns in the primary winding, the electromotive force in the winding is

$$e_1 = -w_1 \frac{\mathrm{d}F}{\mathrm{d}t}. \tag{1.45}$$

The negative sign in the expression (1.45) indicates that the direction of the generated electromotive force is opposite to that of the instantaneous value of the AC voltage u_1 applied to the primary winding. This electromotive force produced in the primary winding is known as the counter-electromotive force. This

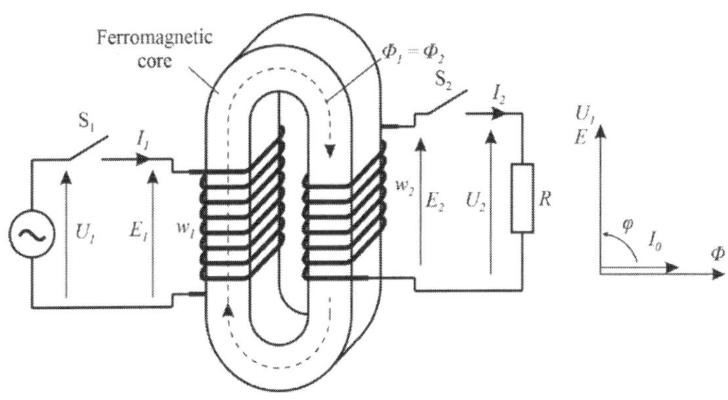

FIGURE 1.39 Design of a transformer and phasor diagram of an idealised no-load condition.

counter-electromotive force significantly reduces the no-load current in the primary winding, as the generated counter-electromotive force e_1 is usually almost as high as the voltage u_1 applied to the primary winding:

$$e_1 = (0,95\ldots0,99)u_1.$$

The changing magnetic flux of the transformer core also passes through all of the turns in the secondary winding and produces an electromotive force e_2 in the winding:

$$e_2 = -w_2 \frac{\mathrm{d}F}{\mathrm{d}t},$$

where w_2 is the number of turns of wire in the secondary winding.

When the switch S_2 is closed (Figure 1.39), a current (called the **load current** I_2) is also generated in the secondary circuit. If the switch S_2 is open, there is no secondary current. This is called a **no-load** state. The primary current I_1 is present even at no load, during which state it is called the **no-load current**, or **magnetisation current** $I_{0.}$

The current I_2 produces a voltage drop in the transformer, and the instantaneous value of the secondary voltage u_2 is slightly lower than the electromotive force e_2. In approximate calculations, we may ignore the voltage drops caused by currents and say that $u_1 = e_1$ and $u_2 = e_2$. Additionally, in a first-order approximation, we may disregard the no-load current as well as the power losses in the windings and the magnetic circuit. If there are no active power losses, then, according to the law of conservation of energy, the active power of the secondary side must be equal to the active power of the primary side: $P_2 = P_1$. A transformer simplified in this way is called an ideal transformer. In an ideal transformer, the following equation holds:

$$U_1 I_1 = U_2 I_2. \quad (1.46)$$

Given that in an ideal transformer the voltage and electromotive force are equal, as well as that both the primary and secondary winding are affected by the same changing magnetic flux, Faraday's law states that the voltage in one turn of wire is the same throughout the entire transformer. As such, the ratio of the number of turns in the windings determines the ratio of the voltages in the windings:

$$\frac{U_1}{U_2} = \frac{w_1}{w_2}. \quad (1.47)$$

If the secondary winding has fewer turns of wire than the primary winding, the transformer will step the voltage down. If, however, the secondary winding has more turns of wire than the primary winding, the transformer will step the voltage up. A transformer enables the magnitude of an AC voltage to be changed, and in a first-order approximation, the secondary voltage can be calculated using the equation

$$U_2 = \frac{w_2}{w_1} U_1.$$

Changing voltage levels is necessary in electrical grids, as we learned in Section 1.1.1. But voltage levels are usually also important for electricity-consuming devices.

As we know, a 100-watt incandescent light bulb outputs more light than a 40-watt bulb. Power is the main parameter that characterises electrical devices.

A power of a hundred watts can be obtained at a voltage of 10 V and a current of 10 A. However, we can also choose a voltage of 100 V and a current of 1 A or, in principle, a voltage of 1 V and a current of 100 A. The power P of an incandescent light bulb is determined by the product of its voltage U and current I, which, for the three options above, can be expressed as

$$P = U_1 I_1 = U_2 I_2 = U_3 I_3. \tag{1.48}$$

This equation can be interpreted as saying that **the same amount of power can be transmitted using different voltage levels.**

In practice, however, the choice of voltage levels is always limited by certain factors. For example, if the voltage of the voltage source is 10 V, while the reference voltage of our light bulb is 100 V, then the bulb will not output any light. We would either have to choose a ten-volt light bulb or use an auxiliary voltage-changing device to adjust the electrical energy parameters in an appropriate manner. Such auxiliary devices are generally called **converters**. In the case above, it would be wiser to choose a ten-volt light bulb, because, first, converters are not exactly cheap and, second, every converter is subject to energy losses.

If, however, our voltage source supplied a voltage of 1,000 V, manufacturing a 100-watt light bulb that is suitable for such a high voltage would be excessively complex and expensive. The filament would have to be extremely long and thin, which would make it difficult to support when heated. Here, using a converter would be a better solution.

If, on the other hand, our voltage source supplied a voltage of 1 V, manufacturing a suitable 100-watt filament would not be too difficult, but difficulties would arise with regard to the connecting wires, as we would need a current of 100 A. Assuming losses of 1% in the connecting wires, the voltage drop in the wires would be $\Delta U \leq 0.01 \cdot 1 = 0.01$ V and the resistance of the wires would thus be R

$\leq 0.01/100 = 0.0001\ \Omega$. Given a distance of, for example, 10 m between the power supply and the light bulb, the power line would need to constructed from copper rods with a diameter of 6.6 cm. That would be an immense waste of copper. Meanwhile, at a voltage of 100 V and with all other conditions being equal, we would only need a wire diameter of 0.66 millimetres.

As we can see, the required diameter of the wire is inversely proportional to the voltage: the higher the voltage, the finer the wires. However, higher voltages are more dangerous and also require that the wires have stronger insulation to prevent dielectric breakdown. As such, excessively high voltages are not good. It turns out that the roughly optimal voltage of a power line can be determined by the distance between the voltage source and the load. A close to optimal voltage can be obtained at one volt per meter, i.e. 1,000 V per kilometre.

1.11.1 Magnetic Circuits of Transformers

Depending on the design of the magnetic core, transformers are divided into core-type and shell-type transformers. A single-phase **core-type transformer** (Figure 1.40a) consists of two **legs**, which hold a winding, and two **yokes**, which provide a closed magnetic circuit. Each winding consists of two parts arranged on the two legs and connected to each other either in series or in parallel. The concentric arrangement of the primary and secondary windings brings them closer together, thereby increasing the electromagnetic linkage between them.

Meanwhile, a single-phase **shell-type transformer** (Figure 1.40b) consists of a single winding-bearing leg and a branching yoke that partially encloses the windings (like a shell).

For a frequency of 50–60 Hz, both the legs and the yokes are constructed from 0.35 to 0.5-mm thick sheets of **electrical steel**. Electrical steel is made by adding 0.8–4.8% **silicon** to iron, which increases the electrical resistivity of the steel and reduces both hysteresis and eddy current losses, but makes the steel sheets hard and brittle. Carbon content degrades the properties of electrical steel and is thus avoided.

In low-power transformers, the cross-section of both the legs and the yoke in the magnetic circuit is square-shaped or rectangular. A rectangular cross-section is also employed for the insulating coil support around which the windings are wrapped (Figure 1.41). In large transformers, the windings placed on a cylindrical support with a ring-shaped cross-section. To make better use of the space inside the windings, the legs of the magnetic circuit are given a stepped cross-section as in Figure 1.42. In very large transformers (one megawatt or higher), the number of steps is higher. In such transformers, the yokes, too, are stepped.

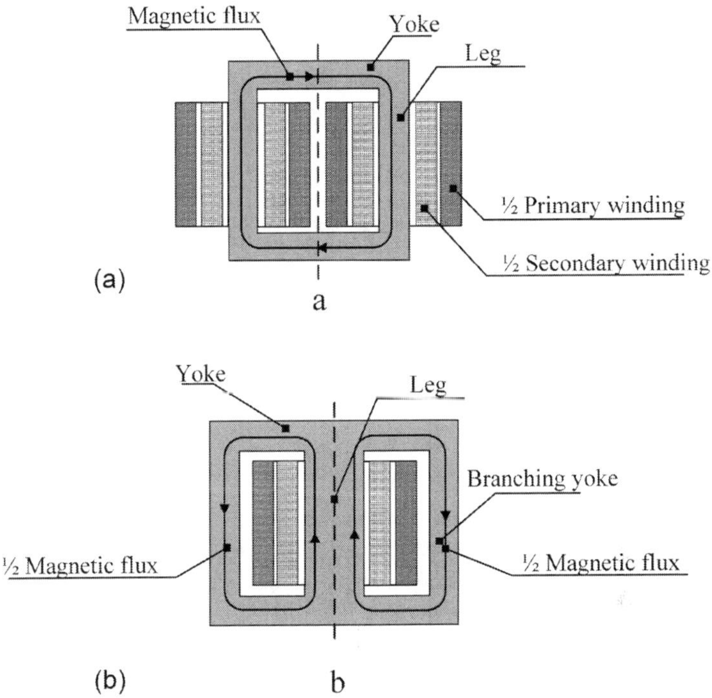

FIGURE 1.40 Cross-section of a core-type transformer (a) and a shell-type transformer (b).

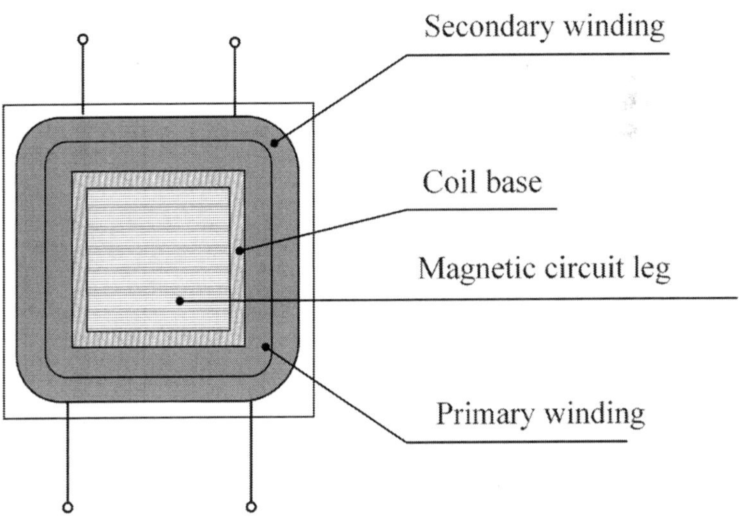

FIGURE 1.41 Cross-section of the leg, coil base, and windings of a small transformer.

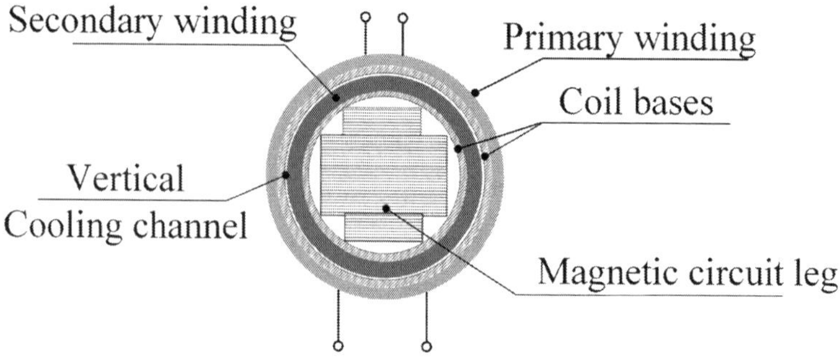

FIGURE 1.42 Cross-section of the leg, coil bases, and winding coils of a power transformer.

1.11.2 Transformer Windings

Windings differ greatly, depending on the rated current and rated voltage. Windings are made either of copper or aluminium wire, more recently also from foil. In an oil-immersed transformer with a copper winding, the current density in the winding is 2.0 to 4.5 MA/m^2 (or A/mm^2), while in a dry-type transformer it is 1.2–3.0 MA/m^2. The upper limit applies to more power transformers. In windings made of aluminium wire, the current densities are 40–45% lower.

Oil-immersed transformers are transformers where the magnetic circuit and windings are placed in a tank filled with oil. The oil provides better cooling, as well as improved resistance of the insulation to the effects of voltage. Most transformers utilised in electrical grids are oil-immersed transformers.

The windings are made of round wire with a cross-section of 0.02–10 mm^2 or profiled wire with a cross-section of 6–60 mm^2. In many cases, the winding coils are put together from several parallel wires. The windings of oil-immersed transformers are made from enamelled and cotton-insulated wire (round wire) or double cable paper and cotton-insulated wire (profiled wire). Meanwhile, the windings of dry-type transformers are made from heat-resistant glass-insulated wire.

Figure 1.43 shows the active component (magnetic circuit and windings) of a three-phase power transformer with concentric windings and the same transformer in its fully assembled form. The active component is placed in an oil tank with cooling ribs. Mounted on the lid of the tank, there are **outlets** for the primary and secondary voltage sides, as well as an **expansion tank** to allow the volume of the oil to respond to temperature changes. There is no air space in the transformer tank. The transformer tank and expansion tank are joined together by a pipe, and normally about half of the expansion tank is filled with oil. In smaller transformers, the expansion tank may be absent. Larger power transformers are usually also equipped with a so-called gas relay, which reacts to the pressure of gases generated in the event of a fault inside the transformer and transmits a signal to switch off the transformer from the high-voltage side by means of a power switch.

FIGURE 1.43 Assembly of the active component of a three-phase power transformer (left) and a fully assembled transformer (right).

2 Electromagnetism

2.1 MAGNETIC FIELDS AND HOW THEY ARE PRODUCED

A magnetic field is a component of an electromagnetic field. Human sensory organs perceive visible light and infrared thermal radiation, both of which are electromagnetic fields that have a very high frequency and short wavelength. The magnetic fields used in electrical machines are either unchanging static fields (e.g. the excitation field in a DC machine) or changing fields of a relatively low frequency (0.1–1 kHz). Such fields are imperceptible to humans.

2.1.1 The Magnetic Field of a Long Straight Wire

Every current-carrying wire has a magnetic field around it. The strength of the magnetic field around a long straight wire can be calculated as follows:

$$H = i / 2\pi r, \tag{2.1}$$

where H – magnetic field strength, A/m;
i – the current in the wire, A;
r – distance from the axis of the wire to the line of force, m.

Equation (2.1) characterises each point of space around the wire. It provides a description of the distribution of the strength of the magnetic field. If we wanted to describe, for example, the distribution of temperature in a room, we could do so using mathematical expressions describing the temperature field. Magnetic fields can be considered an invisible and ever-present component of electric currents. This field that inextricably accompanies a current is also called the H-field. The stronger the current in the wire and the closer the point of space is to the surface of the wire, the stronger the H-field in that point of space. The field is strongest on the surface of the wire.

In addition to strength, the H-field has a **direction**. The H-field is a **vector field**. In the medium it passes through, the H-field generates another field, called the B-field, which we will explore further in the next section. Both the H-field and the B-field are conventionally called magnetic fields. The H-field and the B-field always appear together and are identical in direction, but differ in magnitude. 'Magnetic field' is the common generic name of the H-field and the B-field. Temperature cannot be assigned a direction. A temperature field is a scalar field.

DOI: 10.1201/9781003404088-2

The direction of a magnetic field can be determined, for example, using a small permanent magnet. The compasses used to identify the direction of the Earth's magnetic field contain a magnetic needle, one end of which rotates to point towards the Earth's magnetic north pole and the other end towards the south pole. The direction of the magnetic needle indicates the direction of the magnetic field. The north pole of a compass needle points in the direction of the Earth's geographic north pole, which is near the location of the Earth's magnetic south pole. The Earth's magnetic north pole is located in Antarctica (Figure 1.10g).

In order to visually illustrate the strength and direction of a magnetic field in space, the concept of magnetic field **lines of force** has been adopted. The direction of a line of force indicates the direction of the magnetic field.

Figure 2.1 depicts the field of a straight length of wire, the lines of force of which are in a plane that is perpendicular to the axis of the wire. The direction of the lines of force indicates the direction of the magnetic field. The vectors of magnetic field strength H in points B, C, D, E, F, and G point in the direction of the tangents of the corresponding lines of force. The stronger the field, the more densely the force lines are plotted in a drawing. That is why the lines of force are drawn closer together near the wire. The lines of force of the magnetic field of a long straight piece of wire are circles whose centre is the axis of the wire. Therefore, **all lines of force are closed curves encircling the current-carrying wire.** The strength and direction of the field do not change when moving parallel to the axis of the wire. In different points along all other lines, however, the magnetic field of the wire varies in both magnitude and direction. Such fields are called inhomogeneous. If the direction and strength of a magnetic field are the same in every point of a certain region of space, that field is called **homogeneous**. In the air gaps of electrical machines, the magnetic field is roughly homogeneous.

The relationship between the direction of the current in a wire and the direction of the accompanying magnetic field is given by the corkscrew rule. **When a**

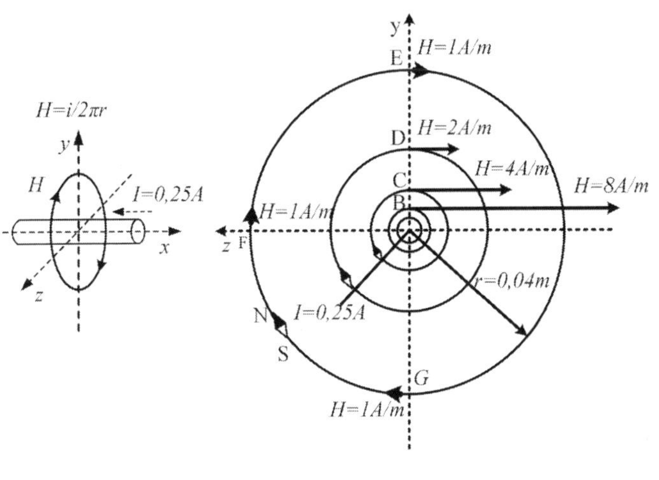

FIGURE 2.1 Magnetic field of a long straight wire carrying a current of 0.25 A.

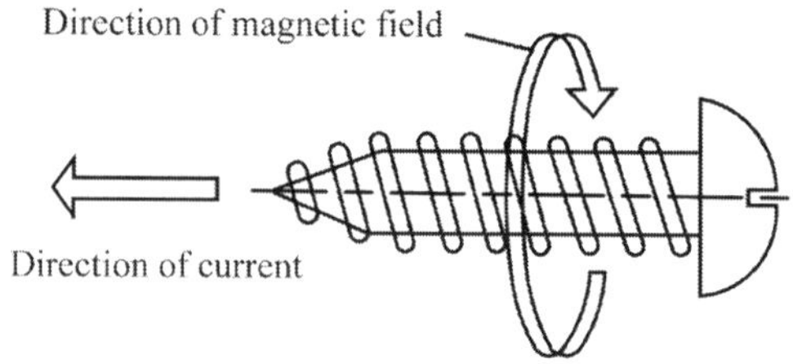

FIGURE 2.2 The right-handed screw rule.

right-handed screw is rotated so that it moves in the direction of the current *i*, the direction of rotation of the screw coincides with the direction of the lines of force of the magnetic field H around the wire. This is illustrated in Figure 2.2.

2.1.2 Magnetic Flux Density Field and Magnetic Flux

When two current-carrying wires are placed close together, a mechanical force arises between them. This is called the induced (caused) force of the magnetic fields of the two currents. It is similar to the force seen between two permanent magnets. Given two currents, one of which is very high and the other very low, the field of the low current will not significantly change that of the high current. This fact enables us to describe the different points of the field of a high current in terms of how strong a force the field exerts on a small wire that is carrying a low current.

By measuring the force at different points in space, we can describe the entire region of space in terms of the mechanical force induced there. This kind of description of space constitutes a specific field (see also Section 1.4). It is called the **magnetic induction field** or **magnetic flux density field**. The two different names refer to different aspects of the same phenomenon. The former indicates that the field is one of an induced force. This field manifests, among other things, as a force.

Meanwhile, the name 'magnetic flux density field' relates to the nature of the field. The *H*-field of a current passing through a wire generates a **magnetic flux** in the space surrounding the wire. Magnetic flux is similar in nature to electric current or fluid flow. Electric current is the movement of electrons. Fluid flow is the movement of molecules. The particles that cause magnetic flux have been called hypothetical magnetic charges. That these charges exist is suggested by the similarity of the properties of magnetic fluxes and electric currents. However, the existence of magnetic charges has neither been proven nor disproven as of yet. Despite the mystery of its inner workings, we know for a fact that magnetic fluxes are real.

The density of a magnetic flux can be expressed via the strength of the magnetic field that induced it:

$$\boldsymbol{B} = \mu \boldsymbol{H}. \tag{2.2}$$

where $\boldsymbol{B}$ is the magnetic flux density vector, which at each point in space is identical in direction to the inducing magnetic field. The unit of measurement of magnetic flux density is **webers per square metre** (Wb/m^2). The unit Wb/m^2 is also called the tesla, denoted by T (1 T = 1 Wb/m^2). 1 T means a magnetic induction in which a wire positioned perpendicularly to the induction field is subjected to a force of 1 N per 1 m of the length of the wire when the wire is carrying a current of 1 A. The factor μ is a quantity characterising the medium, called **magnetic permeability**, or simply **permeability**. The magnetic permeability of vacuum is denoted by μ_0, $\mu_0 = 4\pi \cdot 10^{-7}$ H/m. The magnetic permeability of air is roughly identical to that of a vacuum.

Magnetic field strength H and magnetic flux density B characterise a **point** in space. Magnetic flux Φ, meanwhile, characterises a **surface**.

In the simplest case, if the magnetic flux density field B is homogeneous and the lines of force are perpendicular to a surface A (Figure 2.3), the magnetic flux Φ through that surface can be expressed as the product of the magnetic flux density and the area of the surface.

$$F = BA. \tag{2.3}$$

The unit of magnetic flux is the **weber** (Wb). Magnetic flux is figuratively viewed as the number of lines of force passing through a surface. The stronger the field, the more lines of force per unit of area.

Magnetic flux is not considered to have a direction in space; it is a scalar. However, in magnetic circuits, a magnetic flux can have a positive or negative direction, similarly to electric currents. As a general rule, if the field is irregular, or inhomogeneous, and is not perpendicular to the surface, the magnetic flux can be found by integrating the magnetic flux density over the area A of the surface (Figure 2.4).

$$F = \int_A \boldsymbol{B} \mathrm{d}\boldsymbol{A} = \int_A \boldsymbol{B} \cos\alpha \mathrm{d}\boldsymbol{A}. \tag{2.4}$$

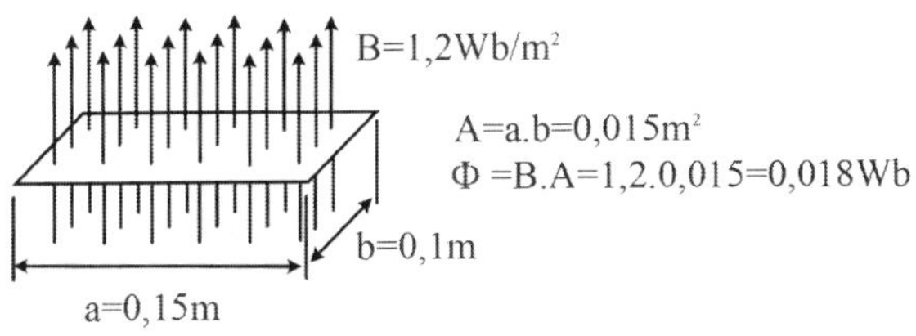

FIGURE 2.3 Magnetic flux through a surface where the magnetic flux density B field is homogeneous and perpendicular to that surface.

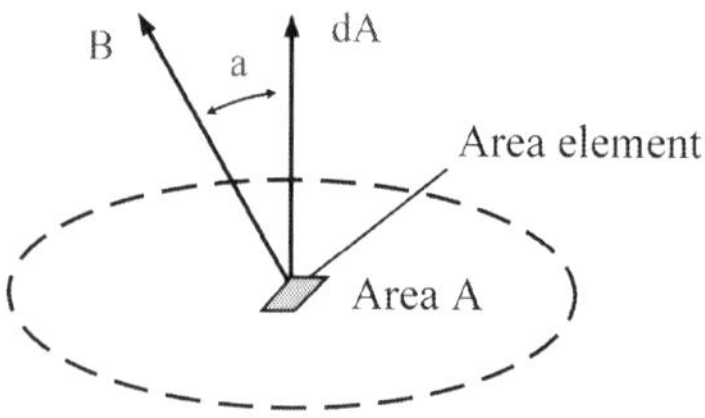

FIGURE 2.4 Calculating magnetic flux by integrating the scalar product of vectors B and dA over the area A.

Here, d$\boldsymbol{A}$ is the vector of an area A element, which is perpendicular to that area element, $\boldsymbol{B}$d$\boldsymbol{A}$ is the scalar product of the vectors $\boldsymbol{B}$ and d$\boldsymbol{A}$, and α is the angle between the vectors $\boldsymbol{B}$ and d$\boldsymbol{A}$.

2.1.3 Analogy between the Generation Processes of a Magnetic Flux, Electric Current, and Fluid Flow

The concepts of H-field, magnetic flux Φ, and magnetic flux density B are easier to comprehend by comparing the process of the generation of a magnetic flux to that of either electric current or fluid flow (Figure 2.5). In the section of the magnetic circuit between areas A_1 and A_2, a magnetic flux Φ (Wb) is generated. Magnetic flux is analogous to electric current I (A) in an electrical circuit and volumetric fluid flow rate Q (m^3/s) in a hydraulic circuit (pipeline).

The magnetic field generating device is called the **source of magnetomotive force**. Similarly, an electric field generating device is called the **source of electromotive force**, or voltage source. In terms of construction, the source of magnetomotive force is an electrical wire wrapped around a magnetic circuit. When a current is passed through the wire, it generates a magnetic field.

To generate a current of 30 A in a cast-iron rod (Figure 2.5 and Table 2.1) with a cross-sectional area of 10 mm^2, an electric field strength of 1.5 V/m is required. Similarly, in our example, to obtain a magnetic flux of 0.014 Wb, you need to apply a magnetic field strength of 200 A/m, and to obtain a volumetric fluid flow rate of 0.6 l/s, you need a pressure field strength of 1,500 Pa/m. Electric field, magnetic field, and pressure field are analogous concepts which describe a **motive effect per unit of length (metre) over a distance**. The motive effects for the entire distance under consideration are, respectively, the voltage U, the magnetisation excitation Hl, or the differential pressure P_{dif}.

A medium impedes motion. In a magnetic circuit, the effect of the medium is described by magnetic permeability μ, and in an electrical circuit, it is described by electrical conductivity γ. The analogue to these in a hydraulic circuit is hydraulic conductivity γ_{hyd} ($m^2/(Pa{\cdot}s)$).

A motive effect produces a current or flux/flow. In a magnetic circuit, this flux is characterised by magnetic flux density B per unit of cross-sectional area (Wb/

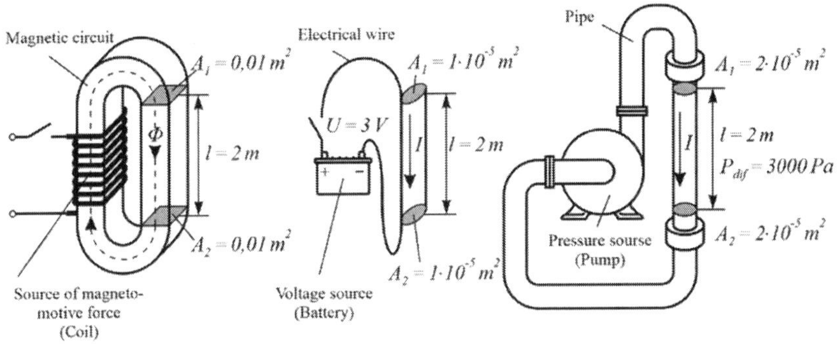

FIGURE 2.5 Generation of magnetic flux, electric current, and fluid flow by means, respectively, of a magnetic field, electric field, and pressure field.

m^2). The analogue to this in an electrical circuit is current density J (A/m^2), and in a hydraulic circuit it is flow velocity (m/s).

In an unbranched circuit, the magnetic flux Φ , electric current I, and volumetric fluid flow rate Q are constant throughout the circuit, but the flux density B, current density J, and flow velocity v may vary between different parts of the circuit, if the magnetic permeability μ, hydraulic/electrical conductivity γ , or cross-sectional area A are different.

2.1.4 Suitability of Fields for Electrical Machines and Introducing Them into the Air Gap

The equations $E = vBl$ and $F = iBl$ (see Section 1.10.1) for the voltage and force induced in a magnetic field indicate that the higher the magnetic flux density B, the greater the voltage E and force F induced in the wire. A higher magnetic flux density B permits electrical machines to be both lighter and smaller. The field of a single straight length of wire is neither very strong nor homogeneous, which makes it inadequate for the air gap of an electrical machine.

To obtain a strong homogeneous field in the air gap, the following methods are used.

- **A coil** is made from **winding wire**. Compared to regular insulated electrical wire, the **insulation layer** of winding wire is very thin (0.01–0.1 mm). The turns of the coil are positioned close together and their fields combine. The field inside the coil becomes significantly stronger.
- **A core made of ferromagnetic material** is placed inside the coil. The magnetic permeability of this core is 2,000–6,000 times higher than that of air. This helps to greatly increase the magnetic flux density in electrical machines, as $B = \mu H$.
- The magnetic field produced by the coil is introduced into the air gap by means of a **magnetic circuit**. The coil core and the magnetic circuit are both usually constructed from the same ferromagnetic material. In fields

TABLE 2.1
Analogy between the parameters of the magnetic circuit, electrical circuit, and hydraulic circuit

Magnetic circuit	Electrical circuit	Fluid circulation system
1. Cross-sectional area of the magnetic circuit $A_1 = A_2 = 0.01$ m^2	1. Cross-sectional area of the cast-iron rod $A_1 = A_2 = 1 \cdot 10^{-5}$ m^2	1. Internal cross-sectional area of the pipe $A_1 = A_2 = 2 \cdot 10^{-5}$ m^2
2. Length of the magnetic circuit between areas A_1 and A_2 $l = 2$ m	2. Rod length $l = 2$ m	2. Pipe length $l = 2$ m
3. Magnetic permeability of the steel material of the magnetic circuit $\mu = 0.007$ H/m	3. Electrical resistivity of the rod $0.5 \cdot 10^{-6}$ Ωm (i.e. 0.5 Ω–m^2/m) and electrical conductivity $\gamma = 2 \cdot 10^6$ Ω^{-1}m^{-1}	3. Hydraulic resistivity of the rod 500 Pa·s/m^2 and hydraulic conductivity $\gamma_{hyd} = 2 \cdot 10^{-3}$ m^2/(Pa·s)
4. The coil was supplied with a current and, as a result, the region between areas A_1 and A_2 was subjected to a magnetisation excitation of $Hl = 400$ A	4. The ends of the rod were supplied, from the battery, with a voltage of $U = 3$ V	4. Starting the pump created a pressure difference between areas A_1 and A_2 in the pipe (i.e. a dynamic differential pressure): $P_{dif} = 3{,}000$ Pa
5. Due to the uniform cross-sectional area and material, a homogeneous magnetic field was generated, with a strength of $H = Hl / l = 400 / 2 = 200\text{A} / \text{m}$	5. Due to the uniform cross-sectional area and material of the rod, a homogeneous electric field was generated, with a strength of $E = U / l = 3 / 2 = 1{,}5\text{V} / \text{m}$	5. Due to the uniform cross-sectional area, a dynamic pressure field with the strength F was generated (the pipe is horizontal): $F = P_{dif} / l = 3000 / 2 = 1500\text{Pa} / \text{m}$
6. The magnetic field caused a magnetic flux. The magnetic flux density is $B = \mu H = 0.007 \cdot 200 = 1.4$ Wb/m^2	6. The electric field generated a current in the rod. The current density is $J = \gamma E = 2 \cdot 10^6 \cdot 1.5 = 3 \cdot 10^6$ A/m^2	6. The pressure field caused the fluid to flow. The average flow velocity of the fluid is $v = \gamma_{hyd} F = 2 \cdot 10^{-3} \cdot 1{,}500 = 3$ m/s
7. The magnetic flux through the cross-sectional area of the circuit is $\Phi = BA = 1.4 \cdot 0.01 = 0.014$ Wb	7. The current is $I = JA = 3 \cdot 10^6 \cdot 1 \cdot 10^{-5} = 30$ A	7. The volumetric fluid flow rate (consumption) is $Q = vA = 3 \cdot 2 \cdot 10^{-5} = 6 \cdot 10^{-5}$ m^3/s = 0.6 l/s

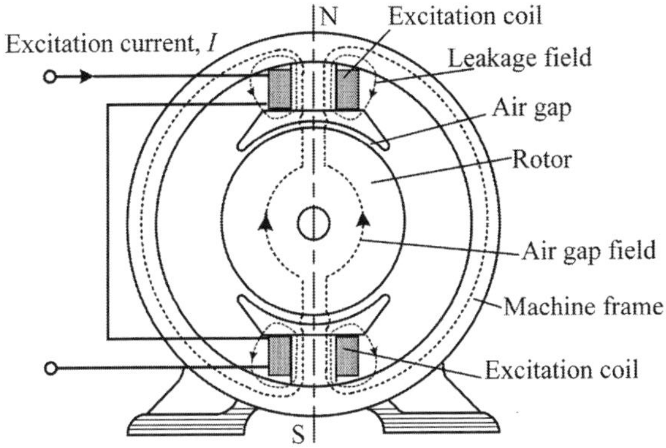

FIGURE 2.6 Magnetic circuit of a DC machine.

> with a frequency of 50 Hz, **electrical sheet steel is used.** A magnetic circuit conducts magnetic fields in roughly the same way as an electrical wire conducts electricity. However, magnetic fields travel in air in a much easier manner than electricity, which can barely pass through air at all. Due to this, a small part of the magnetic field often exits from the magnetic circuit into air in unintended places. This part of the magnetic field is called the **leakage field**.

Figure 2.6 shows the magnetic circuit of the excitation field of a DC machine (the axis of rotation of the machine is perpendicular to the plane of the figure). When a current is run through the two excitation coils, it generates a magnetic field. The cores of the excitation coils form one part of the magnetic circuit. A second part is formed by the turning rotor. A third part is formed by the body of the machine where the magnetic field splits into two. The lines of force of the leakage field encircle the turns of the coil without passing through the air gap. These lines of force do not serve the purpose of the machine. Useful lines of force pass through the air gap.

2.1.5 Ampère's Law and the Magnetic Field of a Coil

The fields of multiple wires combine in space. If multiple wires pass through a surface enclosed by a closed loop, the algebraic sum of the currents of these wires is called the **total current** ΣI passing through that surface. In the case shown in Figure 2.7, $\Sigma I = I_2 + I_4 - I_1 - I_3$. The magnetic field strength H along a closed loop with a length of l and the total current ΣI passing through the loop are related according to Ampère's law.

$$\oint H.dl = \sum I \tag{2.5}$$

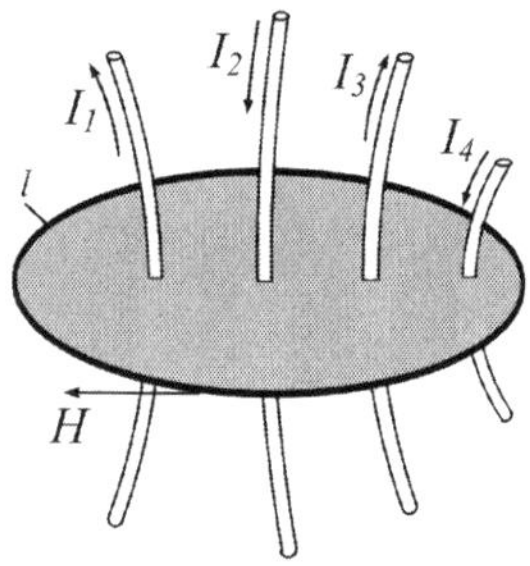

FIGURE 2.7 Determining the strength of a magnetic field produced by multiple current-carrying wires.

where $\boldsymbol{H}$ – vector of magnetic field strength;
d$\boldsymbol{l}$ – loop element.

In the case of a closed loop positioned along a magnetic line of force, $\boldsymbol{H}$ and d$\boldsymbol{l}$ are always identical in direction, which means that their direction may be disregarded. In such a case

$$\int H.\mathrm{d}l = \sum I \tag{2.6}$$

In the case of a single straight current-carrying length of wire, the line of force is circular and the field strength is identical at all points of the line of force. In such a case

$$H\int \mathrm{d}l = I \text{ thus } Hl = I. \tag{2.7}$$

If the line of force is located at a distance r from the axis of the wire, then $l = 2\pi\ r$, which gives us equation (2.1) presented in Section 2.1.1: $H = i/2\pi\ r$.

However, in the case of two lengths of wire that are carrying currents which are equal in intensity but opposite in direction, the total current equals zero $\Sigma\ I = 0$. Nevertheless, the field in the space around these wires is not equal to zero. In such a case, equation (2.6) yields a positive sum for the loop enclosing one wire and a negative sum for the loop enclosing the other wire. Let us consider a field surrounding a circular turn of wire as an example (Figure 2.8 The current I passes through the plane shown in the figure from bottom to top in the right-hand part of the turn and from top to bottom in the left-hand part. Due to this, the lines of force in both the left-hand and right-hand parts of the turn of wire travel in the same direction. The total field (also shown in the figure) is stronger inside the turn than outside of it.

An even stronger field can be obtained by placing a large number of turns carrying a current that proceeds in the same direction close to each other. The resulting construction is called a coil. Figure 2.9 shows the field of a single-layer

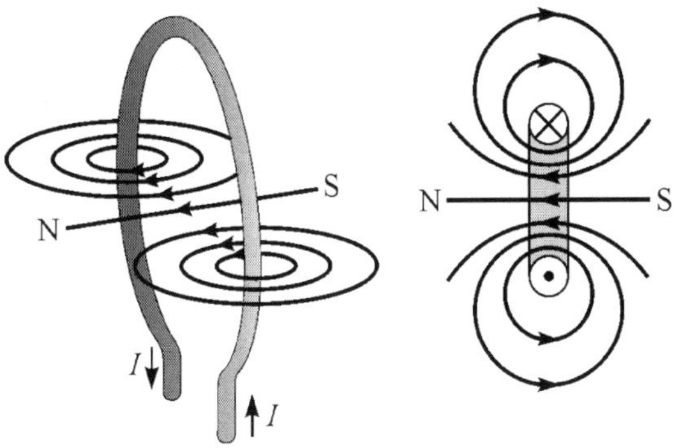

FIGURE 2.8 Magnetic field of a circular turn of wire.

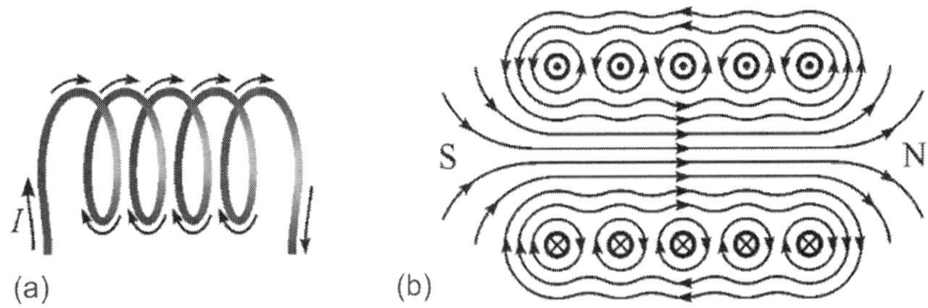

FIGURE 2.9 A coil (a) and the magnetic field in a plane passing through its axis (b).

cylindrical coil, or solenoid. The total current in the coil is equal to the product of the current I and the number of turns of wire w.

$$\sum I = Iw. \tag{2.8}$$

For a coil, Ampère's law as given by equation (2.6) takes the form

$$\int H\mathrm{d}l = Iw. \tag{2.9}$$

2.1.6 Calculation of Magnetic Circuits

Magnetic circuits generally consist of several parts of different lengths l_1, l_2, l_3, … l_k, across each of which the strength of the magnetic field is approximately constant (Figure 2.10).

In such a case, the integral on the left side of equation (2.9) can be replaced with the summation of k

$$\int H.\mathrm{d}l = \sum_1^k H_k l_k \tag{2.10}$$

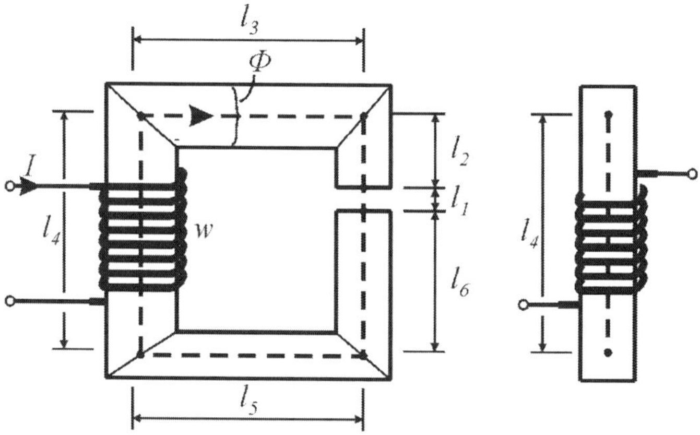

FIGURE 2.10 Application of Ampère's law to a magnetic circuit.

converting Ampère's law to the form

$$H_1 l_1 + H_2 l_2 + H_3 l_3 + \ldots + H_k l_k = Iw \tag{2.11}$$

The product Iw is also called the **magnetomotive force** (denoted as F_{m}), and its unit is the ampere (A), although the ampere-turn was also used in the past. The product $H\,l$ is called the **magnetisation excitation** and is also measured in amperes. Magnetomotive force can be viewed as being the same to a magnetic circuit as electromotive force is to an electrical circuit. The magnetomotive force generated by a coil is the source of the magnetic field in a magnetic circuit.

The longer the magnetic circuit, the higher the magnetomotive force F_{m} needed to produce a field with a certain strength, as

$$H = \frac{Iw}{l} = \frac{F_{\mathrm{m}}}{l}. \tag{2.12}$$

It is often necessary to determine the magnitude of the magnetic flux Φ generated in a magnetic circuit, knowing the magnetisation excitation F_{m}, the cross-sectional area A of the circuit, the length l of the circuit, and the absolute magnetic permeability μ of the material. Given a magnetic circuit with a uniform cross section and an invariable magnetic permeability

$$B = \mu H = \mu \frac{F_{\mathrm{m}}}{l} \tag{2.13}$$

and

$$F = BA = \mu A \frac{F_{\mathrm{m}}}{l} = \frac{F_{\mathrm{m}}}{l / \mu A}. \tag{2.14}$$

The quantity $l/\mu A$ is called the **magnetic resistance**, or **reluctance** R_m:

$$R_m = \frac{l}{\mu A}. \tag{2.15}$$

In order to obtain a higher magnetic flux, it is desirable for the magnetic resistance R_m to also be minimal, as with electrical resistance R. Magnetic resistance decreases as the length of the magnetic circuit decreases, the cross-sectional area of the circuit increases, and the magnetic permeability of the circuit material increases. The magnetic resistance equation (2.15) is analogous to the electrical resistance equation, except with the electrical resistivity of the wire material ρ ($\Omega \cdot$m) replaced with the material's electrical conductivity γ (Ω^{-1}m^{-1}), $\gamma = 1/\rho$. In such a case, the resistance of the wire is given by

$$R = \rho \frac{l}{A} = \frac{l}{\gamma A}, \tag{2.16}$$

where l is the length of the wire in metres and A is the cross-sectional area of the wire in square metres. Comparing equations (2.15) and (2.16) reveals that magnetic permeability μ and conductivity γ are analogous quantities.

Based on equations (2.14) and (2.15), magnetic flux can be calculated as follows:

$$F = \frac{F_m}{R_m}, \tag{2.17}$$

where F_m is the magnetomotive force and R_m is the magnetic resistance (reluctance). Equation (2.17) is analogous to Ohm's law for electrical circuits $I = U/R$. Magnetic flux Φ (Wb) is an analogue of current I (A), magnetomotive force F_m (A) is an analogue of voltage U (V), and magnetic resistance R_m (A/Wb) is an analogue of electrical resistance R (Ω). Equation (2.17) is known as Ohm's law for magnetic circuits.

2.1.7 Magnetic Properties of Ferromagnetic Materials

The magnetic permeability of various materials can be compared to the magnetic permeability of vacuum μ_0 through **relative magnetic permeability** μ_r:

$$\mu_r = \mu / \mu_0, \tag{2.18}$$

where μ is the absolute magnetic permeability of the material.

The magnetic permeability of most materials differs very little from that of a vacuum, in which case it can be said that $\mu_r \approx 1$. Exceptions include iron, nickel, and cobalt, plus some rare metals and their alloys. Their magnetic permeability is much greater. For steels, μ_r = 200–6,000 or even more. These materials are called

ferromagnetic materials. Ferromagnetic properties persist at temperatures up to 800°C, above which they are lost.

While magnetic permeability in a vacuum μ_0 is a constant, it is not so in ferromagnetic materials. Figure 2.11 shows how the magnetic permeability of steel μ_r depends on field strength H. From the curve, we can see that magnetic permeability decreases as the field strength increases. This phenomenon is explained by the structure of ferromagnetic materials.

The atoms of iron and other ferromagnetic materials have a magnetic field. The magnetic fields of neighbouring atoms try to orient themselves in the same direction. As a result, the magnetic fields of the atoms are combined. These regions with a unidirectional field, which are called **magnetic domains**, are extremely small. If we were to break a permanent magnet in half, instead of getting a separate north and south pole, we would get two shorter permanent magnets (Figure 2.12a). This indicates that a permanent magnet consists of numerous small magnets that are uniform in direction. Magnetic domains resemble ultra-small permanent magnets. Typically, magnetic domains are oriented chaotically in space (Figure 2.12b, e), and their combined field is very small.

However, these domains are not always oriented chaotically. When an external magnetic field is applied to a set of magnetic domains (e.g. by using a coil, as shown in Figure 2.12d), the fields of the domains begin to align with the direction of the external field. As a result, the field of the coil and that of the domains are combined (Figure 2.12c, f).

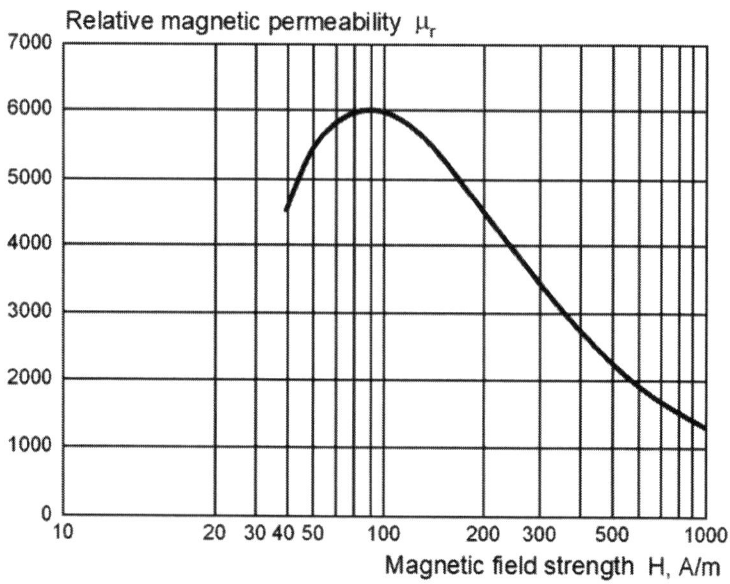

FIGURE 2.11 Relationship between the relative magnetic permeability of steel and magnetic field strength.

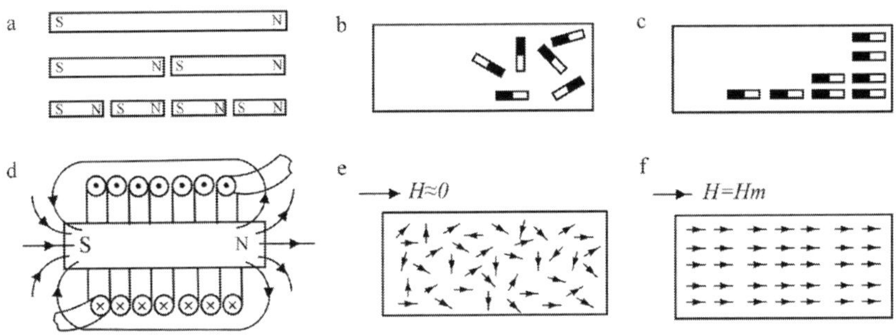

FIGURE 2.12 Splitting of a magnetised steel rod (a), chaotic orientation of domains in non-magnetised steel (b, e), their orientation in the presence of a strong external magnetic field (c, f), and the magnetisation of a steel rod in the magnetic field of a coil (d).

In a certain range of field strengths (50–150 A/m in Figure 2.11), the change in μ_r is quite small. Within this range, the number of domains aligned with the external field and the field of the domains increases roughly in proportion to the external field. The more the external field increases, the more domains will align with the field. Eventually, there will be no more unaligned domains (Figure 2.12f). The field of the domains will stop increasing and μ_r will become small. This phenomenon is called **magnetic saturation**. Saturation occurs at a certain *H*-field strength, which varies between materials.

The magnetisation of materials is more frequently described using magnetisation curves $B = f(H)$ (Figure 2.13a). A higher μ value corresponds to a curve with a steeper slope. In some materials (such as permalloy), the slope of the curve, and thus also μ, increases very sharply. A sharp change in the curve indicates rapid saturation. The region where the slope of a magnetisation curve changes abruptly (200–300 A/m for permalloy) is called the knee point of the curve. The nearly horizontal section of a magnetisation curve is called the saturation region. In electrical steels, saturation occurs between 1.0 and 1.8 T. Higher magnetic flux densities are generally infeasible in electrical machines. Exceptions to this include the fields of machines which utilise superconductor coils, where the magnetic flux density is 2–10 T.

When using an AC field instead of a DC field, the magnetisation curve is replaced by a magnetisation loop, or hysteresis loop (Figure 2.13b). The loop arises as follows. When the current is zero at first and increases for the first time, the accompanying increase in magnetic flux density *B* follows the curve 0A. If we then begin to reduce the current and, with it, the magnetic field strength *H*, the flux density *B* decreases according to the curve AD. **The decrease and increase in current follow different curves.**

In point D, the external field has been reduced to zero, but a **residual magnetic flux** Φ with the flux density *B* is still present in the core. This indicates that not all magnetic domains have returned to their previous random orientation. Some domains maintain the orientation of the previously applied external field (memory effect). Why is that? It is because rotating the atoms in the domains requires

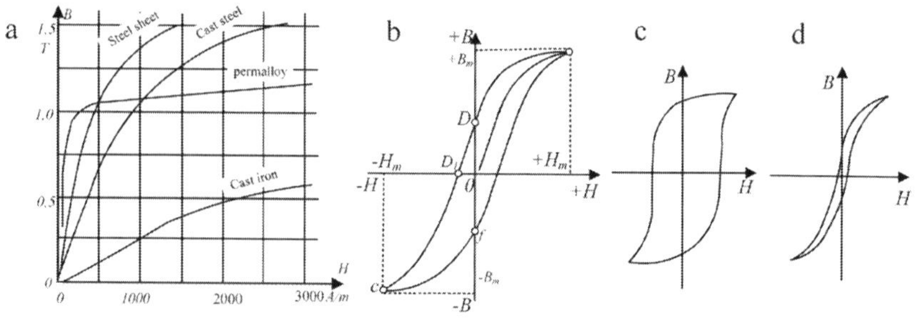

FIGURE 2.13 Magnetisation curves of ferromagnetic materials (a); re-magnetisation of a ferromagnetic material (b); hard magnetic material (c); soft magnetic material (d).

spending energy. If previously the energy of the external magnetic field was used to align the atoms with the field, then after removing the field, there is no energy source available to return all of the domains to their initial orientation. Once the domains have already been aligned one way, a certain proportion of them will maintain this orientation until it is changed by an external energy source. This is why we get a permanent magnet with a residual magnet flux.

In order to get from point D to point D_1 (Figure 2.13b), where the residual magnetic flux is zero, a field that is opposite in direction must be applied. The magnetomotive force required to obtain a residual magnetic flux of zero is called the **coercive force**. The external energy source may also be a strong mechanical shock or input heat. If a permanent magnet falls to the ground, it can partially lose its magnetic properties.

When our AC-generated H-field reaches its negative maximum value $-H_{\mathrm{m}}$ (Figure 2.13b), the magnetic flux density, too, reaches its negative maximum $-B_{\mathrm{m}}$ (in point C). When the AC-generated H-field returns from $-H_{\mathrm{m}}$ to $+H_{\mathrm{m}}$, re-magnetisation occurs according to the curve CFA. Over the course of each AC cycle, re-magnetisation follows the hysteresis loop ADD_1CFA.

For a permanent magnet, it is desirable to have a high residual magnetic flux density and coercivity. Permanent magnetic materials have a wide hysteresis loop (Figure 2.13c). Such materials are said to be **magnetically hard**.

Since re-orienting domains in iron requires energy, energy losses occur in the parts of the magnetic circuits of machines and transformers where alternating magnetic fields are used. The area of the hysteresis loop of steel can be shown to be proportional to the energy spent to re-orient domains in one cycle. The narrower the hysteresis loop, the smaller the re-magnetisation losses. Re-magnetisation losses are also called hysteresis losses. Materials that are designed for use in an alternating magnetic field have a narrow hysteresis loop (Figure 2.13d). Such materials are said to be **magnetically soft**.

In addition, there is another type of loss caused in cores by a variable magnetic field: **eddy current loss**. Hysteresis losses and eddy current losses are collectively referred to as core losses, or **iron losses**.

Permanent magnetic materials

The history of permanent magnetic materials is very long, dating back 4,000 years. The first permanent magnetic material discovered was naturally occurring mineral-rich magnetic iron ore Fe_3O_4. This material was known both in China and in ancient Greece. The first known application for this magnetic material was in land compasses that helped caravans navigate desert landscapes.

The properties of permanent magnets started to be improved in the 1880s, when a thorough study of metal alloys began. Adding tungsten and chromium was found to increase the coercivity of materials. The higher the coercivity, the better. The coercive force shows how large a magnetic field is needed to demagnetise the permanent magnet. The greater the coercive force, the larger the external fields that the permanent magnet can be used in without demagnetising it. Another important parameter of permanent magnets is **energy product**. Energy product is usually measured in kilojoules per cubic metre and it describes the amount of energy needed to magnetise a material. A permanent magnet with a higher energy product, all other properties being the same, will be smaller.

In the 20th century, a breakthrough was made in the field of permanent magnets, when magnets made from aluminium–nickel–cobalt alloys (AlNiCo), with a coercivity of 32 kA/m, were developed. In addition, in 1940–50, a family of permanent magnets known as ferrites was discovered. Ferrites became popular in DC electrical machines used in cars, hand tools, etc.

Magnetic materials made from samarium–cobalt alloys (SmCo) date back to 1960. These magnetic materials are well suited for use in electric motors and generators (coercivity 1,000 kA/m, energy product 100–200 kJ/m^3) but are very expensive due to the content of cobalt, which is a rare raw material. The latest and most important discovery in the field of permanent magnetic materials was made in 1983, when a neodymium–iron–boron (NdFeB) magnetic material was developed, with an energy product of up to 400 kJ/m^3 and a coercivity of 2,000 kA/m. Compared to SmCo permanent magnets, NdFeB permanent magnets have superior properties and are cheaper as well. Their downsides are vulnerability to corrosion, which is why they require a proper protective coating, and low mechanical strength. Compared to SmCo magnets, they also have a lower temperature tolerance (maximum operating temperature ranging from 110 to 180°C).

Properties of permanent magnetic materials

Figure 2.14 shows a typical hysteresis loop for a permanent magnet. The properties of a permanent magnetic material, or a magnetically hard material, are chiefly described by the line in the upper left-hand quarter of its hysteresis loop, which is called the demagnetisation curve of the magnet.

Permanent magnets are described using two curves (Figure 2.14). One of them shows the interdependence of the *B*-field and the *H*-field, which was discussed above. The difference here, however, is that the value of the *B*-field measured in teslas is divided by the magnetic permeability of vacuum μ_0, $B_\mu = B/\mu_0$. The unit for this reduced magnetic flux density B_μ thus becomes A/m, as in the *H*-field.

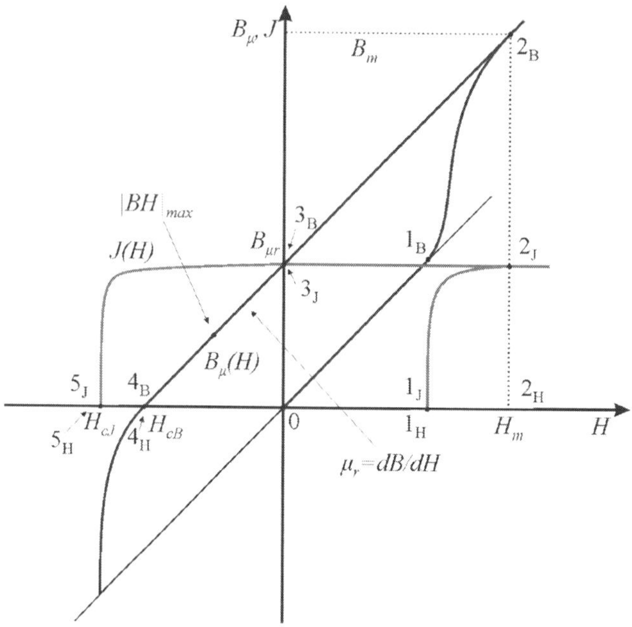

FIGURE 2.14 $B_\mu(H)$ and $J(H)$ curves of a permanent magnet. The curve lines in the upper left-hand quarter are called demagnetisation curves.

Secondly, permanent magnets are characterised by **magnetic polarisation**, or the J-field. The unit of measurement of the J-field is A/m, as in the H-field. All three of the vectors, which characterise the B_μ-field, H-field, and J-field, are identical in direction and related as follows: $\boldsymbol{B}_\mu = \boldsymbol{J} + \boldsymbol{H}$. The J-field describes the total field generated by the ultra-small domains inside the permanent magnet, while the H-field describes the field generated by the electric current in the coil.

Figure 2.14 depicts a situation where an initially non-magnetised coil core made from a permanent magnetic material is placed inside a coil, after which the coil core is made to undergo magnetisation and demagnetisation by application of the field generated by the current in the coil (the H-field on the horizontal axis). During this process, both the B_μ-field (on the vertical axis) and the J-field (also on the vertical axis) begin to change. The J-field shows whether the coil core has developed permanent magnetic properties and how strong they are.

When the core is non-magnetised and no current is present in the coil, all three fields are equal to zero and we are at the starting point, i.e. the origin, of the co-ordinate plane. If we then begin to increase the current in the coil in the positive direction, we will start to move along the curve $B_\mu(H)$ towards point 1_B. Note that the magnetic polarisation J will remain equal to zero up to point 1_J.

When, upon increasing the current in the coil further, the H-field reaches point 2_H from point 1_H, the J-field undergoes a massive change. This field increases from point 1_J to point 2_J. Initially, the *J*-field increases very rapidly, but upon reaching point 2_J, the rate of increase becomes minuscule. This is an expression of the fact that the domains inside the permanent magnet align with the external field quickly at first, until, at point 2_J, a state of saturation is reached (all domains are already aligned with the *H*-field). During the transition from point 1 to point 2, the permanent magnetic material is magnetised. A permanent magnet is formed.

If we were now to reduce the current in the coil to zero (the *H*-field is reduced to zero), the B_μ-field would decrease from point 2_B to point 3_B. Despite the absence of the coil's field, however, we would still have a B_μ-field – generated by the newly formed permanent magnet – with a relatively high flux density. This flux density is called the **residual flux density**. In Figure 2.14, it is denoted by $B_{\mu r}$, because the figure shows the reduced flux density values (indicated by the index μ). However, permanent magnets are described using unreduced residual flux density values, which are denoted by B_r. The quantities B_r and $B_{\mu r}$ are related as follows: $B_r = \mu_0 \cdot B_{\mu r}$. NdFeB-type permanent magnets have the highest B_r values – at up to 1.3 teslas. In ferrite magnets, meanwhile, residual flux density ranges from 0.2 to 0.4 T. The higher the residual flux density, the better, as this allows for a higher magnetic flux density in the air gap of the magnetic circuit in which the permanent magnet is utilised.

If we were to reverse the current in the coil at the origin and begin to increase the current in the negative direction, the *H*-field, too, would increase in the negative direction. This would cause the B_μ-field to decrease. When the *H*-field reaches point 4_H by increasing in the negative direction, the B_μ-field will have become equal to zero. The value of the *H*-field at point 4_H is called **normal coercivity** and is denoted by H_{cB}. The greater this coercivity value, the thinner the magnet can be without becoming demagnetised by the external field. If we were to take the *H*-field back to the origin from point 4_H, the flux density B_μ would recover and return to the value it previously held at point 3_B. This is because the *J*-field imparted by the permanent magnet has remained unchanged (the negative *H*-field was not strong enough to reduce the *J*-field to zero).

If, however, we were to increase the demagnetising *H*-field all the way to point 5_H, the magnetic flux density B_μ would become negative at point 5_H and the *J*-field, which describes magnetic polarisation, would reach a value of zero at point 5_J. This indicates that the magnet has been demagnetised. The strength of the *H*-field at point 5_H is called **intrinsic coercivity** and is denoted by H_{cJ}. H_{cJ} shows how strong an external field is required to demagnetise the magnet.

If we were to take the *H*-field back to the origin from point 5_H, the B_μ-field would also decrease to zero. The *J*-field, meanwhile, would remain at zero for the whole duration, because the magnet has now been demagnetised.

The demagnetisation curve of a permanent magnet shows the magnet's basic parameters, which are residual flux density B_r, intrinsic coercivity H_{cJ}, and normal coercivity H_{cB}. From this curve you can also determine the maximum energy product $|B \cdot H|_{max}$ and the relative magnetic permeability μ_r of the magnet.

The maximum energy product $|B \cdot H|_{max}$ of a magnet corresponds to the point on the demagnetisation curve where the product $B \cdot H$ has the highest value. Energy product is usually measured in joules per cubic metre (J/m^3), but the gauss-oersted (G Oe) is also commonly used.

The relative magnetic permeability μ_r of a permanent magnet shows how sharply the curve $B(H)$ drops in the linear starting region. It usually ranges from 1 to 3.5 in permanent magnets.

Modern permanent magnetic materials

Ferrite magnets are made from either iron oxide and barium carbonate $BaCO_3$ or strontium carbonate $StCO_3$. Ferrite magnets are used widely and are considerably cheaper than other magnetic materials on the market.

AlNiCo magnets are made from alloys of aluminium (Al), nickel (Ni), and cobalt (Co), with minor additions of other elements that improve the properties of the magnetic material. AlNiCo magnets have both excellent temperature tolerance and resistance to corrosion, but are easily demagnetised. Their downsides are that they are expensive and have a low coercivity and energy product.

SmCo magnets belong to a group of rare-earth magnets with excellent resistance to corrosion and higher energy products than ferrite and AlNiCo magnets. They also offer better temperature stability than NdFeB magnets. However, their disadvantages are their high price and low mechanical strength.

NdFeB magnets (neodymium magnets) have better magnetic properties than SmCo magnets, but are not as resistant to corrosion nor as temperature-stable. Due to their high energy product, neodymium magnets have found use in applications where compactness is important. NdFeB magnets are prone to corrosion and therefore require a protective coating, which can be either copper, silver, gold, nickel, zinc, tin, or epoxy.

Table 2.2 presents a comparison of the different types of magnets.

2.2 VOLTAGE INDUCED BY A TIME-VARYING MAGNETIC FIELD

2.2.1 Faraday's Law and Lenz's Law

Above, we looked at how voltage and electromotive force are induced in an electrical conductor that is moving in a magnetic field. Now let us explore a magnetic field that is changing over time.

A time-varying magnetic flux can be obtained by changing the current passing through a winding wrapped around a magnetic circuit. The current in our

TABLE 2.2
Comparison of parameters of permanent magnetic materials

Parameter	Symbol	Unit	Ferrite, pressed/cast	SmCo, pressed/cast	NdFeB, pressed/cast
Residual flux density, 20°C	B_r	T	0.4/0.2	1.1/0.6	1.3/0.7
Max energy product	BH_{max}	kJ/m^3	20–30	100–200	200–300
Temperature coefficient	α_{Br}	%/K	–0.2	–0.03	–0.1/–0.1
Temperature coefficient	α_{HcJ}	%/K	0.4	–0.23	–0.6/–0.4
Coercive force, 200°C	H_c	kA/m	–200	–1,000	–2,000
Electrical resistivity	ρ	Ω	/	/	/1.4
Thermal conductivity	λ	W/(m)	4.5/	10/	9/2
Operating temperature	t	°C	/150	250/110	
Price			Low	High	Medium

winding w_1 can be changed by changing the resistance R of the rheostat (Figure 2.15). A rheostat is a variable resistor. Given a second winding w_2 in the magnetic circuit, changing the magnetic flux Φ produces an electromotive force e in it. This is indicated by the voltmeter. We are interested in the relationship between Φ and e. Measuring a magnetic flux directly is difficult. It is easier to measure the current in the winding with an ammeter. In the presence of an air gap in the magnetic circuit, the magnetic flux changes roughly in proportion to the current.

If we change the current and use a self-recording ammeter and voltmeter, we get the curves shown in Figure 2.16. Curves like these that describe changes over time are also called oscillograms. In our oscillograms, we can see that an electromotive force (emf) is induced in the winding w_2 only when the magnetic flux changes. When the magnetic flux is constant, the emf is equal to zero.

The direction of the induced emf is different for an increasing magnetic flux and a decreasing magnetic flux. The magnitude of the electromotive force depends on the rate of increase of the magnetic flux. The faster the flux increases, the greater the emf. Electromotive force also increases as the number of turns of wire increases. When connected to the starting point 1 of the winding, a voltmeter would report a voltage that is two times higher than when connected to the tapping point 2, because in the latter case, the number of turns would be two times lower.

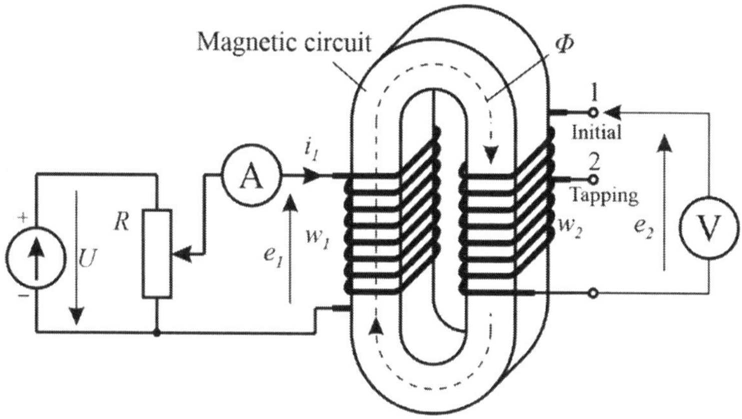

FIGURE 2.15 Generation of a time-varying magnetic flux and induction of an electromotive force by means thereof.

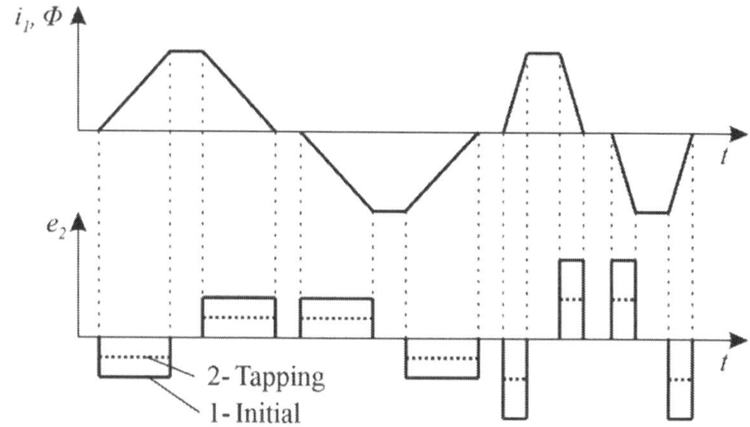

FIGURE 2.16 Electromotive force induced in the winding w_2 by the varying magnetic flux.

The induced electromotive force is generally determined by **Faraday's law**. Faraday's law states that **when a changing magnetic flux passes through a turn of wire, an electromotive force that is proportional to the rate of change of the flux is induced in that turn.** This law is expressed by the following equation:

$$e = -\frac{\mathrm{d}F}{\mathrm{d}t}, \tag{2.19}$$

where e is the induced electromotive force and Φ is the magnetic flux. The minus sign in the equation derives from Lenz's law on the polarity of induced voltages. **Lenz's law** states that the direction of the current induced in a closed loop is such that the magnetic field generated by the current opposes changes in the current-inducing magnetic flux. It follows from this law that **the direction of the voltage induced in a coil is such that the field of the current generated by the voltage works to reduce the initial change in flux** (an initially increasing flux is reduced by the field of the current, while an initially decreasing flux is increased by the field of the current).

To enable a current to be induced in the winding w_2, the resistor R_2 is connected to this winding (Figure 2.17). The direction of the induced voltage and current i_2 is such that the direction of the magnetic flux Φ_2 caused by the current i_2 is the opposite of the direction of the primary flux Φ_1 if the primary flux is increasing. If the primary flux is decreasing, the flux of the induced current is identical in direction to the primary flux. The flux of the induced current opposes changes in the primary flux.

2.2.2 Self-Induction and Inductance

Figure 2.17 shows two coils, w_1 and w_2, in a common magnetic circuit. The magnetic flux in this circuit is changing. The question arises whether the change in flux induces an electromotive force only in the winding w_2 or also in the winding w_1. Upon closer investigation, we will find that an electromotive force is indeed induced in both windings. The fact that the winding w_1 is itself the cause of the changing magnetic flux Φ_1 does not matter. The electromotive force induced in the winding w_1 is called a **self-induced electromotive force**. The resulting self-induced electromotive force may be considered with the voltage source e that is connected in series to the winding (Figure 2.18).

When the current is increasing, the self-induced electromotive force e is opposite in direction to the supply voltage U. When the current is decreasing, U and e are identical in direction. When the current is increasing, the current i enters the voltage source e via the positive terminal. This direction of current corresponds to the supply of energy to the voltage source. In this case, energy is supplied to the magnetic field, since the voltage of the voltage source is the self-induced

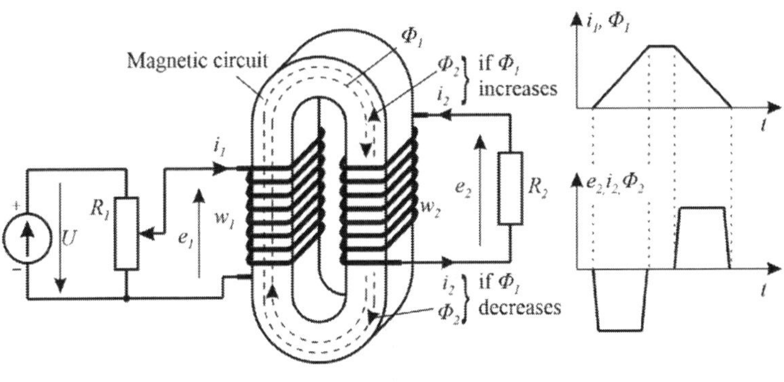

FIGURE 2.17 Emergence, due to induced current, of the magnetic flux Φ_2 in opposition to changes in the primary flux Φ_1.

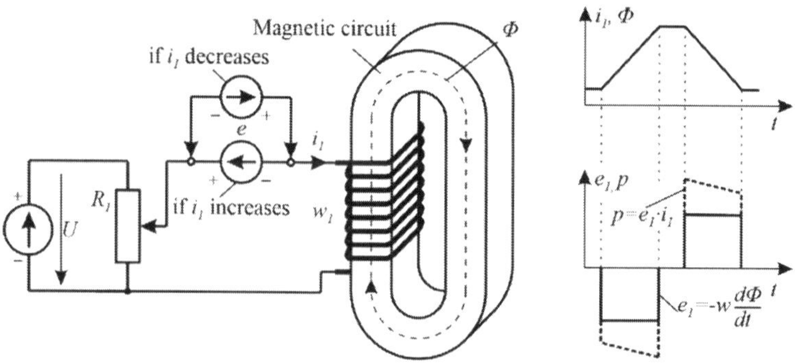

FIGURE 2.18 Generation of self-induced electromotive force by changes in magnetic flux, and the associated exchange of energy between the magnetic field and the electrical circuit.

electromotive force e of the winding. The instantaneous value of the power of the flow of energy to the magnetic field is

$$p = ie. \tag{2.20}$$

When the current is decreasing, the polarity of the voltage source is the opposite of that of the current. The current exits via the positive terminal of the voltage source e (Figure 2.18). This direction of current corresponds to the extraction of energy from the voltage source. In this case, the energy extracted is that stored in the magnetic field, which is returned to the electrical circuit. The exchange of energy between the magnetic field and the electrical circuit occurs by way of the self-induced electromotive force.

The minus sign in equation (2.19) indicates the direction of the self-induced electromotive force relative to the supply voltage. However, this minus sign is often omitted and the correct direction of the electromotive force is later determined on the basis of other criteria.

In a coil with multiple turns of wire, the induced electromotive force is

$$e = -w\frac{\mathrm{d}F}{\mathrm{d}t}. \tag{2.21}$$

The sum of the magnetic fluxes linked with, i.e. passing through, all turns of a coil is called the **flux linkage** Ψ. If all turns are linked with an identical magnetic flux Φ , then

$$\Psi = w\Phi. \tag{2.22}$$

Flux linkage is measured in webers (Wb). Inserting the concept of flux linkage into equation (2.21), the equation becomes

$$e = -\frac{d\Psi}{dt}. \tag{2.23}$$

In the absence of ferromagnetic materials in the medium, the magnetic flux Φ and the flux linkage Ψ are proportional to the current i passing through the coil. The proportionality factor of flux linkage and magnetisation current, i.e. the ratio of the former to the latter,

$$\Psi / i = L \tag{2.24}$$

is called **inductance**, the unit of measurement of which in the SI system is the henry (H); 1 H = 1 Wb/A. Based on equation (2.24), flux linkage is equal to the product of current and inductance. Combining and converting equations (2.24), (2.22), (2.17), and (2.12) reveals the following relationship for inductance:

$$L = \frac{w^2}{R_m}, \tag{2.25}$$

where R_m is the magnetic resistance of the magnetic circuit. The above equation indicates that the inductance of a coil is proportional to the square of the number of turns and inversely proportional to the magnetic resistance of the magnetic circuit. Magnetic resistance depends on the magnetic permeability μ of the medium and the dimensions of the magnetic circuit. A coil can also lack a magnetic circuit. In such a case, the inductance depends solely on the number of turns and the dimensions of the coil. Thus, **inductance is determined by the design of the coil and the properties of the medium.** A conventional straight length of electrical wire that has not been wound into a coil has an inductance too, albeit a low one. Winding a straight length of wire into a coil significantly increases the inductance of the wire.

2.2.3 Inductor as an Energy Storage Component in an Electrical Circuit

It follows from equation (2.24) that flux linkage is equal to the product of current and inductance:

$$\Psi = iL. \tag{2.26}$$

Given two coils carrying an equal current, the flux linkage will be higher in the coil with the higher inductance. **Inductance describes the amount of magnetic flux associated with the coil.**

Self-induced electromotive force can be expressed in terms of inductance. In a non-ferromagnetic medium, μ is constant, as is inductance L. In such a case

$$u = e = \frac{\mathrm{d}\Psi}{\mathrm{d}t} = \frac{\mathrm{d}(iL)}{\mathrm{d}t} = L\frac{\mathrm{d}i}{\mathrm{d}t}. \tag{2.27}$$

The electromotive force e of a coil must be balanced by the circuit's voltage source with voltage U. This is illustrated in Figure 2.19, where $U = e$. The voltage of the voltage source determines the rate of increase of the current

$$\mathrm{d}i / \mathrm{d}t = U / L. \tag{2.28}$$

The rate at which the current in an inductive circuit changes is proportional to the applied voltage and inversely proportional to the inductance. In the absence of resistance in the circuit, the current will become infinitely high. In reality, however, there is always a resistance present, which limits the increase of the current. In an established state, or steady-state operation, the current can be calculated on the basis of Ohm's law $I = U/R$.

A coil is an energy-storage component of an electrical circuit. The inductance of a coil defines its energy storage capacity. The magnetic field energy W_m stored by a coil can be calculated by integrating the instantaneous power p supplied to the coil over time

$$W_\mathrm{m} = \int p\mathrm{d}t = \int ui\mathrm{d}t. \tag{2.29}$$

Based on equation (2.27) and the fact that $u = e$, equation (2.29) can be rewritten as

$$W_\mathrm{m} = \int iL\frac{\mathrm{d}i}{\mathrm{d}t}\mathrm{d}t = L\int i\mathrm{d}i = \frac{Li^2}{2}. \tag{2.30}$$

Figure 2.20a shows for two coils the electromotive force caused by the inductance of the coil, the flux linkage passing through the coil, and the coil's energy and the

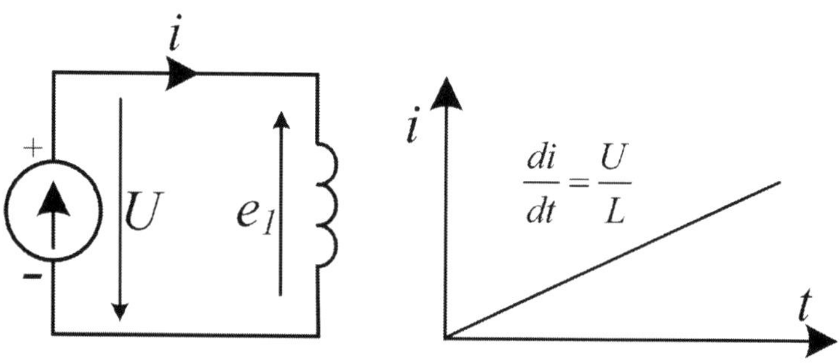

FIGURE 2.19 Explanation of the rate of increase of current.

power of its change. Observe how the curves for the electromotive force (which is essentially a voltage) and the current are shaped differently in the figure. This is due to the fact that the voltage is not proportional to the current, but to the rate of change of the current (equation 2.27). The relationship between flux linkage ψ and current is linear, while that between energy and current is quadratic. If the energy increases linearly, the current increases quadratically.

The equation for calculating the energy of a magnetic field is similar to the equation for the energy W_{el} of an electric field

$$W_{el} = \frac{Cu^2}{2}, \tag{2.31}$$

where C is the capacitance and u is the voltage of the capacitor. Just as the inductance of a coil defines the coil's energy storage capacity, the capacity of a capacitor defines that of the capacitor.

Figure 2.20b shows the charging of two capacitors C_1 and C_2 with a constant current. The oscillograms of an inductive circuit and a capacitive circuit are similar. Current is the equivalent of voltage and vice versa. Flux linkage ψ is the equivalent of charge q. The mathematical equations are analogous. Inductive circuits and capacitive circuits are dual circuits. Inductors and capacitors, or the inductance and capacitance of a component of an electrical device, are energy storage mechanisms. Meanwhile, resistors, or the resistances of the components of an electrical device, are energy converters, as they convert electrical energy into thermal energy.

Resistors, inductors, and capacitors are the fundamental objects of electrical engineering. They are tangible material objects. Current, magnetic flux, magnetic fields, and electric fields are objects as well, but they are not as tangible as material objects. They can be considered a special form of matter.

In addition, space, time, and energy also fulfil the function of objects in the fundamental relations of electrical engineering. For example, the value of a voltage induced in a wire is directly dependent on space and time (the length and travel speed of the wire). The fundamental objects of electrical engineering are grouped together in Table 2.3, which also contains a brief description of each object and their basic parameters together with the units of measurement.

Capacitance and inductance are utilised in the conversion of electrical energy. Electrical energy can be stored in the magnetic field of a coil at a particular voltage U_1, but returned from the magnetic field to the electrical circuit at any other voltage. This fact is used for changing DC voltage levels. In addition to inductance L, converters that perform such functions utilise a switch S (Figure 2.21a). At a certain point in time t_0, the switch S is closed, which switches the coil over to the voltage of the power supply (12 V). The current in the coil then begins to increase at a rate of $\mathrm{d}i_L/\mathrm{d}t = U_1/L$ (see Figure 2.21b). This lasts from the moment t_0 to the moment t_1, when the switch S is opened.

The current i_L in the coil cannot change instantaneously, because it is linked to the energy of the magnetic field. What *can* change instantaneously, however, is the

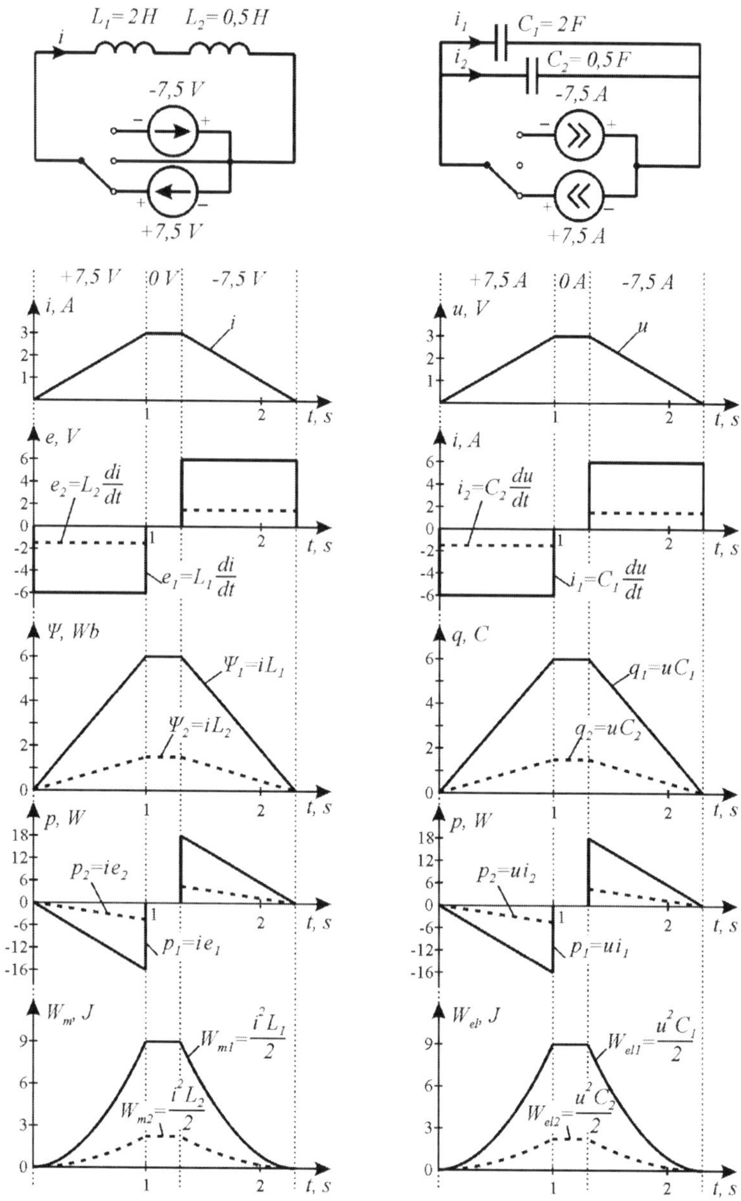

FIGURE 2.20 Storage of energy in inductors and capacitors.

coil's self-induced electromotive force e, which changes according to Faraday's law. When the current i_L tries to start decreasing after the switch S has been opened, the self-induced electromotive force e arises abruptly, shown in Figure 2.21a as a virtual voltage source. This electromotive force is precisely as strong as to cause, when the switch is operated at the moment t_1, the current i_L to remain

TABLE 2.3
Fundamental objects of electrical engineering

Object type	Tangible			Intangible						
Object	**Resistor**	**Inductor**	**Capacitor**	**Current**	**Magnetic flux**	**Magnetic field**	**Electric field**	**Space**	**Time**	**Energy**
Description	Converts electrical energy into thermal energy. Causes a voltage drop when supplied with current.	Stores energy in a magnetic field. Causes a voltage drop when supplied with alternating current.	Stores energy in an electric field.	Movement of electric charges in space. Generates a magnetic field.	Movement of hypothetical magnetic charges in space	Change in the properties of a region of space, manifests in the generation of an induced force and voltage	Change in the properties of a region of space, manifests as a force acting on charged part:cles			
Electronic symbol	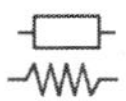									
Basic parameters	Resistance	Inductance Inductive reactance	Capacitance Capacitive reactance	Current Current density	Magnetic flux Magnetic flux density	Magnetic field strength Magnetisation excitation Magnetomotive force	Electric field strength Voltage Electromotive force	Length Area Volume	Time interval Period Frequency Angular frequency	Energy Power

(Continued)

TABLE 2.3 (CONTINUED)
Fundamental objects of electrical engineering

Object type	Tangible				Intangible					
Object	**Resistor**	**Inductor**	**Capacitor**	**Current**	**Magnetic flux**	**Magnetic field**	**Electric field**	**Space**	**Time**	**Energy**
Parameter symbols	R	$L\ x_L$	$C\ x_C$	$I\ J$	$\Phi\ \mathbf{B}$	$\mathbf{H}\ Hl\ F_m$	$\mathbf{E}\ U\ E$	$l\ A\ V$	$\Delta t\ T\ f\ \omega$	$W\ P$
Unit of measurement	Ohm	Henry Ohm	Farad Ohm	Ampere Ampere per square metre	Weber Weber per square metre (Tesla)	Ampere per metre Ampere Ampere	Volt per metre Volt Volt	metre square metre cubic metre	second Hertz Radian per second	Joule Watt
Unit symbol	Ω	H Ω	F Ω	A A/m^2	Wb Wb/m^2 (T)	A/m A A	V/m V V	m m^2 m^3	s Hz rad/s	J W

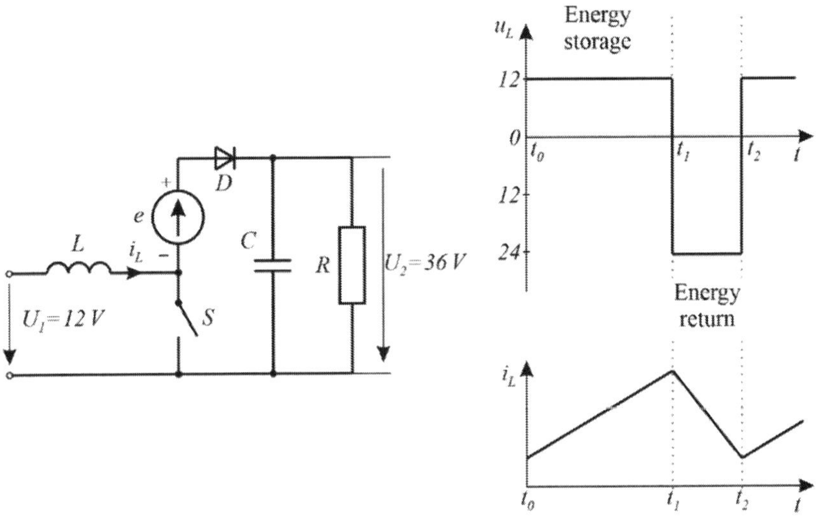

FIGURE 2.21 Use of a magnetic field in a boost converter.

unchanged compared to before the moment of switching t_1. After the moment of switching t_1, however, the current i_L proceeds along a new path, which is as follows: positive terminal of voltage source U_1 – coil L – virtual voltage source e – open diode D – parallel-connected capacitor C and load resistor R – negative terminal of voltage source U_1. The current leaves via the positive terminal of the virtual voltage source, which indicates that energy is being removed from the magnetic circuit of the coil.

According to Kirchhoff's voltage law, the electromotive force of the virtual voltage source is equal to the difference between the voltage of the capacitor C (which equals the converter's output voltage U_2 = 36 V) and the input voltage of the converter (U_1 = 12 V). Thus, in this case, $e = U_2 - U_1 = 24$ V, if we disregard the small voltage drop across the open diode D. The current in the coil decreases at a rate of $di_L/dt = (U_1 - U_2)/L$.

At the moment t_2, the switch is closed again and the cycle repeats. Energy is once more supplied to the magnetic field and the current in the coil starts to increase. Thus, the cyclic supply of magnetic field energy at one moment in time to the coil occurs at one voltage and the return of the magnetic field energy at another moment in time to the electrical circuit occurs at another voltage. Converters that function in this way are known as step-up converters, or boost converters.

The voltage at which energy is returned from the magnetic field to the electrical circuit is determined neither by the inductance of nor the current in the coil. This voltage is determined by the electrical circuit into which the coil's energy is fed. If the voltage in the electrical circuit is very high, then the rate of decrease of the current is very high as well, and the current quickly drops to zero. This is what happens when an electrical circuit is broken by a conventional switch with

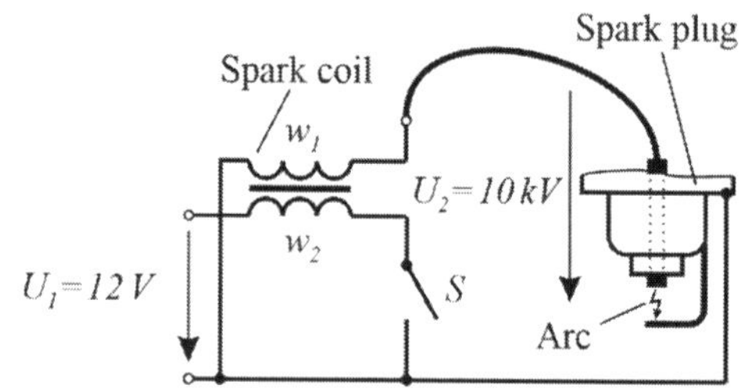

FIGURE 2.22 Generation of a spark in a petrol engine's spark plug by means of energy stored in a magnetic field.

contacts. Every electrical circuit has an inductance, and when a circuit is broken, the voltage in the contacts of the switch becomes so high that a discharge – a momentary electric arc, or spark – occurs in the air between the opening contacts.

In some cases, generating a spark is necessary for a certain function, such as in the petrol engines of motor vehicles. This is achieved with a dual-wound coil (a so-called ignition coil), which has one winding w_2 with a low number of turns and another winding w_1 with a high number of turns (Figure 2.22).

When the switch S, which is a breaker contact or transistor switch, is closed, a voltage of 12 V is applied to the winding w_2 from the battery. Energy is transmitted from the battery to the magnetic field of the ignition coil. When the switch S is opened, a voltage arises in both windings w_1 and w_2. Since the number of turns in the winding w_1 is several hundred times greater than that in the winding w_2, a dielectric breakdown occurs between the electrodes of the spark plug before it can occur in the switch S. Thus, the energy of the magnetic field is removed from the coil via the current in the winding w_1, despite the fact that during the previous time interval, energy was introduced via the winding w_2.

2.2.4 Obtaining a Time-Varying Magnetic Flux by Means of a Body Rotating in a Homogeneous Magnetic Field Sinusoidal Voltage and Sinusoidal Current

Electrical circuits usually utilise sinusoidal voltages. A sinusoidal voltage arises when a coil is rotated in a homogeneous magnetic field at a constant velocity ω . Figure 2.23 depicts the rotation of a coil with a single turn of wire.

As the coil rotates, the magnetic flux passing through it changes. The magnetic flux is highest when the plane of the coil is perpendicular to the lines of force (α = 0°):

$$F_{\mathrm{m}} = BA, \tag{2.32}$$

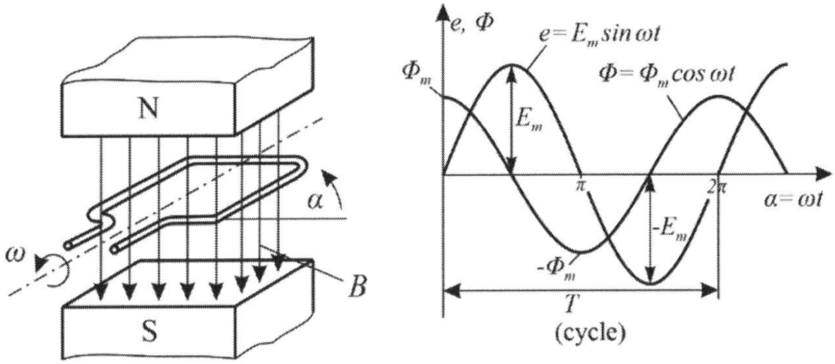

FIGURE 2.23 Generation of a sinusoidal electromotive force.

where

B – magnetic flux density, Wb/m^2;
A – area of the coil, m^2;
Φ_m – maximum magnetic flux, Wb.

When the plane of the coil is parallel to the lines of force of the magnetic field (α = 90°), the magnetic flux passing through the coil is equal to zero:

$$F = 0.$$

In general, the magnetic flux passing through a rotating coil is given by the following equation:

$$F = F_m \cos\alpha. \tag{2.33}$$

When the plane of the coil is perpendicular to the lines of force, then $\alpha = 0°$ and $\cos\alpha = 1$. However, when the plane of the coil is parallel to the lines of force, then $\alpha = 90°$ and $\cos\alpha = 0$.

The angle α increases steadily over time. The general equation for this angle is

$$\alpha = \omega t, \tag{2.34}$$

where ω is the angular velocity of rotation, rad/s.

Over the course of one revolution, the angle α changes by 2π radians. The time it takes to complete one revolution is

$$t_{revolution} = \frac{\alpha}{\omega} = \frac{2\pi}{\omega} = T. \tag{2.35}$$

The time it takes to complete one revolution is also called the **period**, denoted by T. The length of a period is measured in seconds. The inverse of the period is called **frequency**:

$$f = \frac{1}{T}. \tag{2.36}$$

The unit of measurement of frequency is the **hertz** (Hz); 1 Hz = 1 s^{-1}.

Frequency indicates the number of periods per second, which in this case is the number of revolutions completed by the turn of wire in the coil per second. The equation for the magnetic flux passing through the coil can also be given as follows, by substituting time t and angular frequency ω for the angle α :

$$F = F_{\mathrm{m}} \cos \omega t. \tag{2.37}$$

The changing of the magnetic flux passing through the turn of wire induces in the turn an electromotive force

$$e = -\frac{\mathrm{d}F}{\mathrm{d}t}.$$

Since the magnetic flux changes according to equation (2.37), the equation for the electromotive force induced in the turn of wire can be written as:

$$e = -\frac{\mathrm{d}\left(F_{\mathrm{m}} \cos \omega t\right)}{\mathrm{d}t} = \omega F_{\mathrm{m}} \sin \omega t. \tag{2.38}$$

When $\sin\omega t = 1$, the electromotive force is strongest:

$$E_{\mathrm{m}} = \omega F_{\mathrm{m}}, \tag{2.39}$$

E_{m} is the maximum value, or **peak value**, of the sinusoidal electromotive force. The value e of the electromotive force at a particular moment in time is called the **instantaneous value**:

$$e = E_{\mathrm{m}} \sin \omega t. \tag{2.40}$$

If we were to load the source of our sinusoidal electromotive force with a resistance, a sinusoidal current would arise in the circuit, the curve of which can be given by an analogous equation:

$$i = I_{\mathrm{m}} \sin \omega t. \tag{2.41}$$

The instantaneous values of alternating current and voltage are constantly changing between zero and the peak value. This is not the case with direct current. In practice, it is often necessary to compare the effectiveness of direct current and alternating current. An ordinary incandescent light bulb, for example, can work with both direct and alternating current. However, if the rated current of the bulb is, for example, 1 A for direct current, obtaining the same power with alternating current requires that the peak value of the alternating current be higher. In order to take this into account, the concept of the **effective value** of an alternating current has been introduced, defined as follows: **the effective value of an alternating current is that current that will release in the same resistor the same amount of heat as an equivalent direct current.**

It can be shown that the following relationship applies between the effective value I_{ef} and the maximum (peak) value I_m of a sinusoidal current:

$$I_{ef} = I_m / \sqrt{2} = 0,707 I_m. \tag{2.42}$$

Similarly, we can speak about the effective value of a sinusoidal voltage

$$U_{ef} = U_m / \sqrt{2} = 0,707 U_m. \tag{2.43}$$

In European countries, the index 'ef' is usually omitted. The effective value of a current is simply denoted as I, while the effective of a voltage is denoted as U. In the US, the index 'rms' (root-mean-square) is frequently used to indicate an effective value; for example, I_{rms} would be the effective value of a current. The maximum value of an AC mains voltage of 230 V is $U_m = \sqrt{2}\,U = \sqrt{2} \cdot 230 = 325$ V.

Sine waves with the same frequency and an equal initial phase are said to be **in phase**. Meanwhile, sine waves that are shifted in phase by $\pm\pi$ are said to be **in opposite phase**. **A sinusoidal quantity is defined if its maximum value, frequency, and initial phase are known.**

2.2.5 Phase Vector Diagram or Phasor Diagram

Graphically, sinusoidal quantities are depicted as a sine wave, as we saw above, or as a rotating phase vector, or phasor. Drawing a sine wave is more bothersome. Moreover, when there are many sine waves, the graph is harder to follow. That is why electricians mostly use phasor diagrams, or diagrams of phase vectors, which are simpler and clearer than sine wave graphs.

While the name 'phase vector diagram' is accurate, it is long and does not quite roll off the tongue. In the past, these diagrams were simply called 'vector diagrams' – you may come across this in older textbooks, but this name is inaccurate. 'Vector' usually means a spatial vector (vector directed from point A to point B) or a directional physical quantity (such as a vector describing the velocity and direction of an air flow).

Sine waves, on the other hand, are described – in a simplified form – using a phase vector, or phasor. The amplitude and phase of the sine wave are taken to be static, with the amplitude equated with the length of a spatial vector and the phase with the direction of the spatial vector. Placed on a plane and made to rotate

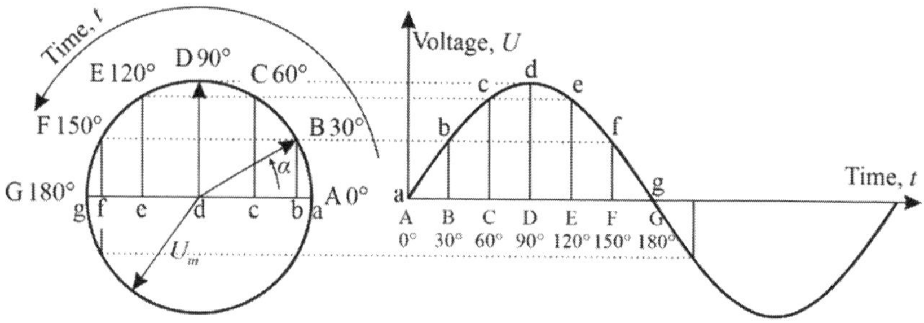

FIGURE 2.24 Substitution of phasors for sine waves.

around its origin, the spatial vector can then be used to determine the instantaneous values of that sine wave. Such a rotating vector is what is known as a phase vector, or phasor. The phase angle and time of the rotating vector will change in unison in the oscillogram of the sine wave. The time in the oscillogram can thus be measured not only in seconds, but also in angular units of measurement: degrees or radians.

A sine wave has a time axis. The passage of time is depicted as translational motion along the horizontal axis. The lines AB, BC, etc. on the time axis in Figure 2.24 represent time intervals. The distances Bb, Cc, etc. between the different points of the sine wave and the time axis indicate the instantaneous value of the voltage.

In a phasor diagram, rotational motion is used in place of translational motion to depict the passage of time. Time is made to 'rotate'. Time intervals are represented by the angles (α in the figure above) formed with the rotating voltage vector, or phasor, Um, as well as by the distances travelled by the end of the rotating voltage vector (arc AB). In the figure, the length of the arc AB is proportional to that of the line AB on the time axis. The instantaneous value of the voltage is indicated by the distance of the end of the rotating vector from the time axis (lines Bb, Cc, etc.). The distance of the end of the rotating vector from the x-axis changes according to the law of sines. For example, based on the geometric design shown in the figure,

$$\textit{the line}\ \mathrm{Bb} = U_{\mathrm{m}} \sin\alpha = U_{\mathrm{m}} \sin 30^\circ = 0.5\,U_{\mathrm{m}};$$
$$\textit{the line}\ \mathrm{Cc} = U_{\mathrm{m}} \sin 60^\circ = 0.866\,U_{\mathrm{m}};$$
$$\textit{the line}\ \mathrm{Dd} = U_{\mathrm{m}} \sin 90^\circ = U_{\mathrm{m}}\ \text{(maximum voltage)}.$$

U_{m} is the length of the phasor in the figure and is related to voltage through the measurement factor, which can be, for example, m_U = 10 V/mm. The phasor is considered to be rotating counterclockwise at a steady speed. However, you can also imagine the phasor as standing still, while the plane of the figure is rotating clockwise. The end result is the same. The rotational speed is determined by the frequency of the alternating current. At a frequency of 50 Hz, the voltage phasor completes 50 revolutions per second, which is equal to an angle of 314 radians, or 18,000°.

In practice, we are usually interested in the phase shift between different quantities. When preparing a phasor diagram, you are free to choose the direction of the first phasor. Other phasors must then be rotated relative to the first one by an angle equal to the phase shift of the relevant quantity.

In a phasor diagram, phasors with different initial phases and amplitudes can be combined using the methods of vector algebra. This is much easier than using two current or voltage curves to construct a third one.

2.2.6 Phase Angle and Phase Shift

When connecting to the grid at time $t = 0$, the mains voltage may not equal zero. In all likelihood, α will hold a non-zero value. This value is usually denoted by ψ and is called the **initial phase angle**, or **initial phase** (Figure 2.25). In such a case,

$$e = E_{\mathrm{m}} \sin(\omega t + \psi). \tag{2.44}$$

The initial phase angle, or initial phase, is the electrical angle ψ traversed in the time from the start of the current cycle to the moment of commencement of observation, which is placed at the origin of the co-ordinate plane.

At time $t = 0$, which is the moment of commencement of observation in the figure above, an angle of 60°, i.e. $\pi/3$, has been traversed since the start of the electromotive force cycle. At this point, $\omega t = 0$ and the **initial instantaneous value** of the electromotive force is

$$e_0 = E_{\mathrm{m}} \sin\psi. \tag{2.45}$$

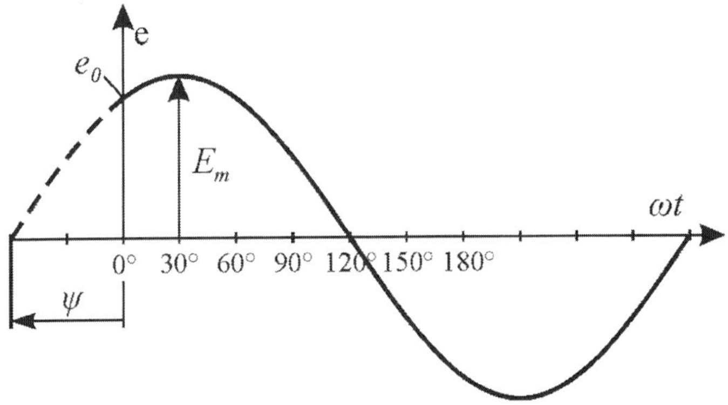

FIGURE 2.25 Initial phase ψ of a sine wave.

Positive initial phases are positioned to the left of the origin of the co-ordinate plane, while negative initial phases are positioned to the right.

When two sine waves with the same frequency, such as the voltages of two different generators, are offset in time relative to each other, there is said to be a **phase shift** and a **phase shift angle**. Figure 2.26 shows a phase shift of 30°.

The sine curve that is **leading in phase** is the curve whose cycle starts earlier, and the sine curve that is **lagging in phase** is the curve whose cycle starts later. In the figure above, the electromotive force e_1 is leading in phase compared to the electromotive force e_2, or in other words, e_2 is lagging in phase relative to e_1.

Phase shift angles between voltages and currents are denoted by the Greek lowercase letter ϕ . This angle may be measured either as the difference between the maximum values or the zero values. More generally,

$$\varphi = \psi_1 - \psi_2$$

where ϕ – phase shift angle;

ψ_1 – initial phase of the voltage sine wave;

ψ_2 – initial phase of the current sine wave.

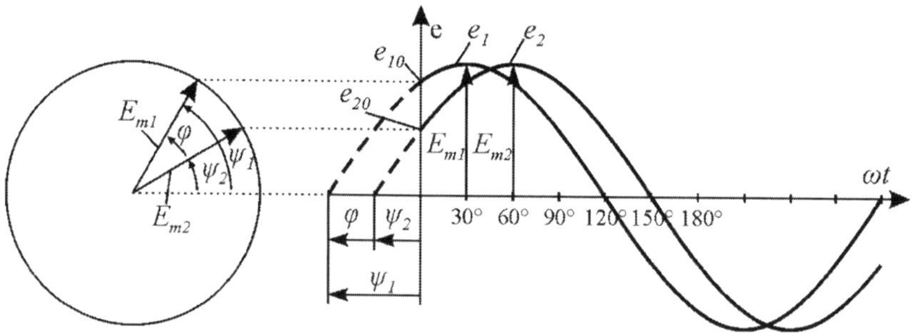

FIGURE 2.26 Phase shift between two sinusoidal electromotive force curves.

3 Basic Concepts of Alternating Current Circuits

AC machines are powered either by a single-phase or three-phase alternating current. Calculations for AC circuits require using a different methodology than for DC circuits. This is because inductance and capacitance impede alternating currents in a very different way than they do direct currents.

Additionally, AC circuits involve a unique type of power: reactive power. While in the case of active power, energy flows only one way (active power enters the electrical circuit and then leaves the circuit), in the case of reactive power, energy bounces back and forth in the electrical circuit. For example, magnetic field energy stored in inductance will pass through an electrical circuit and enter a capacitor, where it is stored as electric field energy, and in the next half-cycle it will return to inductance.

Transformers and AC motors also require reactive power to operate. Understanding AC circuit theory is essential for understanding the processes taking place in these devices.

3.1 INDUCTANCE AND CAPACITANCE IN AN AC CIRCUIT: REACTANCE

Say we have two simple circuits. One is an inductance L and the other is a capacitance C, both driven by a sinusoidal voltage (Figure 3.1).

We want to know how we can determine the current in these circuits. We know from before that the differential dependencies exist between current and voltage. For the inductance, $u = L\,\mathrm{d}i/\mathrm{d}t$, while for the capacitance, $i = C\,\mathrm{d}u/\mathrm{d}t$. Let us assume that the current in the inductance is sinusoidal.

$$i_L = I_{Lm} \sin \omega t \tag{3.1}$$

This current causes, for the inductance, the voltage

$$u_L = L\frac{\mathrm{d}i}{\mathrm{d}t} = L\frac{\mathrm{d}\left(I_{Lm}\sin\omega t\right)}{\mathrm{d}t} = \omega L I_{Lm}\cos\omega t. \tag{3.2}$$

The current for the capacitance can be calculated in a similar way

DOI: 10.1201/9781003404088-3

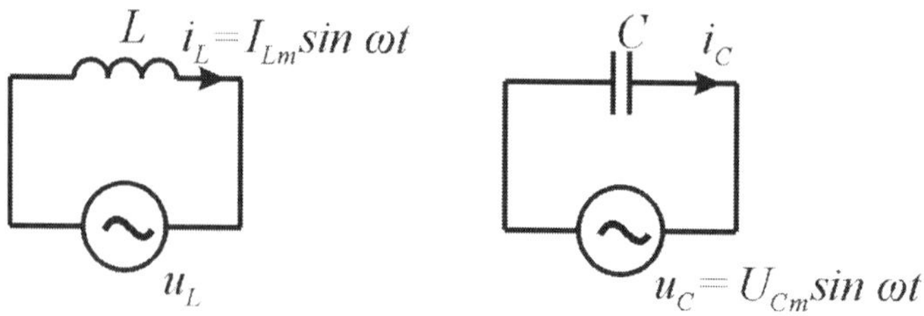

FIGURE 3.1 Inductance and capacitance driven by a sinusoidal voltage.

$$i_C = C\frac{\mathrm{d}u}{\mathrm{d}t} = C\frac{\mathrm{d}\left(U_{\mathrm{Cm}}\sin\omega t\right)}{\mathrm{d}t} = \omega C U_{\mathrm{Cm}}\cos\omega t. \tag{3.3}$$

Equations (3.2) and (3.3) describe sinusoidal quantities. Therefore, a sinusoidal current arises in both the inductance and the capacitance when they are driven by a sinusoidal voltage. The multipliers in equation (3.2) likely determine the maximum value of the voltage in the inductance

$$\omega L I_{L\mathrm{m}} = U_{L\mathrm{m}} \tag{3.4}$$

while the multipliers in equation (3.3) determine the maximum value of the capacitor's current

$$\omega C U_{\mathrm{Cm}} = I_{\mathrm{Cm}}. \tag{3.5}$$

Since $\cos\omega t = \sin(\omega t + \pi/2)$, we may replace the cosine functions obtained by differentiation with sine functions. Thus, we can rewrite equations (3.2) and (3.3) as

$$u_L = U_{L\mathrm{m}}\sin\left(\omega t + \pi/2\right), \tag{3.6}$$

$$i_C = I_{\mathrm{Cm}}\sin\left(\omega t + \pi/2\right). \tag{3.7}$$

It follows from equation (3.6) that the voltage phasor U_L of the inductive circuit is ahead of the current phasor I_L by an angle of $\pi/2$, as $i_L = I_{L\mathrm{m}}\sin\omega$ t. It follows from equation (3.7) that the current phasor I_C of the capacitive circuit is ahead of the voltage phasor U_C by an angle of $\pi/2$. This is illustrated in Figure 3.2.

Rearranging equation (3.4) gives us the ratio between the maximum voltage and maximum current of the inductive circuit, which we can then convert into the ratio of the effective values

$$\frac{U_{L\mathrm{m}}}{I_{L\mathrm{m}}} = \frac{\sqrt{2}U_L}{\sqrt{2}I_L} = \frac{U_L}{I_L} = \omega L. \tag{3.8}$$

Rearranging equation (3.5) in an analogous manner, we get

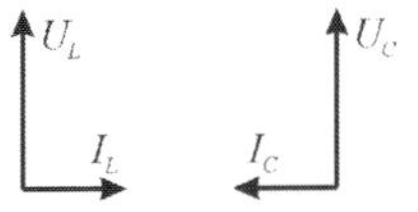

FIGURE 3.2 Position of the current phasor relative to the voltage phasor in the inductive circuit and the capacitive circuit.

$$\frac{U_{Cm}}{I_{Cm}} = \frac{U_C}{I_C} =$$

$$\frac{1}{\omega C}. \tag{3.9}$$

In a DC circuit, the ratio between the voltage and current determines the resistance. Equations (3.8) and (3.9) also describe the ratio between the voltage and current; therefore, the right-hand sides ωL and $1/\omega C$ of the equations represent the AC resistance or reactance. These reactances are denoted as follows:

$$\omega L = x_L, \tag{3.10}$$

$$\frac{1}{\omega C} = x_C. \tag{3.11}$$

Here, x_L is **inductive reactance**, while x_C is **capacitive reactance**. Both inductive and capacitive reactance are generally called **reactance**, denoted by x.

Equations (3.8) and (3.9) may also be written as

$$x_L = U_L / I_L \tag{3.12}$$

and

$$x_C = U_C / I_C. \tag{3.13}$$

Equations (3.12) and (3.13) express Ohm's law for AC circuits.

Inductive reactance x_L increases as frequency f increases (Figure 3.3). Capacitive reactance x_C, on the other hand, decreases as frequency increases. Given an inductance L and a capacitance C, there will always be a certain frequency f_0, at which the inductive reactance and capacitive reactance are equal. This frequency is called the **resonant frequency**.

When the frequency is zero, i.e. in the case of DC voltage, inductive reactance is equal to zero and capacitive reactance is equal to infinity. At high frequencies, the opposite occurs: inductive reactance becomes high and capacitive reactance becomes low.

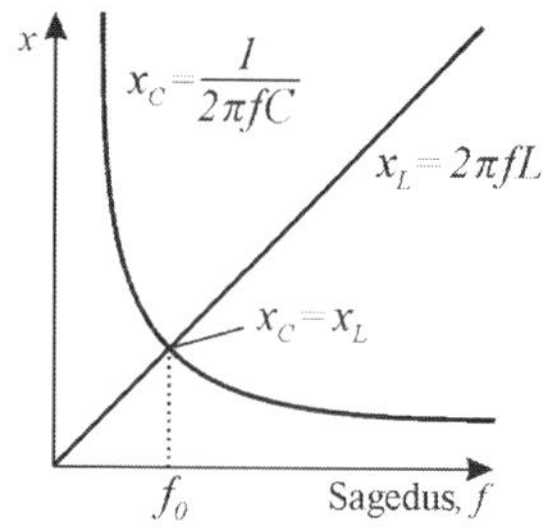

FIGURE 3.3 Dependence of inductive and capacitive reactance on frequency.

3.2 MIXED RESISTANCE IN AN AC CIRCUIT: IMPEDANCE

When a resistance r and an inductance L are connected in series (Figure 3.4a), the same current I passes through these elements.

This current causes a voltage drop $U_r = Ir$ across the resistor r. This voltage drop is in phase with the current (phasors U_r and I in the figure). The current I causes a voltage drop $U_L = Ix_L$ across the inductance. The voltage phasor U_L of the inductance is ahead of the current phasor I by 90°. Therefore, the voltage phasors U_L and U_r of the inductance and the resistance are shifted by 90° relative to each other. At the same time, the geometric sum of these phasors must be equal to the voltage U of the voltage source:

$$\vec{U}_L + \vec{U}_r = \vec{U}. \tag{3.14}$$

The geometric addition of the phasors gives us a **voltage triangle** (Figure 3.4a), where U_L and U_r are the orthogonal and U is the hypotenuse. Based on this triangle,

$$U = \sqrt{U_r^2 + U_L^2}. \tag{3.15}$$

From the voltage triangle, we can derive a similar **resistance triangle** by dividing all three voltages U_r, U_L, and U by the current I. This gives us $U_r/I = r$, $U_L/I = x_L$, and

$$U / I = z, \tag{3.16}$$

where z is the circuit's **apparent resistance**, or **impedance**. Impedance is also measured in ohms. Impedance determines the current generated in the circuit. Impedance can be calculated from the resistance r and the inductive reactance x_L as their geometric sum:

$$z = \sqrt{r^2 + x_L^2}. \tag{3.17}$$

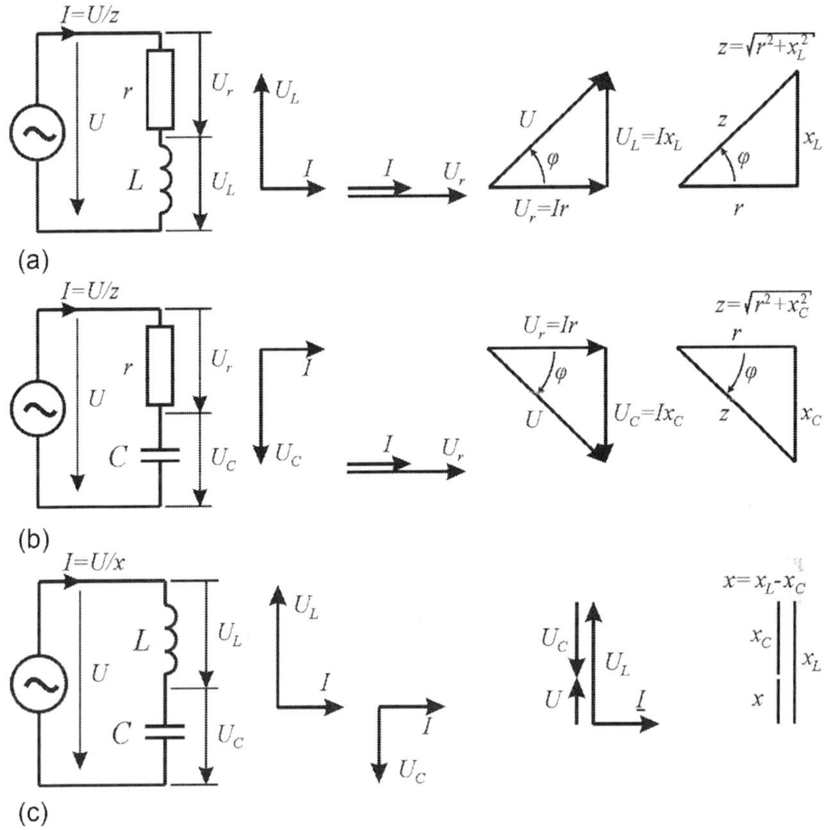

FIGURE 3.4 Resistance and inductive reactance (a), resistance and capacitive reactance (b), and inductive and capacitive reactance (c) connected in series.

In order to determine the current in the circuit shown in Figure 3.4a, we must first calculate the impedance z according to equation (3.17) and then the current I based on equation (3.16).

Example

1. We calculate the inductive reactance

$$x_L = \omega L = 2\pi f L = 2\pi \cdot 50 \cdot 0{,}1 = 31{,}4\,\mathrm{W}.$$

2. We calculate the impedance

$$z = \sqrt{r^2 + x_L^2} = \sqrt{45^2 + 31{,}4^2} = 54{,}9\,\mathrm{W}.$$

3. We calculate the current

$$I = U / z = 230 / 54{,}9 = 4{,}2\text{A}.$$

The phase shift angle φ between the voltage and current can be determined both on the basis of the voltage triangle and the resistance triangle:

$$\cos\varphi = U_r / U = r / z. \tag{3.18}$$

When a resistance r and capacitance C (Figure 3.4b) are connected in series, the same current passes through both of these elements. This current also causes a voltage drop across both elements. In this case, in contrast with the previous one, the voltage phasor U_C of the capacitance is lagging behind the current phasor I by 90°. The voltage phasors U_C and U_r of the capacitance and resistance are shifted by 90° relative to each other, and also must be added together geometrically. The geometric addition of the phasors gives us the voltage triangle (Figure 3.4b), from which we can derive the resistance triangle. The resistance triangle similarly allows us to determine the impedance

$$z = \sqrt{r^2 + x_C^2}. \tag{3.19}$$

When a capacitance C and an inductance L are connected in series (Figure 3.4c), the same current I again passes through these elements. The current causes voltage drops across these elements, and the voltage phasor U_C of the capacitance lags behind the current phasor by 90°, while the voltage phasor U_L of the inductance leads the current by 90°. The angle of shift between the phasors U_C and U_L is therefore 90° multiplied by two, which equals 180°. Therefore, the phasors U_C and U_L are collinear, but opposite in direction. The sum of these phasors must be equal to the supply voltage U:

$$\vec{U}_C + \vec{U}_L = \vec{U}. \tag{3.20}$$

Adding these phasors together does not produce a triangle. The modulus of the resultant phasor is equal to the arithmetic difference between the moduli of the component phasors. By convention, the direction of U_L is considered positive and the direction of U_C is considered negative. Thus, based on the effective values

$$U = U_L - U_C. \tag{3.21}$$

Dividing the expression above by the current I yields

$$x = x_L - x_C, \tag{3.22}$$

where $x = U/I$ is the reactance of the circuit. When the inductive reactance is greater than the capacitive reactance ($x_L > x_C$), then the circuit reactance is positive and the overall circuit is inductive. The current phasor of the circuit will then be lagging behind the voltage phasor by 90°, as shown in Figure 3.4c. In the opposite case – when the capacitive reactance is greater than the inductive reactance ($x_C > x_L$) – then the overall circuit is capacitive. When $x_C = x_L$, then the resistance of the AC series circuit shown in Figure 3.4c is equal to zero.

In general, given a series circuit with a larger number of resistances, inductive reactances, and capacitive reactances, the total impedance is given by

$$z = \sqrt{\left(\Sigma r\right)^2 + \left(\Sigma x_L - \Sigma x_C\right)^2}. \tag{3.23}$$

3.3 AC POWER

Instantaneous AC power means the product of the instantaneous values of voltage and current

$$p = ui. \tag{3.24}$$

The instantaneous values of voltage and current are given by

$$u = U_{\mathrm{m}} \sin \omega t$$

and

$$i = I_{\mathrm{m}} \sin\left(\omega t \pm \varphi\right).$$

Thus, instantaneous power is given by

$$p = U_{\mathrm{m}} I_{\mathrm{m}} \sin \omega t \sin\left(\omega t \pm \varphi\right). \tag{3.25}$$

Since the product of the sine values can be rewritten as

$$\sin \omega t \sin\left(\omega t \pm \varphi\right) = \frac{1}{2}\left[\cos \varphi - \cos\left(2\omega t \pm \varphi\right)\right]$$

and

$$U_{\mathrm{m}} I_{\mathrm{m}} = 2UI,$$

the equation for instantaneous power becomes

$$p = UI \cos \varphi - UI \cos\left(2\omega t \pm \varphi\right). \tag{3.26}$$

Integrating over one period to get the average power P over a period gives us:

$$P = \int_0^T \left[UI\cos\varphi - UI\cos\left(2\omega t \pm \varphi\right)\right]\mathrm{d}t = UI\cos\varphi. \tag{3.27}$$

This average power is called **active power** and is measured in watts (W).

The product UI is called **apparent power** S:

$$S = UI. \tag{3.28}$$

Apparent power is measured in volt-amperes (V ·A).

The ratio of active power to apparent power

$$\frac{P}{S} = \frac{UI\cos\varphi}{UI} = \cos\varphi \tag{3.29}$$

is called the **power factor**. The product

$$S\sin\varphi = Q \tag{3.30}$$

is called **reactive power** and is measured in vars (volt-amperes reactive).

AC circuits thus involve three different types of power, which differ in how the phase shift angle ϕ between the current and voltage curves is taken into account:

$$S = UI,$$

$$P = S\cos\varphi = UI\cos\varphi,$$

$$Q = S\sin\varphi = UI\sin\varphi.$$

To understand the importance of the phase shift angle, let us explore how power behaves for a resistive load, such as a resistor, and a reactive load, such as an inductor or capacitor.

In a resistive load (resistor), the current is in phase with the voltage. This means that $\phi = 0$ and $\cos\phi = 1$. As a result

$$S = P = UI\ , Q = 0. \tag{3.31}$$

In the resistor, all of the power is spent to produce heat. Energy leaves the electrical circuit and is released to the environment. The reactive power is equal to zero. The active power in this case is thermal power, and it is also equal to the apparent power.

The instantaneous power of the resistive load can be calculated based on equation (3.25) by setting ϕ equal to 0. This gives us

$$p = U_m I_m \sin^2 \omega t = 2UI \sin^2 \omega t = 2P \sin^2 \omega t. \tag{3.32}$$

Figure 3.5a shows the curve for the instantaneous power of the resistive load. In both half-periods, the power is positive. The maximum instantaneous value of the power is two times the average power, calculated as $P = UI$, while the minimum instantaneous value equals zero.

In an inductor, the current lags behind the voltage, and ideally $\phi = 90°$. Thus $\cos\phi = 0$ and $\sin\phi = 1$. In such a case

$$S = Q = UI \ , P = 0. \tag{3.33}$$

The active power is equal to zero because no energy leaves the electrical circuit. The power is entirely a reactive power. To understand what this means, let us consider the instantaneous value of the power based on equation (3.25) and an angle ϕ of $-90°$:

$$p = U_m I_m \sin \omega t \sin(\omega t - 90°) = -U_m I_m \sin \omega t \cos \omega t,$$

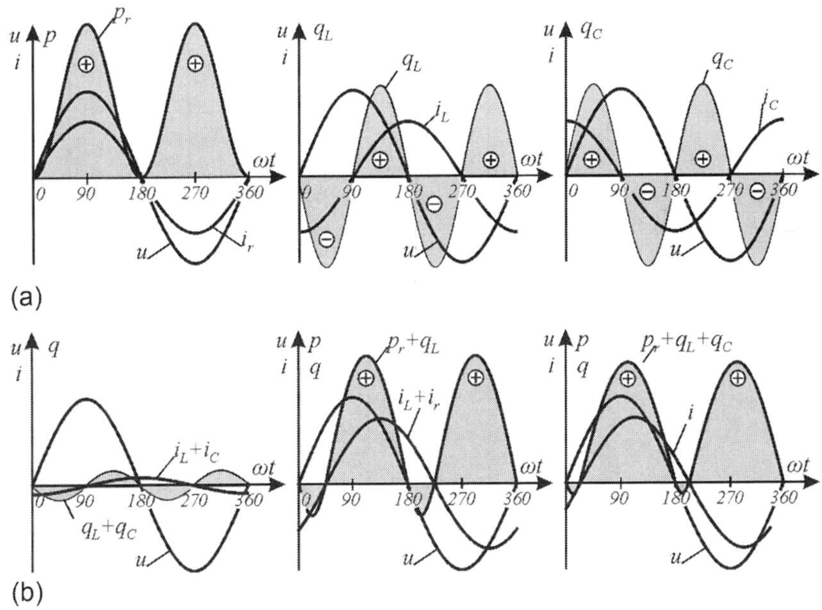

FIGURE 3.5 Instantaneous value curves for the voltage, current, and power of a resistor (a), inductor (b), capacitor (c), parallel-connected coil and capacitor (d), parallel-connected coil and resistor (e), and parallel-connected resistor, coil, and capacitor (f).

which can be further rearranged as

$$p = -UI \sin 2\omega t. \tag{3.34}$$

Figure 3.5b shows the curve for the instantaneous power of the inductor. The instantaneous power fluctuates at a frequency that is twice as high as the mains frequency. In the first half of the first half-period of the voltage (0° to 90°), the instantaneous power is negative, and in the second half of the same half-period (90° to 180°) it is positive. The second half-period is analogous. When the instantaneous power is negative, the current is opposite in direction to the voltage. This corresponds to the voltage source or current source being run in active mode. Instantaneous power is negative when the current in the inductance is decreasing (the current was generated in the previous half-period, which is not shown in the figure). As the current decreases, so does the energy of the magnetic field, and this energy is delivered to the grid. Here, the inductance is acting as an energy source, which manifests as a negative instantaneous power.

In the second half of the half-period (90° to 180°), the current is increasing, as is the energy of the magnetic field. This energy is drawn from the grid (the current and voltage are identical in direction in the oscillogram). Thus, in the case of reactive power, energy does not leave the electrical circuit and is not spent. Energy only swings back and forth between the inductor and the voltage source (electrical grid). The value of the instantaneous power does not reveal whether the power is active or reactive. Determining the reactive power requires observing at least one period.

In electrical grids, it is normally desirable to avoid or reduce reactive power. A voltage source situated in Narva and an inductor in Tallinn, for example, would be separated by an over 200-kilometre-long power line (Figure 3.6) with a total resistance of r_{line}. With the inductor consuming reactive power equal to Q, it is subjected to a current of $I = Q/U$. This current causes a **power loss** P_{loss} in the line, which can be calculated on the basis of the Joule–Lenz law:

$$P_{\text{loss}} = I^2 r_{\text{line}}. \tag{3.35}$$

For larger electrical loads, a reactive energy meter is installed in addition to an active energy meter. The charge for the consumption of reactive energy is proportional to the amount of energy dissipated as heat in the line. Normally, reactive energy charges are about 10 times lower than active energy charges because active energy is transmitted fully to the load, while reactive energy only causes transmission losses. Some appliances, such as induction motors, require reactive power to bring about their magnetic fields.

If, instead of the inductor, we were to connect a capacitor to the grid, the current and voltage would also be shifted in phase by 90°. What would be different, however, is that the capacitor's current would lead the voltage by 90°. This is shown in Figure 3.5c. In the case of the capacitor, the current and voltage are

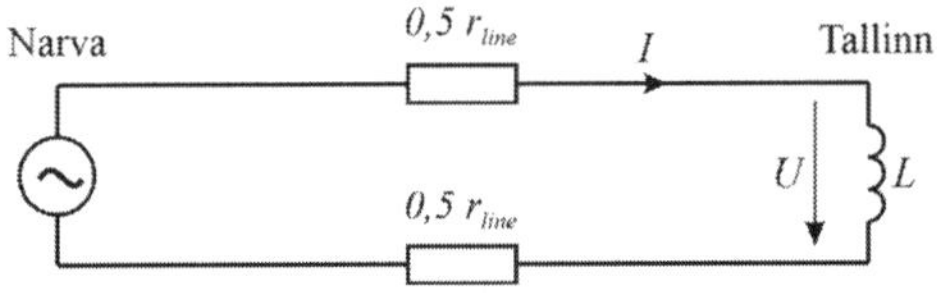

FIGURE 3.6 Line power loss in the transmission of reactive power.

identical in direction in the first half of the half-period (0° to 90°) of the mains voltage. During this time, the capacitor is charged and its energy increases. The instantaneous value of the power is thus positive, and energy flows from the grid to the capacitor. In the second half of the half-period, the capacitor is emptied, and its current is opposite in direction. Energy is delivered from the capacitor to the grid. Thus, the power of the capacitor, too, is reactive power, with energy swinging back and forth between the voltage source and the capacitor.

If we compare the instantaneous value curves for the power and current of the inductor and the capacitor (Figure 3.5b, c), we can see that the curves are opposite in phase (180° phase shift). This makes it possible to use capacitors to free transmission lines from the inductive reactive power generated, for example, by electric motors. The method is called **reactive power compensation**. Reactive power compensation reduces the reactive power drawn from the line, as shown in Figure 3.5d. An inductor and a capacitor exchange energy via the wires connecting them. These wires should be short in order to avoid high energy consumption. Companies are free to choose whether they will compensate for reactive energy themselves by means of a capacitor bank or purchase reactive energy from the grid.

If the active power is much larger than the reactive power (Figure 3.5e), using a capacitor bank may not be the most practical choice. However, a capacitor bank would still enable you to compensate for reactive power (Figure 3.5f).

3.4 THREE-PHASE CIRCUITS

Three-phase circuits are put together from three single-phase circuits. The supply voltages for the single-phase circuits are usually generated by the same AC generator. The voltages of the different phases are shifted in time by one third of a period (Figure 3.7). The phases are usually denoted by the Latin characters A, B, and C.

The instantaneous values of the electromotive forces of the three phases are respectively

$$e_{\mathrm{A}} = E_{\mathrm{m}} \sin \omega t;$$

$$e_{\mathrm{B}} = E_{\mathrm{m}} \sin\left(\omega t - 120°\right); \tag{3.36}$$

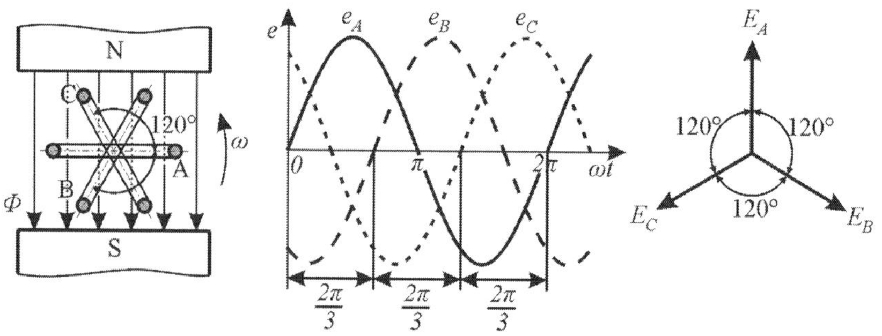

FIGURE 3.7 Model (a), electromotive force curves (b), and phasor diagram (c) of a three-phase generator.

$$e_C = E_m \sin(\omega t + 120°).$$

When three phases are interconnected, they are called a three-phase system.

3.4.1 Star Connection of a Three-Phase System

If we join the ends of the windings of a three-phase generator together, we get a four-wire voltage system with output terminals for the phases A, B, and C and a winding connection point called the neutral point (Figure 3.8).

$$U_A = U_B = U_C = 230\,V;\ U_{AB} = U_{BC} = U_{CA} = 400\,V$$

From the generator, a four-wire line – consisting of three line wires A, B, and C as well as a neutral wire – leads to a load. The currents I_A, I_B, and I_C are called line currents.

If all of the phases have identical loads with equal power, the currents I_A, I_B, and I_C are equal and shifted in phase by 120°. The sum of these currents is equal to zero. In such cases, the load is said to be balanced. In the case of a balanced load, there is no current in the neutral wire, which may thus be omitted. This leaves a three-wire three-phase system, which is what is used in transmission lines.

The low-voltage grids of production facilities and residential buildings generally have five wires: phases A, B, and C, neutral N, and ground (protective earth) PE. The neutral wire is connected to the protective earth at the point of entry to the building, but from there the neutral wire (blue) and protective earth (striped yellow and green) run separately and are isolated. In normal operation, the protective earth does not carry a current. The voltages U_l between the line wires are $\sqrt{3}$ times higher than those between the neutral wire and the line wires. These smaller voltages between the neutral wire and the line wires are called the phase voltages U_{ph}.

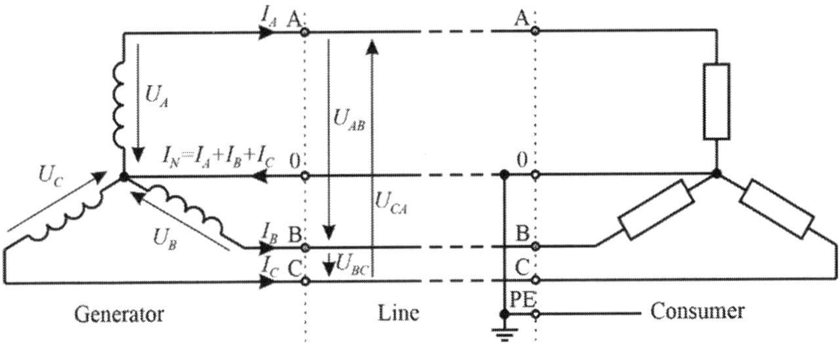

FIGURE 3.8 Star-connected three-phase system.

$$U_l = \sqrt{3}\,U_f. \tag{3.37}$$

The system currently used in European countries utilises a phase voltage of 230 V and a line voltage of 400 V. The voltage values above are so-called rated voltages. The actual mains voltage usually differs from the rated voltage, but the difference must be within certain limits. The maximum permissible difference from the rated voltage is (usually around) ±10%.

3.4.2 Delta-Connected Three-Phase System

In a generator, it is also possible to connect the end of the winding of one phase to the start of the winding of the next phase (Figure 3.9). This yields a so-called delta connection. In a delta-connected system, there is no neutral wire. Loads are attached between the phase wires. The voltages between the line wires are equal to those of the loads. However, the line currents $I_{A,}$ I_B, and I_C are greater than the phase currents I_{AB}, I_{BC}, and I_{CA}.

$$I_l = \sqrt{3}\,I_f, \tag{3.38}$$

where I_l is the line current and I_{ph} is the phase current. The reasons for this are revealed by the phasor diagram in Figure 3.9, where we can see that each line current is obtained by adding together two current phasors that are oriented at an angle of 120° relative to each other.

$$I_{CA} = I_{AB} = I_{BC} = 1\text{ A};\ I_A = I_B = I_C = 1.73\text{ A}$$

The power P of the three-phase system is equal to the sum of the powers of the three phases:

$$P = 3P_f = 3U_f I_f \cos\varphi. \tag{3.39}$$

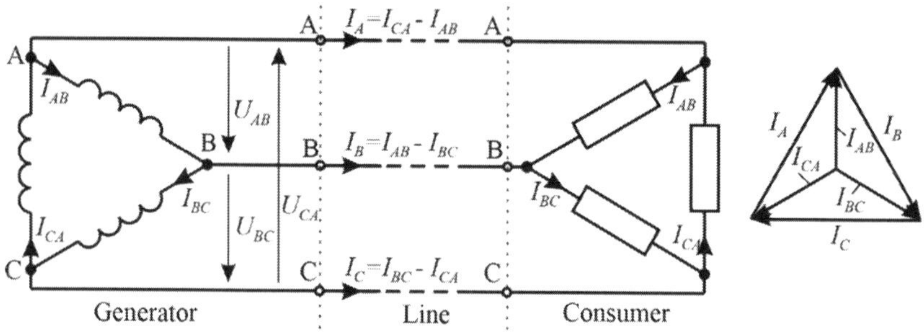

FIGURE 3.9 Delta-connected three-phase system.

The power can also be calculated from the line parameters. For a star configuration, the line and phase currents are equal, but the voltages vary. Hence

$$P = 3\frac{U_l}{\sqrt{3}} I_l \cos\varphi = \sqrt{3} U_l I_l \cos\varphi. \tag{3.40}$$

For a delta configuration, the line and phase voltages are equal, but the currents vary. Hence

$$P = 3U_l \frac{I_l}{\sqrt{3}} \cos\varphi = \sqrt{3} U_l I_l \cos\varphi. \tag{3.41}$$

Thus, the equation is identical for both connection types. Equations (3.40) and (3.41) apply only to balanced loads, where all phase currents are equal.

For balanced loads, the apparent power S and reactive power Q of the three-phase system can be calculated as follows:

$$S = \sqrt{3} UI, \tag{3.42}$$

$$Q = \sqrt{3} UI \sin\varphi. \tag{3.43}$$

4 Transformers

In the first chapter, we explored the design and general principle of operation of transformers. Now let us delve into the different functions of transformers and the processes that each of them involves. We will learn about voltage drops in loaded transformers and how to calculate them on the basis of an equivalent circuit, how to determine the currents in a short-circuited transformer, how to adjust the voltage in a transformer, the types of winding connections in three-phase transformers and how to configure these transformers for parallel operation, specialty transformers, and transient responses in transformers during energising and abrupt short-circuiting.

4.1 LOSSLESS NO-LOAD OPERATION OF A TRANSFORMER

By closing the switch S_1, our transformer is switched to a sinusoidal voltage U_1 and a current I_1 is produced in the winding w_1 (Figure 4.1). This current generates **a magnetic flux** Φ_0 along the core. The changing magnetic flux induces an **electromotive force** in both windings. When the S_2 switch is closed, a current is also produced in the secondary circuit; this current is called the **load current** I_2. When the switch S_2 is open, there is no secondary current. This is known as a **no-load** state. The primary current I_1 is present even at no load, during which state it is called the **no-load current**, or **magnetisation current** I_0.

Let us initially disregard the hysteresis and eddy current losses in the core of the transformer. And let us also ignore the relatively small active power losses that occur when current flows through the primary winding. According to Faraday's law, the magnetic flux Φ_1 and the induced electromotive force e_1 are related as follows:

$$e_1 = -w_1 \frac{dF_1}{dt}, \tag{4.1}$$

w_1 is the number of turns in the primary winding.

If there is no resistance in the primary circuit of the transformer, the instantaneous value of the electromotive force e_1 must always equal the instantaneous value of the primary voltage, which changes according to the law of sines $u_1 = U_{1m}$ $\sin\omega t$. Therefore, we may substitute the voltage u_1 for the electromotive force e_1 in equation (4.1), which gives us

$$U_{1m} \sin \omega t = -w_1 \frac{dF_1}{dt}. \tag{4.2}$$

DOI: 10.1201/9781003404088-4

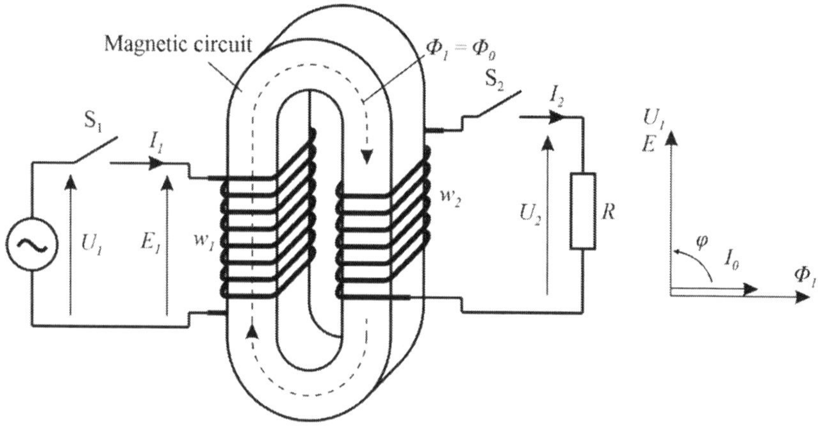

FIGURE 4.1 Transformer operated at no load, and the phasor diagram for the idealised no-load operation.

The left side of equation (4.2) describes the changing voltage, while the right side describes the magnetic flux balancing this changing voltage. While Faraday's law describes the electromotive force generated by a changing magnetic flux, equation (4.2) describes the opposite process, i.e. how a magnetic flux changes due to a changing voltage. We are interested in understanding the patterns the changing of the magnetic flux follows. To find out, let us rearrange equation (4.2) as

$$\mathrm{d}F = -\frac{\sqrt{2}U_1}{w_1}\sin\omega t\mathrm{d}t,$$

which, when integrated, gives us

$$F = -\frac{\sqrt{2}U_1}{w_1}\int\sin\omega t\mathrm{d}t = \frac{\sqrt{2}U_1}{\omega w_1}\cos\omega t = \frac{\sqrt{2}U_1}{2\pi f w_1}\cos\omega t + K, \tag{4.3}$$

since $-\int\sin\omega t\mathrm{d}(\omega t) = \frac{1}{\omega}\cos\omega t + K$ ($K = 0$ for a flux without a DC component) and $\omega = 2\pi f$, where f is the frequency of the supply voltage.

Therefore, **when the voltage changes according to the law of sines, the change in the magnetic flux follows the law of cosines.** The curve of the magnetic flux Φ lags behind the voltage curve by 90°, as shown in the phasor diagram in Figure 4.1. The idealised no-load current I_0 is also sinusoidal and is in phase with the magnetic flux. Therefore, the phase shift angle between current and voltage is $\phi = 90°$. The active power consumed by a lossless transformer at no load is

$$P = 3U_1 I_0 \cos 90° = 0, \tag{4.4}$$

while the reactive power is given by

$$Q = U_1 I_0 \sin 90^\circ = U_1 I_0. \tag{4.5}$$

A transformer always needs reactive power from the grid for operation. Without reactive power, the transformer is unable to generate the magnetic flux needed to induce the secondary voltage.

In the event of losses, ϕ is slightly lower than 90°, and the active power given by equation (4.4) is greater than zero. Equation (4.3) may also be written as

$$F = F_m \cos \omega t,$$

where Φ_m is the maximum value of the magnetic flux.

$$F_m = \frac{\sqrt{2} U_1}{2\pi f w_1}. \tag{4.6}$$

Equation (4.6) tells us two things. First, the amplitude of the magnetic flux decreases as the frequency f increases. A lower magnetic flux enables us to reduce the cross-sectional area of the ferromagnetic core (i.e. the magnetic circuit). It also accommodates shorter winding turns, resulting in a smaller and lighter transformer. If we used a frequency of 500 Hz instead of 50 Hz for a transformer with the same power, the transformer would be 10 times lighter. Why, then, is electrical energy transmitted and distributed at a frequency of 50 Hz? It is because as frequency increases, so does reactive power. Back in the year 1900, a frequency of 50 Hz (or 60 Hz in the USA) was roughly optimal. Today, it would be optimal to use a frequency of ~ 400 Hz, but we cannot swap out the existing power system and loads. This would be extremely costly.

Second, equation (4.6) shows that magnetic flux decreases as the number of turns in the primary winding increases. Due to this, given the same voltage and frequency, the primary winding of a transformer that has a core with a small cross-sectional area will always hold many more turns of wire than that of a transformer that has a core with a large cross-sectional area.

Since, with the simplifications made above (ignoring resistive losses), the primary voltage U_1 is equal to the counter-electromotive force induced in the winding, i.e. $U_1 = E_1$, we can conclude from equation (4.6) that

$$E_1 = \sqrt{2}\pi f w_1 F_m = 4,44 f w_1 F_m. \tag{4.7}$$

The electromotive force is not dependent on the resistance of the circuit. As such, equation (4.7) also applies to real-life transformers, where losses are present, and is known as the transformer electromotive force equation. It is one of the fundamental equations for transformers in the case of a sinusoidal voltage.

Equation (4.7) can also be used to find the electromotive force E_2 for the secondary winding by plugging the number of turns w_2 in the secondary winding into the equation:

$$E_2 = 4,44 f w_2 F_m. \tag{4.8}$$

The ratio between the electromotive forces of the primary winding and the secondary winding is called the **transformation ratio** (or transformation coefficient) k_{12}:

$$E_1 / E_2 = k_{12}. \tag{4.9}$$

Equations (4.7)–(4.9) reveal the fact that

$$k_{12} = w_1 / w_2, \tag{4.10}$$

i.e. that a transformer's transformation ratio is equal to the ratio between the number of turns, i.e. the *turns ratio.* The linear relationship between the number of turns and the electromotive force also follows directly from Faraday's law $e = -w\, \mathrm{d}\Phi / \mathrm{d}t$. **In order to reduce the electromotive force and, with it, the voltage, the number of turns in the secondary winding must be less than that in the primary winding.**

The transformer electromotive force equation may be transposed as follows:

$$E_1 = 4,44 f w_1 F_m = 4,44 f w_1 B_m A, \tag{4.11}$$

where B_m – maximum magnetic flux density in the transformer core,
A – cross-sectional area of the metal material of the transformer core.

The maximum magnetic flux density is limited by the saturation of the transformer core and ranges from 1.0 to 1.7 Wb/m^2 depending on the material of the core. Transformers are designed in a way that ensures that the maximum magnetic flux density is reached at the rated voltage of the transformer. At such a maximum magnetic flux density, the no-load current of the transformer is relatively low, usually under 10% of the rated current. If a voltage that is higher than the rated voltage is applied to the transformer, the magnetisation current increases abruptly. This is illustrated in Figure 4.2.

According to equation (4.11), the magnetic flux density B in the transformer core increases in proportion to the electromotive force E. And, approximately, $E = U$. At the same time, Ampère's law tells us that the relationship between the transformer no-load current I_0 and the strength of the core's magnetic field H is linear. That is why the curve for the relationship between the maximum values of the transformer no-load current and the voltage $I_{0m} = f(U_m)$ in Figure 4.2 is similar to the transformer core magnetisation curve $B_m = f(H_m)$. At the rated voltage, the maximum voltage is U_{m1} and the maximum no-load current is I_{0m1}. If the voltage rises above the rated voltage, for example, by 20% (U_{m2}), the no-load current becomes about five times higher (I_{0m2}), at which point it already equals

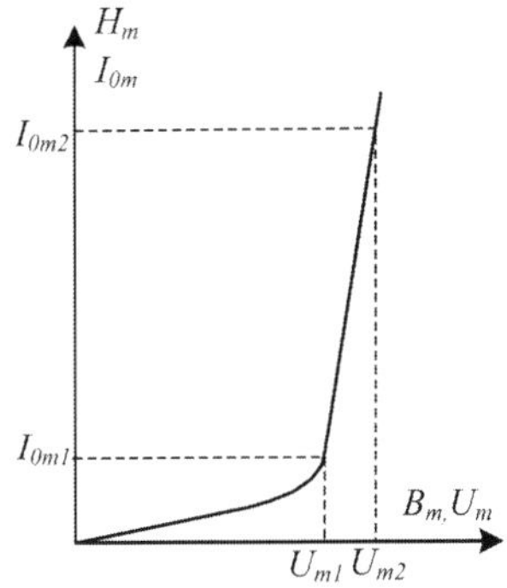

FIGURE 4.2 Sharp increase in the maximum no-load current I_{0m} of the transformer when the voltage rises above the rated voltage.

approximately half the rated current. Consequently, when a voltage that is 1.5–2 times higher than the rated voltage is applied to a transformer, the no-load current will significantly exceed the rated current, and the insulation of the transformer will break down due to overheating even if no load is applied. Below the rated voltage, transformers will function without any problems.

Equation (4.11) can also be used as the starting point for determining the number of turns for transformer windings, if we have, for example, an old set of magnetic circuit sheet metal that we want to fit with new windings. For the primary winding, the number of turns is given by

$$w_1 = U_1 / 4{,}44 f B_m A. \tag{4.12}$$

If we do not know the exact composition of our sheet metal, we can assume that $B_m = 1.0$ Wb/m², then we can measure the cross-sectional area of the magnetic circuit, and, if we know the characteristics of the supply voltage, we have everything we need to complete the calculation.

Problem 4.1

We have reclaimed some transformer core sheet metal from an old shell-type transformer. After assembling the pieces of metal, the cross-sectional area of the metal in the leg is 12 cm². We now want to place windings on the leg in order to construct a 230/36 V, 50 Hz transformer. Find the number of turns of wire required for each winding.

Solution

Since we do not know the exact material of the pieces of sheet metal, we will assume that the maximum magnetic density is $B_m = 1.0$ Wb/m². The number of turns required for the primary winding is

$$w_1 = \frac{U_1}{4{,}44 f B_m A} = \frac{230}{4{,}44 \cdot 50 \cdot 1{,}0 \cdot 12 \cdot 10^{-4}} = 863 \text{ turns.}$$

The number of turns required for the secondary winding is

$$w_2 = \frac{U_2}{4{,}44 f B_m A} = \frac{36}{4{,}44 \cdot 50 \cdot 1{,}0 \cdot 12 \cdot 10^{-4}} = 135 \text{ turns.}$$

The transformers used in converters may also be driven by a rectangular-wave voltage instead of a sinusoidal voltage. Figure 4.3 shows how the magnetic flux changes in the presence of such a wave.

For simplicity's sake, let us assume that our transformer has no losses and leakages. During a rectangular-wave pulse, voltage is constant. Thus, we can rearrange equation (4.2) as

$$\mathrm{d}F = -\frac{U_1}{w_1}\mathrm{d}t. \tag{4.13}$$

The minus sign need not be taken into account in this case, and integrating gives us

$$F = \frac{U_1}{w_1}t + K. \tag{4.14}$$

Here, K is the initial magnetic flux at the start of the voltage pulse. Over the course of a voltage pulse, the magnetic flux increases linearly. If the voltage pulses are unipolar and there is a pause between them (Figure 4.3a), the voltage equals zero during each pause. The magnetic flux does not change during this time. Hence, the magnetisation current must also be constant. This is possible if both the resistance of the electrical circuit and the internal resistance of the voltage source are zero. In reality, this is usually not the case, and the magnetic flux will decrease slightly between pulses.

During the next voltage pulse, the magnetic flux resumes increasing in the previous direction. Eventually the magnetic flux always becomes high enough to cause the core to become saturated. In order for the magnetic flux to increase

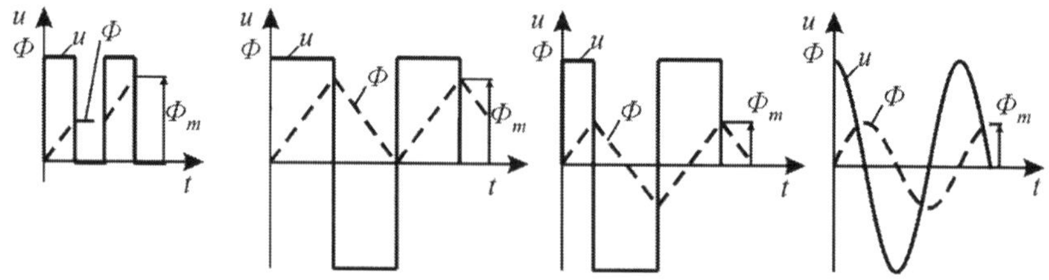

FIGURE 4.3 Changing of magnetic flux in a transformer: unipolar voltage pulses (a); bipolar voltage pulses (b); bipolar voltage pulses with the first pulse being two times shorter in duration (c); sinusoidal voltage (d).

further, the magnetisation current will need to begin to increase very sharply. In an idealised case (rectangular magnetisation curve), the flux will become infinitely high. In reality, however, the current will always reach some final value, at which point the magnetic flux will stop increasing. When this happens, the induced electromotive force and secondary voltage will disappear completely. From this we may conclude that with unipolar pulses and direct current, a transformer will quickly become saturated and will shut down. And not only that. The primary current will become several times higher than the rated current, the transformer winding will overheat, and the insulation will break down. That is, of course, unless the transformer has been equipped with a safety device that switches it off when the current gets too high.

In the case of voltage pulses with an alternating polarity, the magnetic flux decreases during negative pulses, thereby preventing the transformer from reaching a state of saturation. Thus, the transformer stays operational. In the case of a voltage that is alternating in polarity, the magnetic flux (as well as the magnetisation current) may be unipolar, as in Figure 4.3b. However, for a magnetic circuit, it is better to have a magnetic flux with an alternating polarity (Figure 4.3c). Given the same voltage, the maximum magnetic flux Φ_m will then be twice as low. An alternating magnetic flux can be obtained by shortening the duration of the first voltage pulse by half. A sinusoidal voltage immediately produces a magnetic flux without a DC component if the transformer is switched on when the mains voltage is at its peak (Figure 4.3d). If the switch-on occurs at any other time, a DC component is formed, which on real-life transformers, however, decays within a few dozen supply voltage periods (in the presence of a DC component, the positive and negative half-period of the magnetic flux are not equal).

In the case of a rectangular-wave voltage, the shapes of the curves of the voltage and magnetisation current are different: the curve of the magnetisation current is triangular, while that of the voltage is rectangular. In the case of a sinusoidal voltage and an idealised case, these curves are similar. In real-life transformers, however, the curve of the magnetisation current is distorted due to partial saturation.

4.2 FORMATION OF PRIMARY CURRENT LOAD COMPONENT DUE TO SECONDARY CURRENT

Now let us study the primary current – or more precisely, the load component of the primary current – caused by the secondary current. For simplicity, we will assume that all of the lines of force of the magnetic field pass through both winding coils in an identical manner and that there is an equal number of turns in the windings. Additionally, we will ignore the leakage fields that occur in real life. Note that when a current passes through a winding, there is a certain voltage drop. We may consider the windings to have no resistance, but that, instead, two resistors – r_1 and r_2 – are connected to the windings in series, as in Figure 4.4.

Assume that the primary winding of the transformer is driven by a rectangular-wave voltage u_1 (see the oscillogram in Figure 4.5). At time t_1, the load resistor

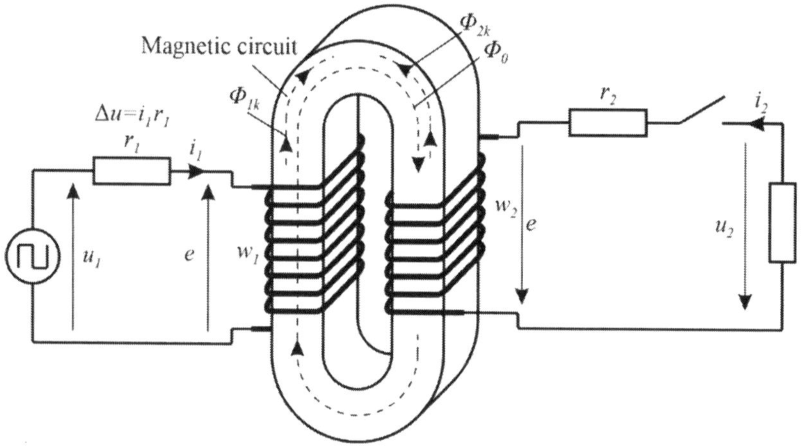

FIGURE 4.4 Loading the transformer with current.

R is switched on and a secondary current i_2 is generated. The load resistor is stepped down until time t_2. Thereafter, it is kept constant, and at time t_3 the load is switched off. The load current of the secondary circuit causes a series of interactions to arise in the transformer, indicated in Figure 4.5 by arrows numbered from one to six.

The process of the increasing of the primary current I_1 due to the secondary current I_2 may also be described using symbols as follows:

$$I_2 \uparrow \Rightarrow F_2 \uparrow \Rightarrow (F_1 - F_2 = F_0) \downarrow \Rightarrow E_1 \downarrow \Rightarrow (U_1 - E_1) \uparrow \Rightarrow \Delta U \uparrow \Rightarrow I_1 \uparrow$$

In short, the field of the secondary current destroys the field of the magnetisation current, for the restoration of which additional current is drawn from the power supply. This energy exchange process is mediated by a magnetic field. The energy of the load current drawn from the primary side is transformed into field energy, and the field energy is transformed into the energy of the secondary current. While, in Section 2.1, we saw that an electric current always produces a magnetic field and inserts energy into it, here, the opposite process occurs as well: energy is transmitted from the magnetic field of the primary winding to the secondary winding, where it manifests as a current.

Normally, there is a certain delay between a cause and its effect. In other words, the cause and effect are temporally separated. The generation of a current and the generation of its magnetic field are not temporally separated in the slightest. In this respect, the relationship between a current and its magnetic field is unique. Thus, the question of which is primary – the current or the field – becomes meaningless. The field may be deemed primary just as well as the current. The relationship between currents and magnetic fields serves as the basis for the operation of all transformers. The processes described above do not occur in

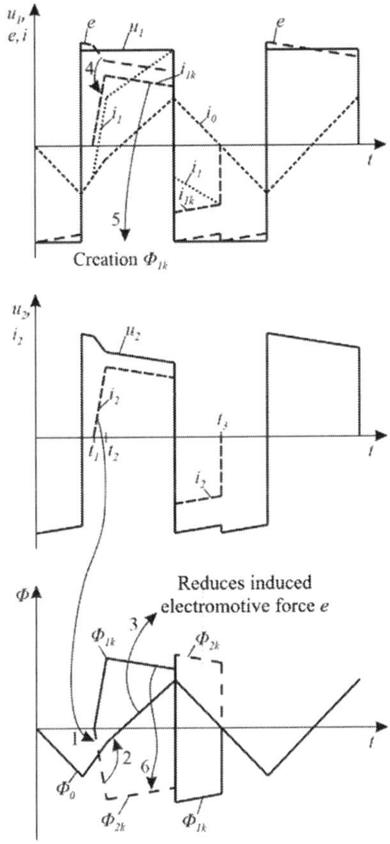

FIGURE 4.5 Formation of the primary current load component $i_{1\text{load}}$ due to the secondary current i_2.

succession, but all at the same time. However, in order to understand what is happening, it is expedient to consider the processes as six successive steps.

The magnetisation current i_0 and the load component of the primary current $i_{1\text{load}}$ combine in the primary winding of the transformer, resulting in the total current i_1 (dotted line in Figure 4.5).

The curve of the electromotive force e in the transformer differs from the curve of the primary voltage, because a voltage drop occurs across the resistance in the primary winding (r_1 in Figure 4.4) At no load, the electromotive force is greater than the voltage at the start of the rectangular-wave pulse, because, at this time, energy stored in the magnetic field of the magnetic circuit is returned to the power supply (the direction of the current is the opposite of that of the voltage). When the instantaneous value of the magnetisation current is zero and there is no load current, the electromotive force is equal to the primary voltage (see the positive voltage pulse on the right in Figure 4.5). The electromotive force is reduced by both the magnetisation current and the load current, which both cause a voltage drop across the resistance in the winding. The secondary voltage further differs from the electromotive force by the voltage drop across the resistance r_2 of the secondary winding, but its curve is similar in this case to that of the electromotive force. The decrease caused by the load current is clearly visible in the curve of the secondary voltage.

Interactions:

1. The current i_2 produces a magnetic flux Φ_{2k} in the magnetic circuit.
2. According to Lenz's law, the magnetic flux Φ_{2k} is opposite in direction to the magnetic flux Φ_0 of the magnetisation current i_0 and slightly reduces it (see also the directions of the magnetic flux in Figure 4.4).
3. The decrease in the magnetic flux Φ_0 causes the induced counter-electromotive force e to decrease (compare with the counter-electromotive force during the next positive voltage pulse, where there is no load current i_2).
4. The decreaseg in the counter-electromotive force e causes the voltage across the resistor r_1 to increase. As a result, the current increases, as given by Ohm's law. The primary current load component i_{1load} is formed.
5. The primary current load component i_{1load} produces its own magnetic flux Φ_{1load} in the magnetic circuit.
6. The magnetic flux Φ_{1load} of the primary current load component is opposite in direction to the magnetic flux Φ_{2load} of the secondary current, as a result of which the currents balance out. Thus, the magnetic flux Φ_0 decreases only due to the increasing of the voltage drop Δu, and this decrease is small.

The current and voltage curves of a real transformer will differ somewhat from those in Figure 4.5, because, in real transformers, the magnetic fluxes of the primary winding and the secondary winding are always unequal due to leakage fluxes.

4.3 COMMON APPLICATIONS AND DESIGNS OF TRANSFORMERS

Transformers are used most commonly in electricity supply systems. Electrical energy is transmitted from power plants to end loads by passing it multiple times through transformers installed in substations. Electricity supply transformers, as a rule, are three-phase machines and are operated at a frequency of 50 Hz or 60 Hz. Since electrical grids contain various rated voltages (0.4 kV, 10 kV, 110 kV, etc.), transformers are designed specifically for those voltages. Transformers may be loaded up to the rated current. Higher currents should not be used over an extended period of time, as this will cause the transformer to overheat and break down. To avoid this, transformers are equipped with fuses or circuit breakers. In regards to the overheating of a transformer, it makes no difference whether the load current of the transformer is in phase with the voltage (such load currents are called resistive load currents) or whether it is shifted in phase with respect to the voltage. This is why the power of transformers is measured in kilovolt-amperes (kVA), which is the unit of measurement of apparent power. Apparent power is given by the product of voltage and current without taking into account the phase shift angle.

Transformers are also quite frequently placed at the loads, as the mains voltage is often unsuitable for loads. Transformers used at loads include ones with a frequency higher than the mains frequency. Transformers operated at a higher frequency are lighter, smaller, and cheaper. However, in order to obtain a higher frequency, the alternating mains voltage must first be rectified and then, using semiconductor switches, converted back to an alternating voltage of a different frequency.

Transformers are also used in the measurement of high voltages and high currents, because they enable measuring instruments to be designed for the optimal rated current (usually 5 A) and rated voltage (usually 100 V).

An overview of the various applications of transformers is given in Figure 4.6.

4.3.1 Construction of Magnetic Circuits for Transformers

For a frequency of 50 Hz or 60 Hz, transformer magnetic circuits are constructed from 0.35- to 0.5-mm thick sheets of electrical steel. The sheets of steel need to be thin because massive magnetic circuits are prone to high eddy currents and high eddy current losses.

With an alternating magnetic flux, an electromotive force is induced in all loops that the magnetic flux passes through (Faraday's law). If the loop is conductive, an electric current is also generated in it. The left half of Figure 4.7 shows a section of a massive magnetic circuit positioned in the path of an alternating magnetic flux. An electromotive force and current are induced both in the short-circuited turn around the magnetic circuit (the short-circuited turn is shown to explain the arising of the losses) as well as in every conductive loop inside the magnetic circuit. In massive magnetic circuits, the current loops have a large cross-sectional area and low resistance, and are subject to high eddy currents. The current causes the magnetic circuit to heat up. This heating occurs at the expense of the energy of the alternating flux passing through the magnetic circuit. To replace the energy lost from the magnetic field, additional energy is drawn from the power supply. The resulting energy loss is called eddy current loss.

In a magnetic circuit constructed from steel sheets that are insulated from each other, as shown in the right half of Figure 4.7, current can only arise in loops that are located inside the individual sheets of steel. Since the sheets are thin, the current loops have a small cross-sectional area and high resistance, and the resulting eddy currents are low. The steel sheets are insulated from each other by a coat of lacquer, a layer of oxide, or a thin sheet of treated paper. Eddy current loss P_e can be calculated using the equation

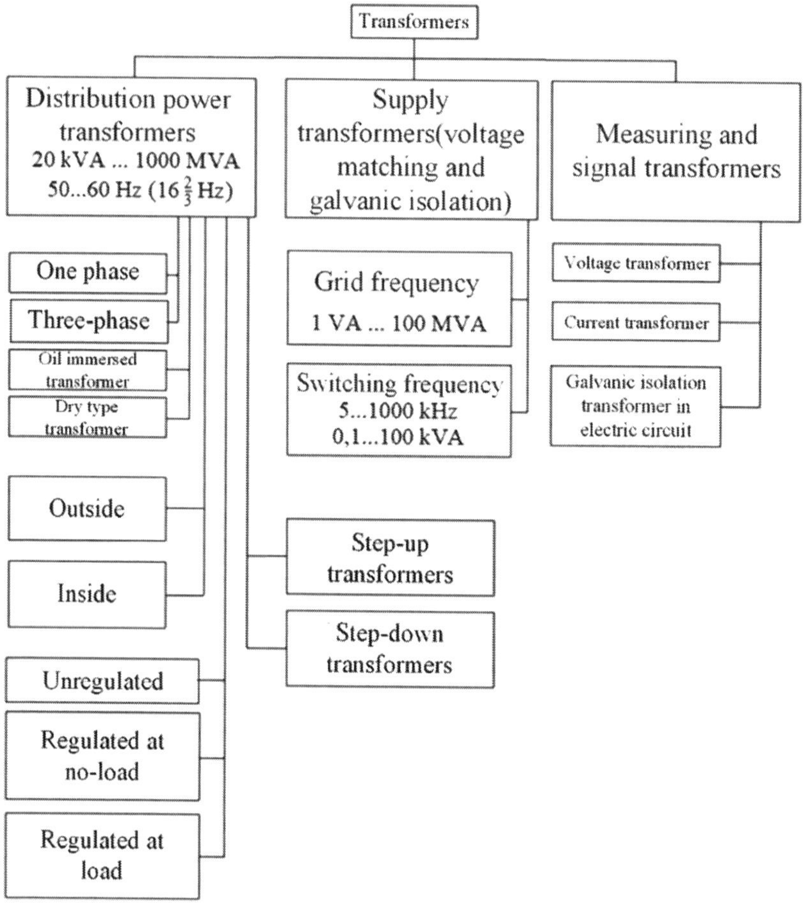

FIGURE 4.6 Types of transformers.

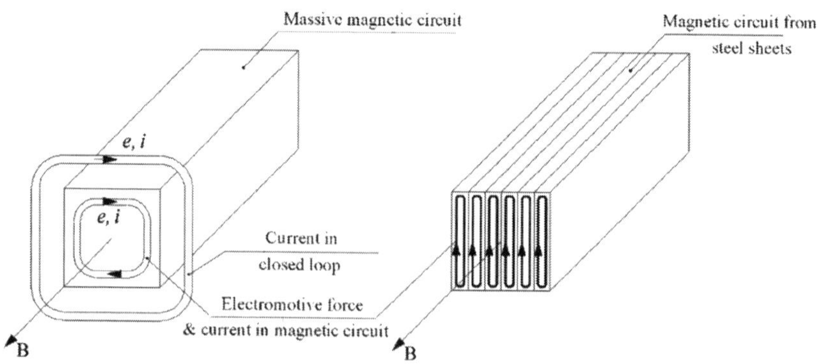

FIGURE 4.7 Eddy current loops in a massive magnetic circuit and a magnetic circuit constructed from steel sheets.

$$P_e = VK_e f^2 d^2 B_m^2, \tag{4.15}$$

where V – volume of the magnetic circuit;
K_e – coefficient for considering the units of measurement and the electrical resistivity of the material of the magnetic circuit;
f – frequency;
d – thickness of steel sheet;
B_m – maximum magnetic flux density.

Eddy current losses increase in proportion to the frequency, the thickness of the steel sheet, and the maximum magnetic flux density squared. At higher frequencies, engineers use either a thinner steel sheet or magnetic circuits constructed from a non-conductive material (**ferrite**, etc.) where iron is supplemented by various substances to reduce the electrical conductivity of the material.

Transformer cores can be assembled from sheet metal in a number of ways. For low-power transformers, the sheet metal may be stamped, yielding monobloc yokes and legs (Figure 4.8).

Additionally, **cold-rolled steel sheets** may be used, if the transformer metal needs to have superior properties. The properties of cold-rolled steel sheets are better in the direction of rolling than in the direction perpendicular to it. That is why the transformer magnetic circuits are wound from **steel strips**. After winding, the magnetic circuit is cut in half along the centre line for installing the windings (Figure 4.9). To minimise the air gap, the surface of the cut is polished smooth.

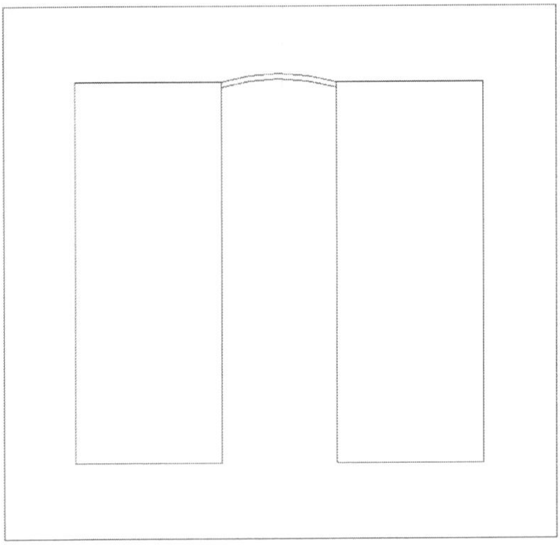

FIGURE 4.8 Stamped steel sheet magnetic circuit for a small transformer.

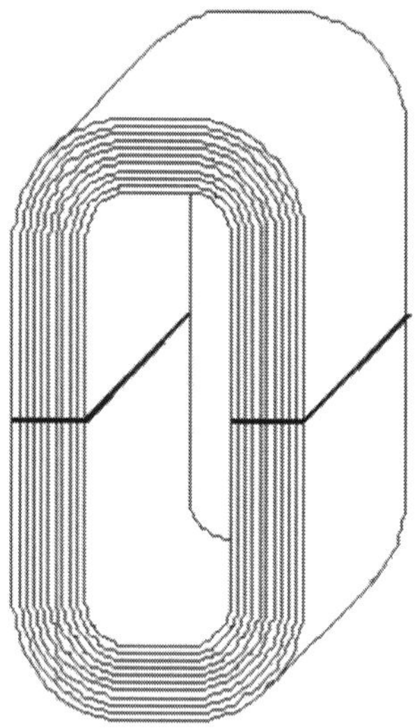

FIGURE 4.9 Magnetic circuit wound from steel strips for a single-phase core-type transformer.

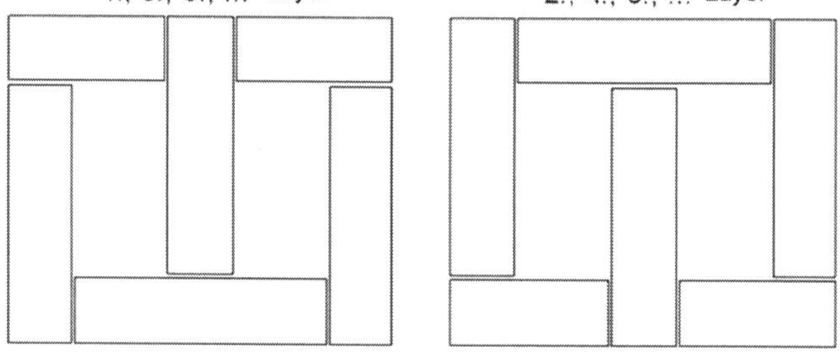

FIGURE 4.10 Assembly of a transformer core using the step-lap method.

For large transformers (starting from a few kilovolt-amperes), the magnetic circuit is assembled from rectangular pieces of steel sheet metal. In a core assembled in this way, there will inevitably be small air gaps at the joints, which are positioned perpendicularly to the path of the magnetic flux. The larger these air gaps, the greater the magnetisation excitation required by the magnetic circuit and the greater the no-load current of the transformer. Air gaps can be minimised by assembling the magnetic circuit in such a way that the gaps between two layers positioned on top of each other do not overlap (Figure 4.10). These cores are known as **step-lapped cores**. When simplicity is required, however, core layers may still be stacked in a way that leaves the gaps lined up across all layers. Such transformer cores are called **butt-lapped cores**.

In small transformers, the stack of sheet metal is held together by a coil support. Meanwhile, in large transformers, keeping the steel core from falling apart requires making holes in the sheet metal for **bolts** that hold the stack of metal sheets in one piece. In order to keep the individual sheets of metal from becoming electrically linked, which could lead to the formation of loops of short-circuit current, the bolts are placed inside insulating casings, and the nuts are seated on insulating washers.

4.3.1.1 Magnetic Circuits of Three-Phase Transformers

Three-phase current can be transformed using three single-phase transformers. The three transformers would then form a three-phase transformer group. However, it is more common to use a single three-phase transformer with a common core for all three phases. This is made possible under a condition of balance in three-phase voltage systems. The assembly of a three-phase transformer is explained in Figure 4.11.

The magnetic flux Φ in the core of each transformer is proportional to the voltage of the primary winding of that transformer. With one transformer per phase in a balanced three-phase system, the alternating magnetic fluxes of the different

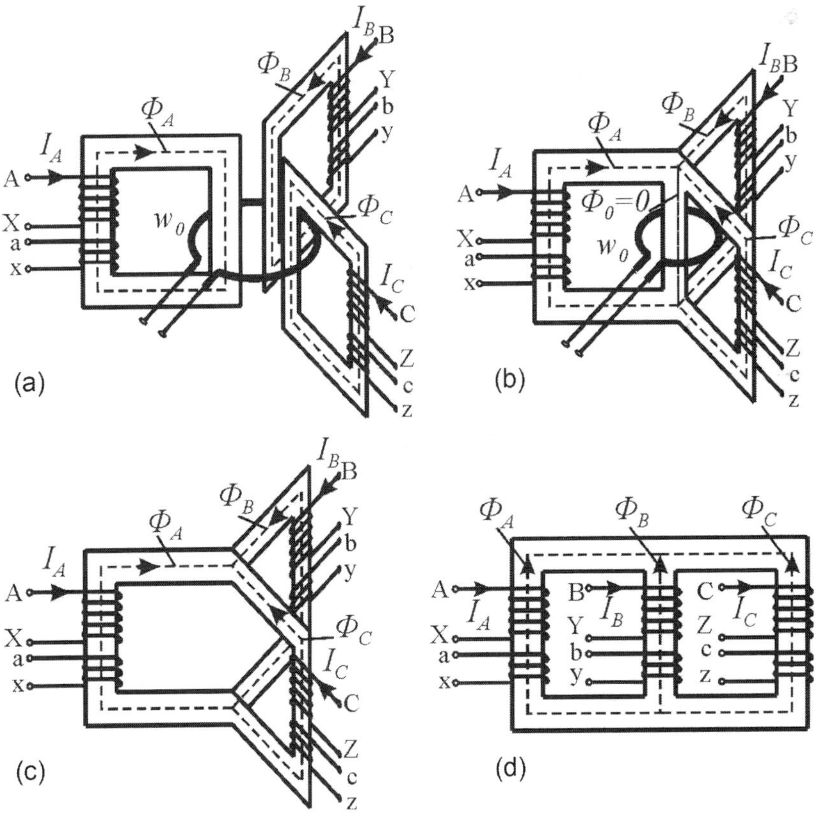

FIGURE 4.11 Obtaining a three-phase transformer by merging three single-phase transformers.

transformers are shifted in time in such a way that the sum of the instantaneous values of their magnetic fluxes is zero at all times (in a phasor diagram, the sum of three equal phasors shifted in phase by 120° is zero). This is considered in Figure 4.11a, where the magnetic circuits of the three single-phase transformers are shown as consisting of a single sheet of metal for simplicity. The transformers of the phases A, B, and C have windings on only one leg. The side yokes of all three transformers are facing each other and are surrounded by a shared winding w_0. No voltage is induced in the winding w_0, because the sum of the magnetic fluxes Φ_A, Φ_B, and Φ_C is zero at all times (the magnetic flux of each phase is always equal and opposite to the sum of the magnetic fluxes of the two other phases, despite the fact that all of the magnetic fluxes are constantly changing). If the three parallel side yokes were replaced with a single side yoke shared by all three transformers (Figure 4.11b), the sum of the magnetic fluxes in this shared side yoke would also be zero. Since there would essentially be no magnetic flux in that side yoke, we could just as well omit it (Figure 4.11c). Omitting the shared side yoke does not change anything. In addition, the length of the yoke of the phase B transformer may be reduced to zero and the magnetic circuits of phases A and C may be repositioned so that they align. While this does lead to an imbalance in the three-phase magnetic circuit because the phase B magnetic circuit is shorter, it does not affect the operation of the transformer in any meaningful way – the only effect is a slight difference between the no-load currents of the three phases. Meanwhile, we get a flat magnetic circuit that is easy to manufacture. Only in cases where a high-powered transformer is difficult to transport are separate transformers used for each phase.

4.3.2 Types of Transformer Windings

Transformer windings need to be constructed in a way that ensures their electrical and mechanical strength and the necessary temperature tolerance. In addition, the winding manufacturing technology must be as simple and inexpensive as possible, and the electrical losses in the windings must be within the prescribed limits.

Depending on how they are positioned on the core and relative to each other, the high-voltage and low-voltage windings (HV and LV windings) of a transformer are divided into **concentric windings**, or **cylindrical windings** (see Figures 1.40, 1.42), and **disc windings** (Figure 4.12).

Cylindrical windings are wrapped around insulating pipes, with the low-voltage winding placed closer to the core. A vertical 5–8-mm wide cooling channel is left between the HV winding and the LV winding. Due to the large number of turns in the HV winding, this winding is divided into several coils stacked on top of each other, in each of which the voltage may be up to 1,000 V. These coils are multi-layered and the inter-layer voltage may be up to 100 V.

In disc windings, the HV and LV discs are installed alternately along the core (Figure 4.12). The inner diameter of the LV discs is slightly smaller than that of the HV discs, which, due to their higher inner diameter, are positioned slightly further away from the leg. The uppermost and lowermost disc is a low-voltage

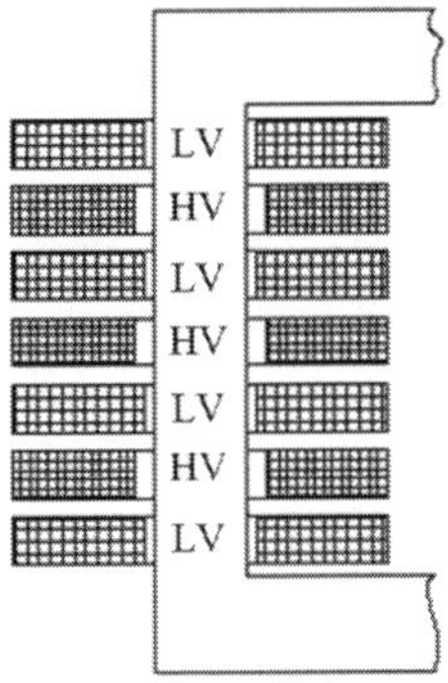

FIGURE 4.12 Disc windings on a transformer leg.

disc. Disc windings are harder to manufacture, but their advantage is a lower electromagnetic leakage. Minimising leakage is essential at very high voltages (110 kV and higher), where the windings need to be spaced farther apart to prevent dielectric breakdown.

4.4 PARAMETERS AND MODEL OF A TRANSFORMER

Figure 4.13 shows a schematic representation of a loaded real-life transformer together with the key parameters to consider in regards to it.

The transformer has two isolated electrical circuits operating at different voltages. The two circuits are interconnected by a common magnetic flux Φ. In addition to the common magnetic flux Φ, each winding has its own magnetic flux, Φ_{1S} and Φ_{2S}, respectively, the loop of which closes in air and does not connect with the other winding. These fluxes are called leakage fluxes. The actual positioning of the lines of force of the leakage field is more complex than that in Figure 4.13, which only shows a single line of force, but this is not important in this case. Considering electrical and magnetic quantities together is difficult. In order to simplify matters, an equivalent circuit may be used, i.e. a theoretical circuit of the transformer that contains only the parameters of the electrical circuit. The equivalent circuit can be created through an intermediate step, which is the transformer model shown in Figure 4.14.

The model has been drawn up with the following equivalent conversions.

- The two components of the current in the primary winding, which are the magnetisation current I_0 and the load current I_{1load} have been split into separate circuits. For this purpose, the primary winding has been imagined as consisting of two parallel halves, with one half w_{10} positioned on the magnetic circuit and generating a magnetic flux, and the other half, or the primary winding w_1, linked in an ideal manner with the secondary winding w_2.

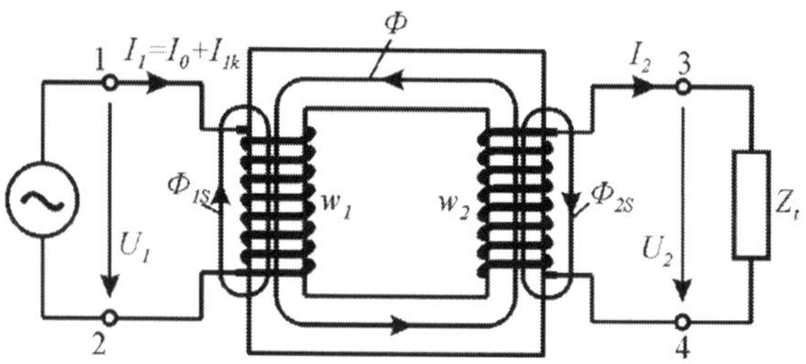

FIGURE 4.13 Real-life transformer and its parameters.

- The leakage flux of the windings has been separated from the windings and inserted into imaginary inductors L_{1S} and L_{2S}. Additionally, electrical resistance, too, has been separated from the windings and inserted

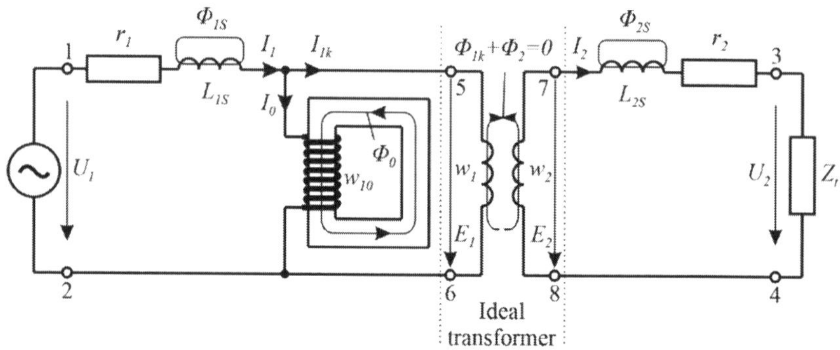

FIGURE 4.14 Transformer model where the two components of the current in the primary winding are split into separate circuits, and the parasitic inductances and winding resistances are separated from the windings.

into separate resistors r_1 and r_2.
- The ideal linkage between the primary winding and the secondary winding has been achieved by means of an imaginary device called an ideal transformer. An ideal transformer has no leakage field or winding resistance. The ideal transformer transfers the voltage applied to the primary terminals 5 and 6 to the secondary terminals 7 and 8. As it does this, the ideal transformer also transfers the current generated in the secondary circuit to the primary circuit. The voltage and electromotive force in the ideal transformer are equal at all times. The electromotive force in the ideal transformer in the model shown in Figure 4.14 is equal to the electromotive force that would occur in the real-life transformer.

The magnetic flux $Φ_0$ of the magnetisation circuit in the transformer model can in turn be transposed into an electrical circuit according to Figure 4.15b or 4.15c.

The resistor $r_{\mu\ \mathrm{p}}$ (Figure 4.15b) has been given a value whereby the power dissipated by it is equal to the iron losses P_{Fe} in the core, which is presumably a known

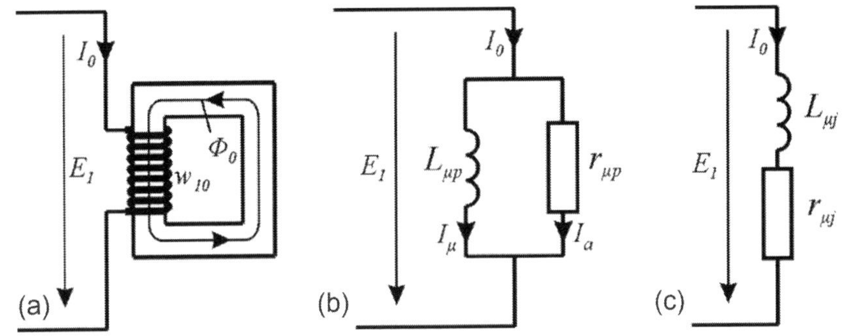

FIGURE 4.15 Replacement of the transformer magnetic circuit (a) with an equivalent electrical circuit connected in parallel (b) or in series (c).

quantity. The index μ suggests here that the magnitude of the resistance is not a constant value, but depends on the operating point on the magnetisation curve. Based on the Joule–Lenz law, we can say that

$$P_{\mathrm{Fe}} = \frac{E_1^2}{r_{\mu \mathrm{p}}} \text{ thus } r_{\mu \mathrm{p}} = \frac{E_1^2}{P_{\mathrm{Fe}}}. \tag{4.16}$$

The lossless inductance $L_{\mu\ \mathrm{p}}$ describes the energy stored in the magnetic field of the transformer core. The inductance $L_{\mu\ \mathrm{p}}$ can be calculated from the magnetic flux Φ_0, the magnetisation current I_0, and the number of turns in the primary winding w_1. Based on the flux linkage equation (2.22) $\psi = \omega\ \Phi$ and the inductance equation (2.24) $L = \psi / i$ presented in the second chapter, after some rearrangement and adding the corresponding indices, we get

$$L_{\mu \mathrm{p}} = \frac{w_1 F_0}{I_\mu},$$

where Φ_0 is the effective value of the magnetic flux. Generally, magnetic circuit calculations are performed using the maximum magnetic flux Φ_{m}. Hence, we get

$$L_{\mu \mathrm{p}} = \frac{w_1 F_{0\mathrm{m}}}{\sqrt{2} I_\mu}. \tag{4.17}$$

If necessary, it is also possible to use an equivalent circuit where the resistor that is accounting for losses and the inductance are connected in series (Figure 4.15c). The resistance in the equivalent series circuit can be calculated as follows (based, once again, on the Joule–Lenz law):

$$r_{\mu \mathrm{j}} = p_{\mathrm{Fe}} / I_0^2. \tag{4.18}$$

The inductances L_{1S} and L_{2S} of the coils accounting for leakage fluxes can be calculated from the magnetic resistance R_{mS} along the path of the leakage flux, after determining the resistance based on the design of the transformer. In a fully assembled transformer, the resistances for parasitic inductances can also be determined experimentally by means of a short-circuit test, which we will discuss in a later section.

4.4.1 Current, Voltage, and Power in the Ideal Transformer

Since there is no leakage flux in the ideal transformer, the magnetic flux Φ_2 of the secondary winding is exactly equal to the opposite magnetic flux Φ_1 of the primary winding. Due to this, the magnetisation excitation of the secondary winding $F_{m2} = I_2 w_2$ must be equal to that of the primary winding $F_{m1} = I_1 w_1$. Therefore, we may say that

$$I_1 w_1 = I_2 w_2$$

or

$$w_1 / w_2 = I_2 / I_1. \quad (4.19)$$

Since voltage and electromotive force are equal in the ideal transformer, we may conclude from equations (4.9) and (4.10) above that

$$U_1 / U_2 = w_1 / w_2. \quad (4.20)$$

Based on equations (4.19) and (4.20), we may say that

$$U_1 / U_2 = I_2 / I_1 = w_1 / w_2 \quad (4.21)$$

or

$$U_1 I_1 = U_2 I_2. \quad (4.22)$$

Equation (4.21) tells us that the **secondary voltage in the ideal transformer is directly proportional to the turns ratio, while the secondary current is inversely proportional** to the turns ratio. **Decreasing the voltage increases the current.** Equation (4.22) tells us that the apparent power on the primary side $S_1 = U_1 I_1$ is equal to that on the secondary side $S_{2} = U_2 I_2$:

$$S_1 = S_2. \quad (4.23)$$

Since there are no losses, the active power does not change:

$$P_1 = P_2. \quad (4.24)$$

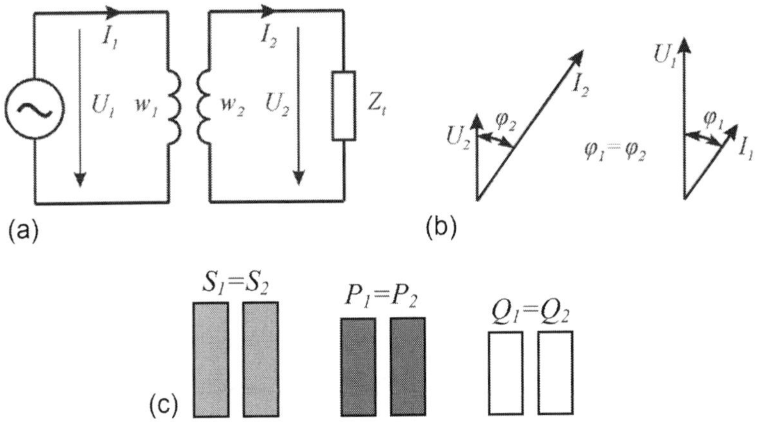

FIGURE 4.16 Equivalent circuit of a loaded ideal transformer (a), diagram of current and voltage phasors when ωl > w_2 (b), and the powers transmitted via the transformer (c).

Equations (4.23) and (4.24) reveal that the reactive power does not change either (Figure 4.16c):

$$Q_1 = Q_2. \tag{4.25}$$

On the secondary side, the phase shift angle ϕ_2 between the current and voltage is determined by the impedance Z_t of the load. An equal phase shift angle can be seen in the primary circuit (ϕ_1 in Figure 4.16b).

In conclusion, the ideal transformer transmits active, reactive, and apparent power without any losses (Figure 4.16). During transmission, only the voltage-to-current ratio is changed. High voltage and low current are changed into low voltage and high current, or vice versa. The no-load current of the ideal transformer is equal to zero.

Transformers are used precisely for the purpose of changing the voltage-to-current ratio. An ideal transformer would be the best tool for the job. Unfortunately, it is impossible to manufacture an ideal transformer, and so we must accept the limitations of real-life transformers. Real-life transformers are subject to resistive losses, and – due to leakage fluxes – they also consume reactive power. Consequently, equations (4.23)–(4.25) and (4.21) are not directly applicable to real-life transformers. Nevertheless, the differences between ideal transformers and real-life transformers are not too great (usually less than 10%). Thus, in approximate calculations, a real-life transformer can be considered an ideal one. When greater accuracy is required, however – such as for maintaining a certain voltage level in electrical grids – one must also take into account losses, leakage, and no-load current. This is usually done by preparing an equivalent circuit for the transformer, which we will discuss in a later section.

4.4.2 Changing the Impedance with an Ideal Transformer

In the case of an impedance-type load Z_t connected to a power supply with the voltage U_1 (Figure 4.17a), the no-load current is given by Ohm's law as $I = U_1/Z_t$. Now, say we are not satisfied with the current intensity I for some reason. We can correct this by connecting the load Z_t to the supply voltage via a transformer.

Say we are using an ideal transformer with the transformation ratio $k_{12} = w_1/w_2$ (Figure 4.17b). We want to find the impedance when the load is combined with the transformer. This new impedance Z_t' is called apparent impedance. We can calculate it from the transformer's primary current and voltage.

$$Z_t' = U_1 / I_1. \tag{4.26}$$

Since, in an ideal transformer, $U_1 = k_{12}U_2$ and $I_1 = I_2/k_{12}$, we get

$$Z_t' = \frac{U_1}{I_1} = \frac{k_{12}U_2}{\frac{I_2}{k_{12}}} = k_{12}^2 \frac{U_2}{I_2}. \tag{4.27}$$

Based on Ohm's law, $U_2/I_2 = Z_t$, which gives us

$$Z_t' = k_{12}^2 Z_t. \tag{4.28}$$

From equation (4.28), we can conclude that **when a load is connected to a power supply via an ideal transformer with a transformation ratio of k_{12}, the load's new resistance is directly proportional to the square of the transformation ratio.** Connecting the load Z_t to the power supply via a transformer is effectively the same as connecting the equivalent load Z_t' directly, without a transformer (Figure 4.17c). A transformer enables to reconcile the power supply voltage with the resistance of a load.

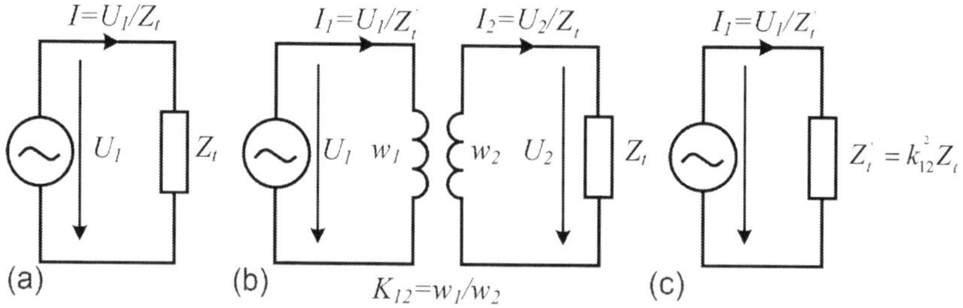

FIGURE 4.17 Connecting the same impedance to a power supply directly (a) and via a transformer (b), and connecting an equivalent impedance to the power supply without a transformer (c).

4.4.3 Calculations for a Circuit Containing an Ideal Transformer

The primary and secondary circuits of a transformer operate at different voltages, which makes performing calculations complicated. In order to simplify matters, either the primary or secondary circuit of may be replaced with an **equivalent circuit**. In this equivalent circuit, the voltages in the different sides of the transformer are set as equal (resulting in a transformation ration of 1 : 1 and an equal number of turns in each winding). This kind of substitution is called **reflecting**, or **referring**, one side of the transformer to the other side. In order to maintain equivalence, the values of the resistances in the reflected circuit must be changed according to equation (4.28) presented in the previous section.

In an ideal transformer (Figure 4.18a) with a transformation ratio of 1 : 1, the voltages of both sides are equal ($u_1 = u_2$), as are the currents ($i_1 = i_2$). Such a transformer may be replaced with the equivalent circuit shown in Figure 4.18b. This equivalent circuit contains only two connections. These are the connections between the adjacent terminals on the primary side and the secondary side (in this case, the **marked terminals** 5 and 7 and, respectively, two unmarked terminals 6 and 8). In diagrams, marked terminals are designated by filled in dots. Transformer terminals are marked to indicate the relative polarity. Marking the terminals in this way is also called transformer winding phase designation, or **phasing**.

For example, if a positive voltage is applied to the terminal 5 in Figure 4.18a at some point and the magnetisation current enters the transformer winding through this terminal, the question arises as to which terminal is positive at that moment in the secondary winding – terminal 7 or terminal 8? Terminal marking is used in order to eliminate this uncertainty. One winding terminal, usually that of the primary winding, is supplied with the marking arbitrarily (e.g. terminal 5 in Figure 4.18a). Which terminals are marked for other windings is determined using a measuring instrument (oscilloscope or voltmeter). The markings are assigned in such a way that the polarity of all marked terminals is identical during each half-period of the AC voltage. The phasing of transformer windings with the help of a voltmeter is illustrated in Figure 4.19.

During phasing, one freely chosen terminal in the primary winding (terminal b in Figure 4.19) is connected to a freely chosen terminal in the secondary winding. A voltmeter is inserted between the other two terminals (a and c). If the voltages of terminals a and b are in phase, the voltmeter will indicate the difference between the primary voltage U_1 and the secondary voltage U_2 (Figure 4.19a). In such a case, we may mark either the terminals connected to the voltmeter in Figure 4.19a or the other two terminals. The formation of the difference between the primary and secondary voltages $U_1 - U_2$ is shown in the phasor diagram. However, if we connect the randomly chosen terminals together and the voltmeter indicates the sum of the primary and secondary voltages, the terminals must be marked as in Figure 4.19b.

If, in an ideal transformer whose windings each contain an equal number of turns, we were to connect the marked terminals of the two windings (terminals 5

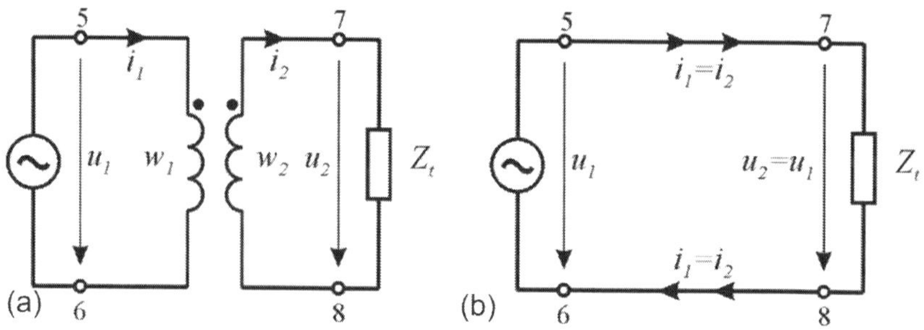

FIGURE 4.18 An ideal transformer (a), given an equal number of turns in the windings, may be replaced by an appropriate connection of the corresponding terminals of the primary and secondary sides (b).

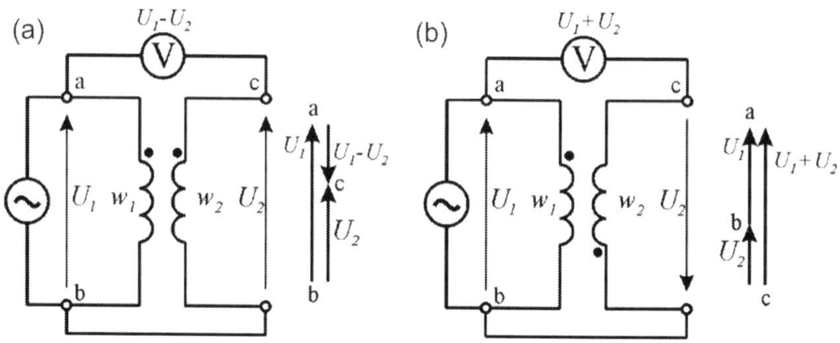

FIGURE 4.19 Determining the phase of transformer windings with a voltmeter connected to adjacent terminals (a) or opposite terminals (b).

and 7 in Figure 4.18a), the voltage between the remaining terminals 6 and 8 would equal zero, because $U_1 - U_2 = 0$. Therefore, we may also connect terminals 6 and 8. In addition, we may remove the transformer windings altogether, which would yield the equivalent circuit shown in Figure 4.18b. The current and voltage in the secondary and primary sides of the transformer remain the same. The load resistor Z_t, which is connected to the secondary terminals 7 and 8, may now also be connected directly to the primary terminals 5 and 6 without anything changing as a result. Now let us look at an example of how to perform calculations for a circuit that contains an ideal transformer.

Problem 4.2

A single-phase AC motor with a rated power P_n of 10 kW, a power factor of 0.8, and an energy conversion efficiency factor of 0.9 is supplied with a voltage of 230 V via a long power line. The resistance of the power line is $rl = 0.3\ \Omega$ and the inductive reactance is $xl = 0.4\ \Omega$ (Figure 4.20). Find the voltage across the motor and the power loss in the transmission line.

Solution

Calculate the rated apparent power S_n and the consumed active power P_t for the motor:

$$S_n = \frac{P_n}{\cos\varphi \cdot \eta} = \frac{10}{0,8 \cdot 0,9} = 13,89\text{kVA},$$

$$P_t = \frac{P_n}{\eta} = \frac{10}{0,9} = 11,11\text{kW}.$$

Calculate the motor's impedance z_t:

$$z_t = \frac{U_n^2}{S_n} = \frac{230^2}{13,89 \cdot 10^3} = 3,81\text{W}.$$

The angle of this complex impedance is determined by the power factor, $\phi = \arccos 0.8 = 36.87°$.

Calculate the real and imaginary part of the complex impedance of the motor that is a load on the power line:

$$r_t = z_t \cos\varphi = 3,81 \cdot 0,8 = 3,05\text{W},$$

$$x_t = z_t \sin\varphi = 3,81 \cdot 0,6 = 2,29\text{W}.$$

The line current equals

$$I_l = U_l / (r_l + \text{j}x_l + r_t + \text{j}x_t) = 230—0° / (0,3 + \text{j}0,4 + 3,05 + \text{j}2,29)$$

$$= 230\angle 0° / (3,35 + \text{j}2,69) = 230\angle 0° / 4,30\angle 38,76° = 53,53\angle -38,76°\text{A}.$$

The voltage across the motor equals

$$U_t = I_l z_t = 53,53\angle -38,76° \cdot 3,81\angle 36,86° = 204\angle -1,9°\text{V}$$

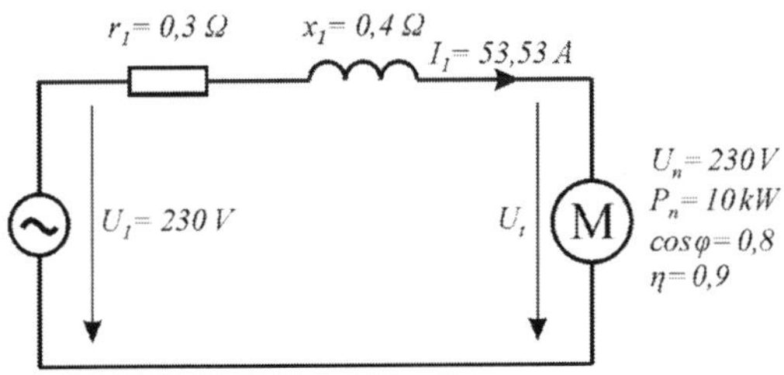

FIGURE 4.20 Supply of power to an electric motor from a low-voltage power line.

and the line loss equals

$$P_{\text{loss}} = I_l^2 r_l = 53,53^2 \cdot 0,3 = 860\,\text{W}.$$

Current regulations generally allow supply voltages to be up to 10% lower than the rated voltage. Thus, the minimum permissible voltage for the motor would be 230 · 0.9 = 207 V. This means that the voltage drop in the line needs to be reduced. We can do this with a transformer, which is a standard method for reducing voltage drops. To do this, we will place a transformer T1 (Figure 4.21a) at the beginning of the line to step the voltage up from 0.23 kilovolts to 6 kilovolts, and another transformer T2 at the end of the line to step the voltage back down to 0.23 kilovolts. The line resistance will remain unchanged, and we will consider the transformers to be ideal transformers.

Now, let us calculate the voltage across the motor and the losses in the line under the new conditions. We will perform the calculations using a referred equivalent circuit. The referral can be done in two different ways. For example, we may shift the entire circuit to a voltage level of 6 kV (Figure 4.21b). In such a case, the line resistance remains unchanged, because the power line operates at precisely this voltage, but the resistance of the load must be referred from 0.23 kV to 6 kV. Using the referral equation (4.28), we get

$$Z_{\text{t}}^{'} = k_{12}^2 Z_{\text{t}} = \left(\frac{6000}{230}\right)^2 \cdot (3,05 + \text{j}2,29) = 2075,6 + \text{j}1558,4\,\text{W}.$$

Now, let us calculate the current in the line

$$I_{l\,6\text{kV}} = \frac{U_t k_{12}}{r_l + \text{j}x_l + r_t^{'} + \text{j}x_t^{'}} = 230 \cdot \frac{6000}{230} / \left(0,3 + \text{j}0,4 + 2075,6 + \text{j}1558,4\right)$$

$$= 6000 / \left(2075,9 + \text{j}2558,8\right) = 6000\angle 0° / 2596\angle 36,90° = 2,31\angle -36,90°\,\text{A}.$$

The voltage across the motor at a voltage level of 6 kV is

$$U_{\text{t}}^{'} = I_{l\,6\text{kV}} z_{\text{t}}^{'} = 2,31\angle -36,90° \left(2075,6 + \text{j}1558,4\right) = 5995,6\angle 0°\,\text{V}$$

and at a level of 230 V it is

$$U_{\text{t}} = k_{21} U_{\text{t}}^{'} = \frac{230}{6000} \cdot 5995,6 = 229,8\,\text{V}.$$

The line loss equals

$$P_{\text{loss}} = I_{l\,6\text{kV}}^2 r_l = 2,31^2 \cdot 0,3 = 1,6\,\text{W}.$$

Alternatively, we may perform this calculation by referring the entire circuit to a voltage level of 230 volts (Figure 4.21c). In such a case, resistance of the load remains unchanged, because the load runs at precisely this voltage, but the line resistance must be referred, because the line voltage is 6 kV. Now we need to refer the transformer T2 from the primary side to the secondary side, wherein the referral ratio is k_{21} = 230/6,000. When performing referrals, it is important to keep in mind that when transitioning from a higher voltage to a lower voltage, the resistance decreases, and vice versa. The resistance of the 6 kV power line at a voltage level of 230 V is

$$r_l' = k_{21}^2 r_l = \left(\frac{230}{6000}\right)^2 \cdot 0,3 = 0,00044\text{W},$$

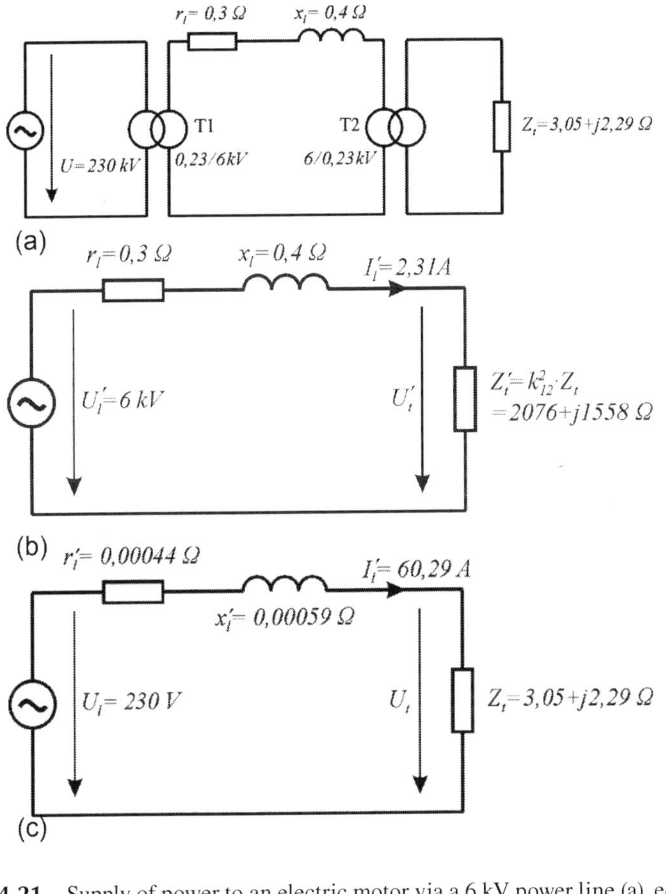

FIGURE 4.21 Supply of power to an electric motor via a 6 kV power line (a), equivalent circuit referred to the 6 kV voltage level (b), and equivalent circuit referred to the 230 V voltage level (c).

$$x_l^{'} = k_{21}^2 x_l = \left(\frac{230}{6000}\right)^2 \cdot 0,4 = 0,00059\text{W}.$$

The line current equals

$$I_l = \frac{U_l}{r_l^{'} + \text{j}x_l^{'} + r_\text{t} + \text{j}x_\text{t}} = \frac{230Ð0°}{0,00044 + \text{j}0,00059 + 3,05 + \text{j}2,29}$$

$$= \frac{230\angle 0°}{3,815\angle 36,90°} = 60,29\angle -36,90°\text{A}.$$

The voltage across the motor equals

$$U_\text{t} = I_l z_\text{t} = 60,29\angle -36,90° \cdot 3,81\angle 36,86° = 229,7\angle -0,04°\text{V}.$$

The line loss equals

$$P_\text{loss} = I_l^2 r_l^{'} = 60,29^2 \cdot 0,00044 = 1,6\text{W}.$$

Both equivalent circuits will yield the same result within the range of precision of the calculation. The equivalent circuit referred to 230 volts (Figure 4.21c) shows that when the line voltage is increased, the equivalent resistance decreases significantly. As a result, the voltage drop and losses across the transmission line become very small compared to when power is transmitted at a low voltage.

The optimum voltage level is considered to be about 1 V per metre, or 1,000 V per kilometre. For example, passenger cars use a voltage of 12 V, while buses use 24 V. Unnecessarily high voltages cause problems with insulation and electrical safety. Electrical energy transmission grids utilise a variety of voltage levels separated by transformers.

4.5 EQUIVALENT CIRCUIT OF A TRANSFORMER

The equivalent circuit of a transformer is a circuit diagram that describes all of the key parameters of a real-life transformer, as well as the relationships between. For a dual-wound transformer it is possible to create two equivalent circuits: one at the primary voltage level (Figure 4.22a) and another at the secondary voltage level (Figure 4.22b).

These equivalent circuits have been created on the basis of the transformer model shown in Figure 4.14. The following changes have been made. The magnetic circuit has been replaced with a parallel electrical circuit. In one case, the resistances in the secondary winding have been referred to the primary side (Figure 4.22a), and in the other case, the resistances in the primary winding and the magnetisation circuit have been referred to the secondary side (Figure 4.22b). The ideal transformer with the transformation ratio 1 : 1 after referral has been

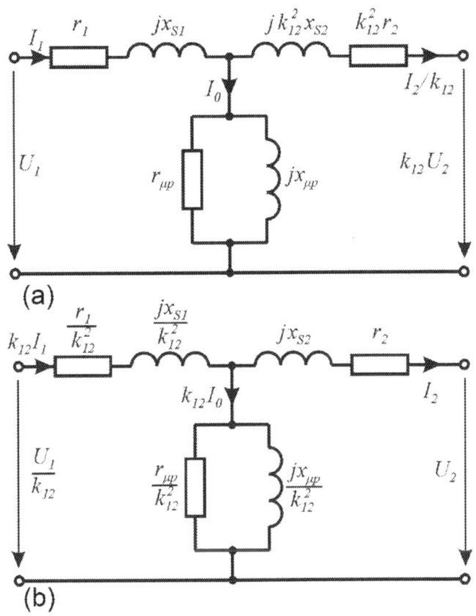

FIGURE 4.22 Transformer equivalent circuit referred to the primary side (a) and transformer equivalent circuit referred to the secondary side (a).

replaced with connections between terminals 5 and 7 and terminals 6 and 8, respectively. The inductances in the equivalent circuit have been replaced with inductive reactances. Thus, the frequency of the supply voltage has been indirectly accounted for.

Referral to the primary side is appropriate when studying the processes taking place in the supply grid. Referral to the secondary side is appropriate when studying the processes taking place in the load circuit. For example, it is possible to study how the intensity of the load current I_2 affects the secondary voltage U_2. The equivalent circuit provides a sufficiently accurate description of the effect of the windings' resistances and leakage-related inductive reactances, as well as the effect of iron losses in the core. The equivalent circuit is least accurate in describing the current I_0 in the magnetisation branch, because it does not take into account the non-linearity of the magnetic circuit. The current through the inductive reactance in the equivalent circuit is sinusoidal and proportional to the voltage. In reality, the magnetisation current is not sinusoidal, and increases quite sharply when the voltage exceeds the rated voltage. This is due to the non-linearity of the magnetisation curve of the electrical steel in the core of the transformer, the effect of which is explained by the graphs in Figure 4.23.

If the maximum magnetic flux is low (Φ_1 in Figure 4.23), the operating point moves along the near-linear part of the magnetisation curve, and the curve of the magnetisation current i_0 is near-sinusoidal. As the maximum magnetic flux increases, the operating point moves into the region around the knee point of the curve and over it (Φ_2 and Φ_3, respectively). In this region, the magnetisation current increases rapidly and becomes non-sinusoidal. The magnetic flux changes in

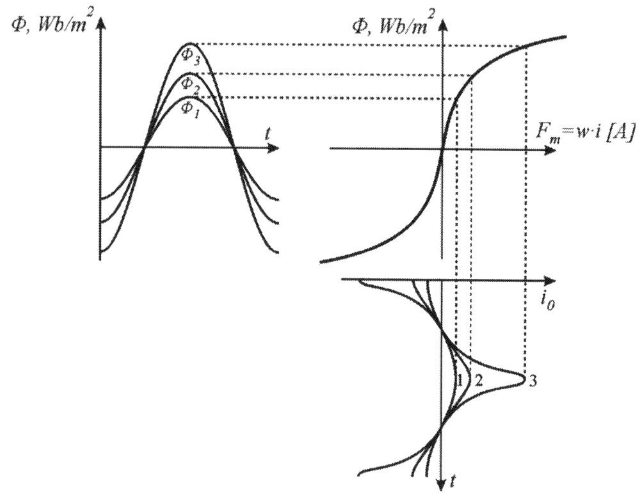

FIGURE 4.23 Effect of the maximum magnetic flux of the transformer core on the magnetisation current.

proportion to the voltage. The voltage rising a mere 20% above the rated voltage can be enough to overload the primary winding of the transformer with magnetisation current. The equivalent circuit, however, does not reflect this fact, because it treats the inductance as linear. Nonetheless, at and below the rated voltage, the equivalent circuit is sufficiently accurate.

The magnetisation current I_0 is relatively low compared to the load current (approximately 10%). The voltage drop caused by this current at the elements r_1 and jx_{S1} is very small (usually less than one percent of the rated voltage) and may usually be disregarded. Thus, we may use the simplified equivalent circuit shown in Figure 4.24, where the primary and secondary winding's resistances and parasitic inductive reactances have been aggregated in consideration of the referral ratio, yielding the equivalent resistances r_{eq} and x_{eq}.

When referring to the primary side,

$$r'_{eq} = r_1 + k_{12}^2 r_2, \tag{4.29}$$

$$x'_{eq} = x_{S1} + k_{12}^2 x_{S2}.$$

When referring to the secondary side,

$$r''_{eq} = \frac{r_1}{k_{12}^2} + r_2, \tag{4.30}$$

$$x''_{eq} = \frac{x_{S1}}{k_{12}^2} + x_{S2}.$$

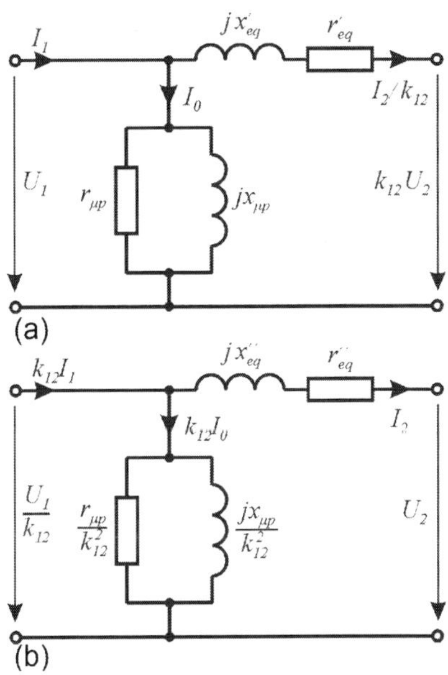

FIGURE 4.24 Simplified equivalent circuit of transformer, with aggregated primary and secondary winding resistances and parasitic inductive reactances: (a) referred to the primary side; (b) referred to the secondary side.

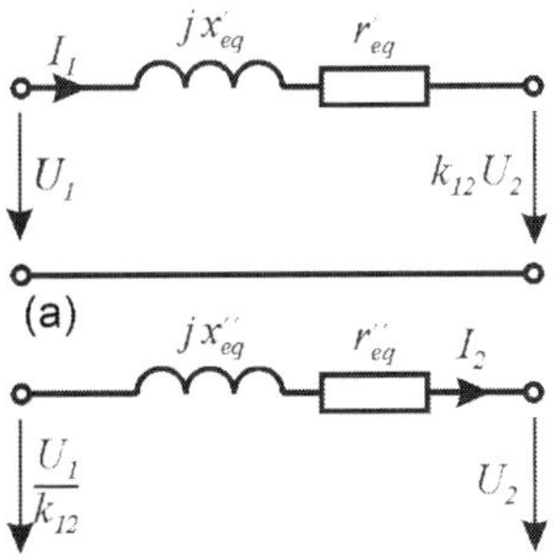

FIGURE 4.25 Simplified equivalent circuit of transformer, with no magnetisation branch: (a) referred to the primary side; (b) referred to the secondary side.

In some cases, the magnetisation current along with the magnetisation branch in the equivalent circuit (Figure 4.24) may be ignored. This gives us an even simpler equivalent circuit (Figure 4.25).

4.6 USE OF A PER-UNIT SYSTEM

If we want to perform calculations for an electrical circuit containing a single dual-wound transformer, we have to choose whether to refer the entire circuit to the higher voltage level or the lower voltage level. With a greater number of transformer windings or transformers, we would have to choose a voltage from a single winding to which to refer every other winding. In some cases, this laborious and time-consuming task can be avoided by adopting a per-unit system. With such a system, the actual values of the voltages, currents, and resistances are replaced with their ratios to established base values. The base voltage, base current, base impedance, and base apparent power form a system of base values. When creating a system of base values, it is necessary to first choose two base values, leaving the remaining two to serve as derived base values. For transformers, it is appropriate to select the rated voltage U_n of one winding and the rated power S_n of one transformer to be the base values. The derived base values will then be the rated current

$$I_n = \frac{S_n}{U_n} \tag{4.31}$$

and the rated impedance

$$Z_n = \frac{U_n}{I_n} = \frac{U_n^2}{S_n}. \tag{4.32}$$

The rated impedance is sometimes simplistically also called the rated resistance.

The actual values and base values are used to find the per-unit voltages, currents, powers, and impedances.

Per-unit voltage

$$u^* = U / U_n. \tag{4.33}$$

Per-unit impedance

$$z^* = z / Z_n. \tag{4.34}$$

Per-unit resistance

$$r^* = r / Z_n. \tag{4.35}$$

Per-unit inductive reactance

$$x^* = x / Z_n. \tag{4.36}$$

Per-unit quantities are all unitless, and are usually denoted by capital letters plus ‘pu’ (per-unit) in subscript, as in $Z_{pu} = 0.3$ pu. Note that for a transformer, the per-unit rated power equals one ($S_{pu} = S_n / S_n = 1$ pu), and the resistance, reactance, and impedance values do not depend on the selected base voltage. However, the base voltage does determine the rated impedance.

Example

We choose U_{1n} = 230 V as the base voltage. The base resistance at the primary voltage is

$$Z_n^{'} = \frac{U_{1n}}{I_n} = \frac{U_{1n}^2}{S_n} = \frac{230^2}{400} = 132,3\text{W}.$$

If we choose the secondary voltage U_{2n} = 12 V as the base voltage, the base resistance is

$$Z_n^{''} = \frac{U_{2n}^2}{S_n} = \frac{12^2}{400} = 0,36\text{W}.$$

The per-unit value of the secondary-side resistance of 1 Ω is

$$\overset{*}{r}_{\text{load}} = \frac{r_{\text{load}}}{Z_n^{''}} = \frac{1}{0,36} = 2,78\text{p.u.}$$

If we now change the load resistance so that $\overset{*}{r}_{\text{koormus}} = 1\text{s}\ddot{u}$, the corresponding actual resistance on the secondary side is, in ohms,

$$r_{\text{load}} = \overset{*}{r}_{\text{load}} Z_n^{''} = 1 \cdot 0,36 = 0,36\text{W}$$

and the power dissipated by a resistor with such a resistance at a voltage of 12 V is

$$P = \frac{U^2}{r_{\text{load}}} = \frac{12^2}{0,36} = 400\text{W}.$$

If the primary side contains a load resistor with a per-unit resistance, again, of $\overset{*}{r}_{\text{load}}$ = 1 pu its resistance in ohms is

$$r_{\text{load}} = \overset{*}{r}_{\text{load}} Z_n^{'} = 1 \cdot 132,3 = 132,3\text{W}$$

and the power dissipated by this resistor at a voltage of 230 V is

$$P = \frac{230^2}{132,3} = 400\text{W}.$$

The example above tells us that when a resistance is expressed as a per-unit value, the power dissipated by the resistor does not depend on the voltage level (secondary or primary voltage). Instead, that power depends on the base apparent power, which is usually the rated apparent power of the transformer. When a load resistance of $r^*_{load} = 1$ pu is connected to the terminals of whichever winding of a transformer with however high the power, the power dissipated by the load is equal to the rated power of the transformer. This enables per-unit values to also be used for comparing the resistances, inductive reactances, losses, and other similar parameters of transformers with different power ratings.

4.7 TRANSFORMER SHORT CIRCUIT AND SHORT-CIRCUIT TEST

4.7.1 Why Is It Necessary to Investigate the Short-Circuit Operation of a Transformer?

Most buildings today are equipped with electricity. Electrical wires can be found in nearly every room of a building. Damage to the insulation of electrical wires can result in a so-called short circuit. This involves bare electrical wires coming into contact with each other and the resistance between the two live wires becoming very small (close to zero). According to Ohm's law, when the resistance becomes zero, the current becomes infinitely high. In reality, currents cannot become infinitely high, but they can become much higher than the normal current level. High currents are dangerous, as they cause the wires to heat up, which can, in turn, cause a fire to start.

Although damage to the insulation of wires is not a very common occurrence, strict requirements have been established to protect all electrical installations against short circuits. Short-circuit protection consists in switching off the voltage as quickly as possible when a dangerously high current occurs in the wires. This prevents the wires from overheating. The switching is achieved by means of a circuit breaker or fuses. However, neither circuit breakers nor fuses are able to switch off extremely high currents. Choosing the appropriate electrical safety device requires knowing how high the potential short-circuit currents in the electrical grid can get. An electrical safety device must not trip at the normal operating current, but must trip in the event of a short circuit. The trip currents of electrical safety devices are known quantities, and we must be sure that when a short circuit occurs, the current exceeds the trip current of the device.

Figure 4.26 shows part of the electricity supply system of a building and the corresponding circuit diagram. The circuit diagram also indicates the short-circuit currents for cases where there is a short circuit in the power line or the building.

Short-circuit currents never become infinitely high, because power lines and transformers always have a resistance and an inductive reactance. When it comes to limiting short-circuit currents, the resistances of transformers play a major role, as they are both relatively high and stable. A short circuit can occur right at the beginning of a power line, where the length of the wires and, therefore, the line resistance are still negligible. The short-circuit impedance of a transformer, along with the active and reactive components of the impedance, can be found both mathematically and experimentally. The experimental method is known

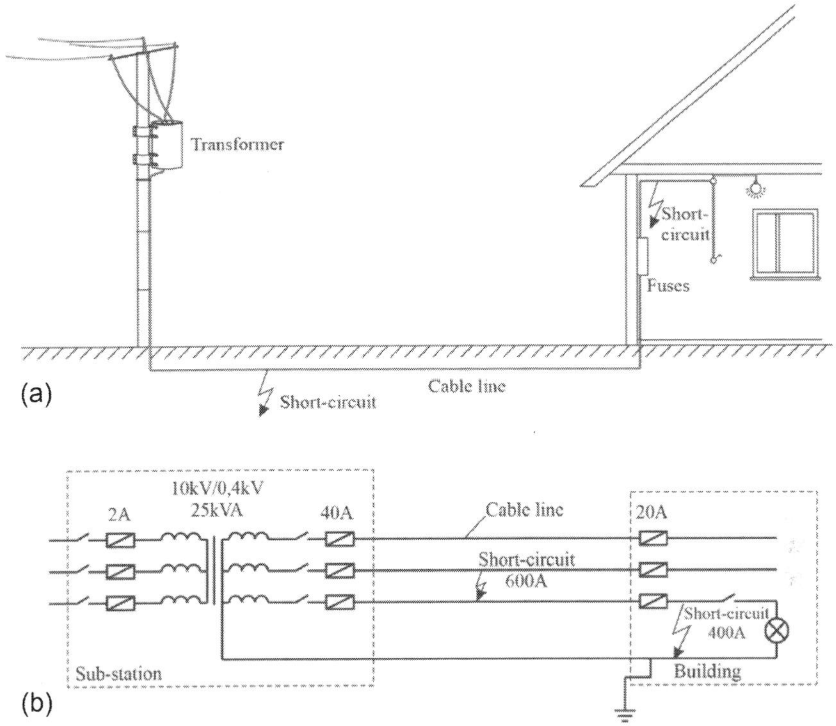

FIGURE 4.26 Supply of electricity to a building (a) and the corresponding circuit diagram with the potential short-circuit currents (b).

as short-circuit testing. When a short circuit occurs at the rated voltage of the transformer, the resulting short-circuit currents will exceed the calculated rated voltage of the transformer by about 20 times. Such conditions are not suitable for conducting tests, because electrical safety devices need to immediately switch off any currents that high, making it impossible to perform measurements. In order to allow measurements to be performed, short-circuit tests are conducted at voltages 10 to 20 times below the rated voltage.

4.7.2 Conducting a Short-Circuit Test

Figure 4.27 shows the operating points of a transformer with a rated voltage of U_n = 400 V, rated current of I_n = 90 A, and a short-circuit impedance of z_k = 0.33 Ω for short-circuit testing, short-circuit operation, rated operation, and no-load operation. Figure 4.28 shows the diagram for the short-circuit testing of a single-phase transformer and an equivalent circuit for the short-circuited transformer.

The secondary winding of the transformer being tested is short-circuited; a voltmeter and the voltage winding of a wattmeter are connected to the primary winding in parallel, and an ammeter and the current winding of the wattmeter are

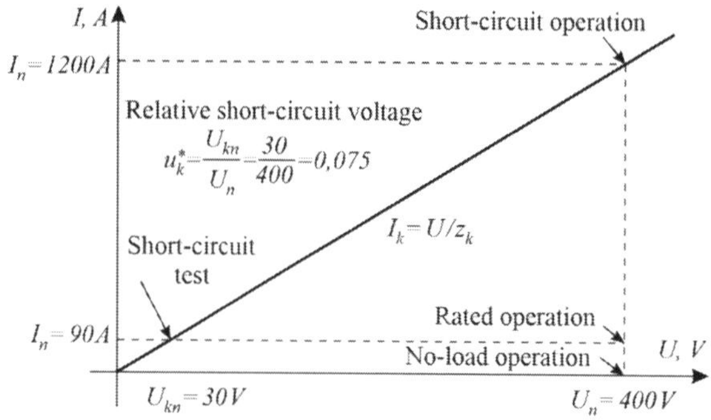

FIGURE 4.27 Different operating statuses of a transformer and the operating point for short-circuit testing.

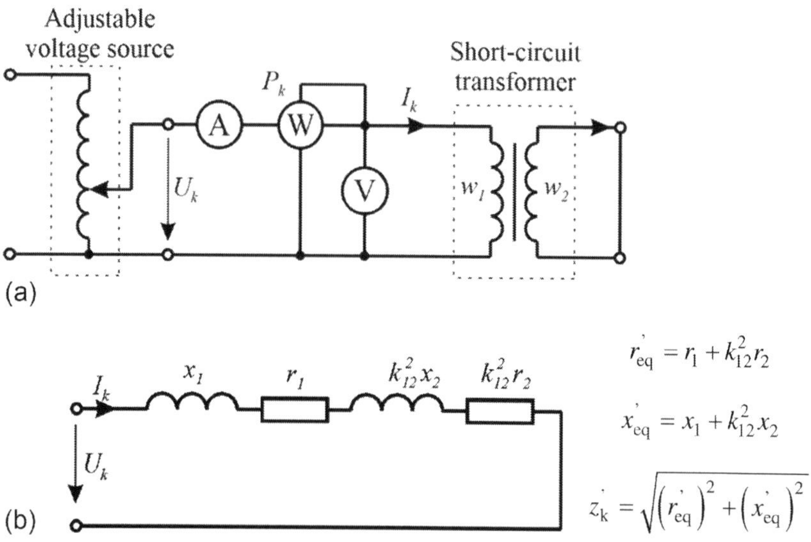

FIGURE 4.28 Diagram for the short-circuit testing of a single-phase transformer (a) and an equivalent circuit for the short-circuited transformer (b).

connected to the primary winding in series (Figure 4.28a). The primary winding is supplied with a voltage from a variable voltage source, which in this case is an autotransformer, via the windings of the ammeter and the wattmeter. At the start of the test, the voltage of the variable voltage source is zero; then it is increased while observing the ammeter reading. The voltage is increased in steps, and the readings of the instruments are recorded at each step, in order to later establish the short-circuit characteristics as a function of the voltage. The voltage is stepped up until the ammeter reads the rated current of the primary winding. When this happens, the transformer being tested has reached the rated operating point for short-circuit testing. Figure 4.27 shows the rated operating point for short-circuit testing for a specific transformer. At the rated current of 90 A, the voltage supplied to the primary winding of this transformer is $U_{kn} = 30$ V. Quantities relating to short-circuit operation are denoted by the index k. The same figure also shows the rated operating point, where the current is 90 A and the voltage is 400 V. Note that under rated operation, the secondary winding of the transformer is not short-circuited, but contains a load that is consuming power at 90 A. If, however, a short circuit were to occur on the secondary side of the transformer at the rated voltage, the current on the primary side would rise to $I_k = 1{,}200$ A (Figure 4.27). This current value is not a measured value, but has been calculated from the short-circuit impedance z_k. The short-circuit impedance z_k, in turn, is determined by a short-circuit test, where, according to the above, the currents and voltages are significantly lower. This resolves the problem of how to experimentally determine high and dangerous currents which occur during a short circuit.

The equivalent circuit of the short-circuited transformer shown in Figure 4.28b has no magnetisation branch. During short-circuit testing, the voltage across the primary winding is much lower than the rated voltage. Due to this, the magnetisation current is negligible and need not be taken into account. During a real short circuit, the voltage across the magnetisation branch would be about twice as low as the rated voltage, as the high current would cause a large voltage drop across the primary winding. Due to this, the magnetisation current would be several times lower than during rated operation (the AC resistance of the magnetisation branch is non-linear). Ultimately, the contribution of the magnetisation current to the short-circuit current is quite insignificant even during a real short circuit, and therefore need not be taken into account.

4.7.3 Use of Data Obtained from a Short-Circuit Test

During a transformer short-circuit test, the readings of the ammeter, wattmeter, and voltmeter are recorded. These readings are then used to calculate the impedance z_k, reactance x_k, and resistance r_k. The calculation equations for single-phase transformers are as follows:

$$z_k = U_k / I_n; r_k = P_k / I_n^2; x_k = \sqrt{z_k^2 - r_k^2}. \tag{4.37}$$

The resistance r_k calculated on the basis of the measurement results corresponds to the equivalent resistance $r_{eq}^{'}$ of the equivalent circuit shown in Figure 4.28b. The short-circuit resistance is related to the resistances of the equivalent circuit presented in Section 4.5 as follows:

$$r_k = r_1 + k_{12}^2 r_2 = r_1 + r_2^{'}. \tag{4.38}$$

A similar situation applies to the reactance

$$x_k = x_1 + k_{12}^2 x_2 = x_1 + x_2^{'} \tag{4.39}$$

and the impedance

$$z_k = \sqrt{r_k^2 + x_k^2}. \tag{4.40}$$

Thus, resistances determined on the basis of a short-circuit test constitute the sums of the corresponding resistances of the primary side and the secondary side, with the resistances of the secondary side referred to the primary side. A short-circuit test cannot be used to determine the resistances of the primary winding and the secondary winding separately. If, however, it is necessary to know the approximate values of these resistances, for electrical resistance we may say that

$$r_1 \approx r_2^{'} \approx 0,5r_k. \tag{4.41}$$

In the case of three-phase transformers, the currents and voltages are measured for all phases, and then the average values of the line current Il_n and the line voltage for short-circuit testing Ul_k are determined. The power is determined as the total short-circuit power P_k for all three phases.

In the case of star-connected primary windings, the phase-specific short-circuit parameters are calculated as follows:

$$z_k = U_{lk} / \sqrt{3}I_{ln}; r_k = P_k / 3I_{ln}^2; x_k = \sqrt{z_k^2 - r_k^2} \tag{4.42}$$

In the case of a delta connection:

$$z_k = \sqrt{3}U_{lk} / I_{ln}; r_k = P_k / I_{ln}^2; x_k = \sqrt{z_k^2 - r_k^2}. \tag{4.43}$$

The short-circuit resistance values z_k, x_k, and r_k found on the basis of the test results can be used to draw a short-circuit resistance triangle (Figure 4.29a). Multiplying the short-circuit resistance values by the rated current yields the short-circuit voltage $U_k = z_k I_n$, the reactive component of the short-circuit voltage $U_{kr} = x_k I_n$, and the active component of the short-circuit voltage $U_{ka} = r_k I_n$. These voltages can be used to draw up a phasor diagram of the short-circuit test (Figure 4.29b).

The angle φ_k between the short-circuit voltage phasor U_k and current phasor I_k is the short-circuit phase shift angle. In the triangle of short-circuit resistances, it is also possible to separately show the resistance triangles of the primary winding and the secondary winding (Figure 4.29a). The voltages, currents, and resistances U_{kn}, I_n, z_k, r_k, and x_k measured and calculated during a short-circuit test can also be expressed as per-unit values. The base voltage of the per-unit system would be the rated voltage U_n and the base current would be the rated current I_n. This means that the base impedance is $Z_n = U_n / I_n$. The per-unit short-circuit voltage u_k^* is thus given by:

$$\overset{*}{u}_{\text{k}} = U_{\text{kn}} / U_{\text{n}}. \tag{4.44}$$

The per-unit short-circuit voltage $\overset{*}{u}_{\text{k}}$ can be shown to equal the per-unit impedance:

$$\overset{*}{u}_{\text{k}} = U_{\text{kn}} / U_{\text{n}} = z_{\text{k}} I_{\text{n}} / U_{\text{n}} = z_{\text{k}} / Z_{\text{n}} = \overset{*}{z}_{\text{k}}. \tag{4.45}$$

In practice, per-unit short-circuit voltages are often expressed as a percentage:

$$u_{\text{k}\%} = \frac{U_{\text{kn}}}{U_{\text{n}}} \cdot 100 = 100\overset{*}{u}_{\text{k}}. \tag{4.46}$$

The per-unit short-circuit resistance $\overset{*}{r}_{\text{k}}$ and per-unit short-circuit reactance $\overset{*}{x}_{\text{k}}$ are also used:

$$\overset{*}{r}_{\text{k}} = r_{\text{k}} / Z_{\text{n}}; \overset{*}{x}_{\text{k}} = x_{\text{k}} / Z_{\text{n}}. \tag{4.47}$$

The per-unit active component $\overset{*}{u}_{\text{ka}}$ of the short-circuit voltage can be shown to equal the per-unit resistance $\overset{*}{r}_{\text{k}}$, and a similar relationship exists between the reactive component of the short-circuit voltage and the reactance

$$\overset{*}{u}_{\text{ka}} = \overset{*}{r}_{\text{k}}; \overset{*}{u}_{\text{kr}} = \overset{*}{x}_{\text{k}}. \tag{4.48}$$

Short-circuit voltage is a key parameter of transformers. The per-unit short-circuit voltage is indicated as a percentage on the nameplate of a transformer. In power transformers, $u_{\text{k}}\% = 5$–15%, with the higher end of the range applying to high-voltage transformers, in which the windings are spaced farther apart, resulting in higher leakage fluxes. Often, the active component $u_{\text{ka}}\%$ and the reactive

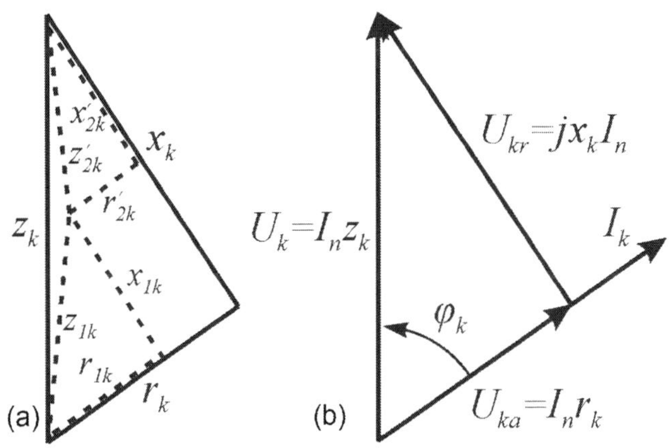

FIGURE 4.29 Triangle of transformer short-circuit resistances (a) and phasor diagram of short-circuit test (b).

component $u_{kr}\%$ of the per-unit short-circuit voltage are also indicated on the transformer's nameplate.

The voltages that occur during transformer short-circuit tests are about 20 times lower than the transformer's rated voltage. Due to this, the magnetic flux density is also low, and the iron losses in the transformer core are several hundred times lower than during rated operation. In other words, they are negligibly small. The active short-circuit power P_k is mainly used to cover the copper losses in the windings. A much smaller but still significant part of the power P_k is lost as hysteresis and eddy current losses caused by leakage fluxes travelling along the transformer tank and fasteners. These leakage fluxes are as large during a short-circuit test as they are during rated operation, because the currents are equal to the rated currents. The short-circuit resistance r_k calculated on the basis of the short-circuit power P_k also takes into account the above additional losses due to leakage fluxes, and is, as a result, slightly higher than the ohmic resistance of the windings. Ohmic resistance is measured by an ohmmeter.

Say we know the magnitudes of the active and reactive components of the short-circuit voltage of a transformer (from, e.g., the data on the transformer's nameplate) and we want to know the sum P_{Cu} of the transformer's copper losses and losses caused by leakage fluxes at a given load current I. We can then calculate this sum from the short-circuit resistance r_k:

$$P_{Cu} = r_k I^2. \tag{4.49}$$

Problem 4.3

Solution

$$Z_n' = \frac{U_{1n}^2}{S_n} = \frac{230^2}{630} = 83,97\mathrm{W}.$$

Calculate the short-circuit resistance of the primary side:

$$r_k' = \overset{*}{u}_{ka} Z_n' = 0,07 \cdot 83,97 = 5,88\mathrm{W}.$$

Calculate the rated current of the primary side:

$$I_n = S_n / U_{1n} = 630/230 = 2,74\mathrm{A}.$$

The power loss in copper at half the rated load is

$$P_{Cu} = r_k' \left(0,5I_{1n}\right)^2 = 5,88 \cdot 0,5^2 \cdot 2,74^2 = 11\mathrm{W}.$$

The per-unit short-circuit voltage and other ratings indicated on the nameplate of the transformer can be used to find the short-circuit currents of this transformer.

If a short circuit occurs at the rated primary voltage, the short-circuit current in the primary side is

$$I_{1k} = U_{1n} / z_k^{'}, \tag{4.50}$$

where I_{1k} – short-circuit current in the primary winding;
U_{1n} – rated voltage of the primary winding;
$z_k^{'}$ – short-circuit impedance referred to the primary side.
Equation (4.50) can also be given with per-unit quantities:

$$\overset{*}{I_k} = 1/\overset{*}{z_k} = 1/\overset{*}{u_k}. \tag{4.51}$$

From the per-unit short-circuit current $\overset{*}{I_k}$, we can switch to the real short-circuit current

$$I_k = \overset{*}{I_k} I_n = I_n / \overset{*}{u_k}. \tag{4.52}$$

Problem 4.4

Solution

$$I_{1k} = \frac{I_{1n}}{\overset{*}{u_k}} = \frac{I_{1n}}{u_{k\%}} \cdot 100 = \frac{6 \cdot 100}{7} = 85,7\text{A}.$$

The rated current of the secondary side is

$$I_{2n} = \frac{U_1 I_{1n}}{U_2} = \frac{400 \cdot 6}{36} = 66,67\text{A}.$$

The short-circuit current on the secondary side is

$$I_{2k} = \frac{I_{2n}}{u_{k\%}} \cdot 100 = \frac{66,67 \cdot 100}{7} = 952,4\text{A}.$$

If we know the values of the active and reactive components of the per-unit short-circuit current in the transformer, we can use them to ascertain the secondary voltage in the transformer under load, as well as the change in the secondary voltage due to the load.

Problem 4.5

Solution

$$Z_n'' = \frac{U_{20}^2}{S_n} = \frac{115^2}{2000} = 6,6125\text{W}.$$

Based on equation (3.49) and the fact that $\overset{*}{u}_{ka} = 0,01u_{ka\%}$ and $\overset{*}{u}_{kr} = 0,01u_{kr\%}$, calculate the short-circuit resistance and reactance referred to the secondary side:

$$r_k'' = \overset{*}{u}_{ka} Z_n'' = 0,07 \cdot 6,6125 = 0,4629\text{W},$$

$$x_k'' = \overset{*}{u}_{kr} Z_n'' = 0,05 \cdot 6,6125 = 0,3306\text{W}.$$

Calculate the secondary current I'' using an equivalent circuit for the secondary side (Figure 4.30):

$$I'' = \frac{U_{20}}{\sqrt{\left(r_k'' + r\right)^2 + \left(x_k''\right)^2}} = \frac{115}{\sqrt{\left(0,4629 + 10\right)^2 + 0,3306^2}} = 10,99\text{A}.$$

Calculate the voltage across the load resistor r, which constitutes the secondary voltage of the transformer under load:

$$U_2 = I''r = 10,99 \cdot 10 = 109,9\text{V}.$$

The change in the secondary voltage of the transformer under load is

$$\Delta U = U_{20} - U_2 = 115 - 109,9 = 5,1\text{V}.$$

4.8 TRANSFORMER NO-LOAD TEST

A transformer short-circuit test reveals one part of the parameters of the equivalent circuit of the transformer, but not the parameters of the magnetisation branch of the equivalent circuit. These can be determined by conducting a no-load test (also known as an open-circuit test). Additionally, a no-load test can be used to determine the transformer's transformation ratio and no-load losses.

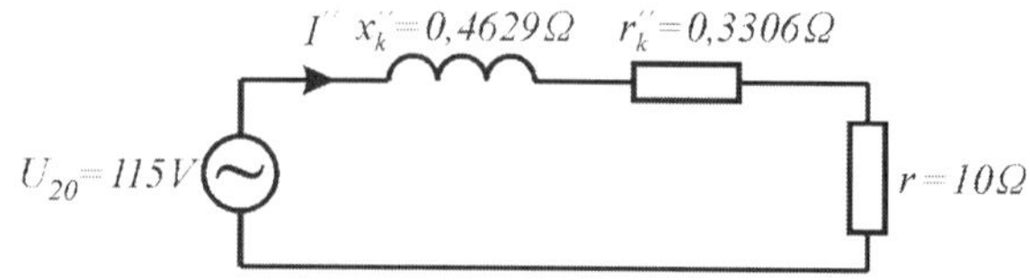

FIGURE 4.30 Equivalent circuit of transformer referred to the secondary side for Problem 4.6.

The circuit diagram for a no-load test is illustrated in Figure 4.31a. An ammeter, wattmeter, and voltmeter are inserted into the primary side, while only a voltmeter is inserted into the secondary side. The primary voltage is increased from zero to 1.1 times the rated voltage, and the primary voltage U_{10}, secondary voltage U_{20}, current $I_{10,}$ and power P_0 are recorded at regular intervals. Based on the measurement results, the no-load characteristics are established as a function of the primary voltage.

Based on the readings of the instruments recorded at the rated voltage $U_{10} = U_{1n}$, the no-load impedance z_0, resistance r_0, and inductive reactance x_0 are calculated (no-load quantities are denoted by the index 0):

$$z_0 = \frac{U_{10}}{I_{10}}\ ;r_0 = \frac{P_0}{I_{10}^2}\ ;x_0 = \sqrt{z_0^2 - r_0^2}\ . \tag{4.53}$$

The calculated resistance r_0 and inductive reactance x_0 are approximately equal to the series-connected resistance and inductive reactance in the magnetisation branch of the equivalent circuit of the transformer (Figure 4.31b, c).

Based on the test data, we can then determine the transformation ratio

$$k_{12} = w_1 / w_2 \approx U_{10} / U_{20},$$

the power factor

$$\cos\varphi_0 = \frac{P_0}{U_{10} I_{10}} \tag{4.54}$$

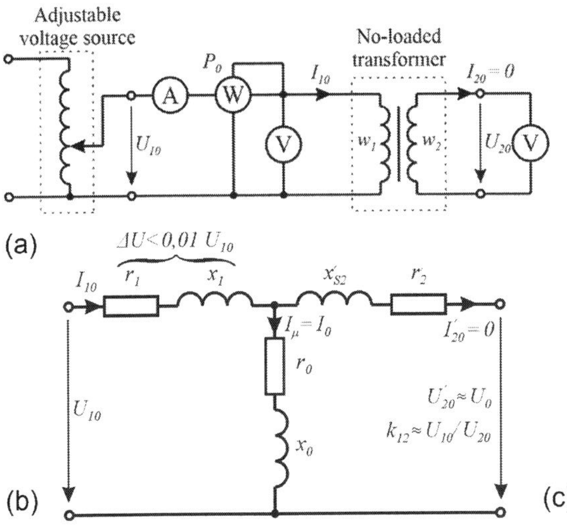

FIGURE 4.31 Circuit diagram for the no-load test of a single-phase transformer (a), equivalent circuit of the transformer at no load (b), and simplified equivalent circuit for no-load testing (c).

and the per-unit no-load current i_0^*

$$i_0^* = \frac{I_{10}}{I_{1n}}.$$

The per-unit no-load current usually falls between 3 and 8% of the rated current, with values in the upper end of this range corresponding to smaller transformers. The no-load inductive reactance x_0 is usually much higher than the resistance r_0, and thus the no-load power factor $\cos\phi_0 \leq 0.1$. The no-load loss P_0 consists virtually only of iron losses P_{Fe} (combined hysteresis loss and eddy current loss) in the transformer core

$$P_0 = I_0^2 r_0 \approx P_{Fe}, \tag{4.55}$$

because, at no load, copper losses only occur in the primary winding and are very small.

No-load losses, or iron losses, occur at all load currents in a transformer. Their magnitude does not depend on the transformer's load current. What it does, however, depend on is the voltage. At a voltage of, say, 10% above the rated voltage, the no-load loss of a transformer will be around 30% higher. Conversely, the copper losses discussed above, which can be determined by a short-circuit test, depend heavily on the load current. The power loss of a transformer is the sum of the power loss in iron and power loss in copper.

4.9 VOLTAGE REGULATION OF TRANSFORMER

The voltage regulation of a transformer is defined as the arithmetic difference between the no-load voltage across the transformer's secondary winding and the voltage at the rated load across the same winding. It is assumed that the primary voltage and frequency of the transformer do not change when the voltage regulation is determined.

The definition does not specify the nature of the rated load, i.e. whether it is a resistive, inductive, capacitive, or some kind of mixed load. The magnitude of the voltage regulation, however, depends greatly on the nature of the load, which must be borne in mind. The changing of the voltage of a substation transformer causes a change in voltage at the loads. This is undesirable. The change in voltage at loads must be within established limits. In order to comply with established requirements, it is necessary to know a transformer's voltage regulation. A transformer provides a significant proportion of the overall change in voltage.

The voltage regulation of a transformer can be determined with sufficient precision based on its simplest equivalent circuit (Figure 4.32). For this, it is necessary to find the secondary voltage U_2 when the secondary current I_2 is equal to the rated current I_{2n}. This should be done in the case of mixed loads (e.g. variants 1, 2, and 3 in Figure 4.32).

A transformer's voltage drops the most in the case of an inductive load. In such a case, the voltage drop may be equal to the transformer's short-circuit voltage u_k (see the phasor diagram in Figure 4.33a). If, in a low-voltage grid with a rated voltage of 400 V, the per-unit short-circuit voltage of a transformer is, for example, $u_k^* = 5,6\%$, the maximum voltage regulation is 22.4 V (this change in voltage is a voltage drop). Note that usually the permissible voltage deviation from the rated voltage in a low-voltage grid is ±10%. Therefore, the maximum voltage regulation of our transformer is about half the voltage deviation permitted in this grid.

For a resistive load, the voltage regulation of a transformer is lower than for an inductive load. If a reactive power compensator, such as a capacitor bank, is used, this compensator reduces the inductive load of the transformer and thereby also the voltage regulation of the transformer. When the number of capacitors is high enough to cause the reactive power to become capacitive (the current phasor leads the voltage phasor, as in Figures 4.33c, d), the voltage regulation of the transformer is either zero (Figures 4.33c) or even negative (Figures 4.33d). A negative voltage regulation means that the transformer's voltage under load is higher than its no-load voltage.

From the phasor diagram, it is also possible to derive the mathematical equation for calculating the transformer's voltage regulation

$$\Delta U_{\%} = u_{ka\%} \cos\varphi_2 + u_{kr\%} \sin\varphi_2 + \frac{\left(u_{kr\%} \cos\varphi_2 - u_{ka\%} \sin\varphi_2\right)^2}{200}. \quad (4.56)$$

Here, $\Delta U_{\%}$ – voltage regulation of the transformer as a percentage value;
$u_{ka\%}$ – active component of short-circuit voltage as a percentage value;
$u_{kr\%}$ – reactive component of short-circuit voltage as a percentage value;
ϕ_2 – phase shift angle between secondary current and secondary voltage.

The last component of equation (4.56) is relatively small and may therefore be omitted, leaving

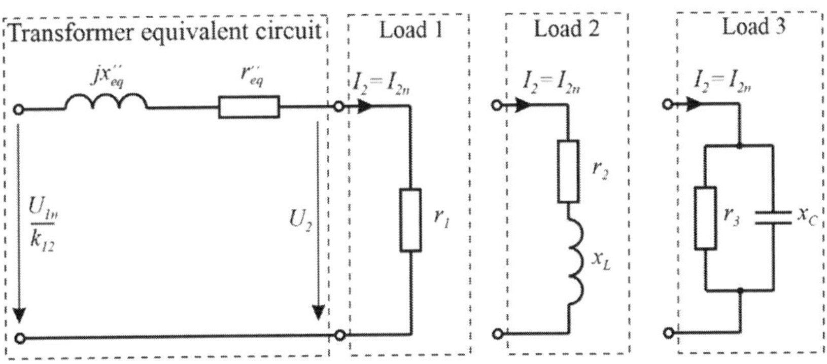

FIGURE 4.32 Circuit diagram for determining the voltage regulation of a transformer.

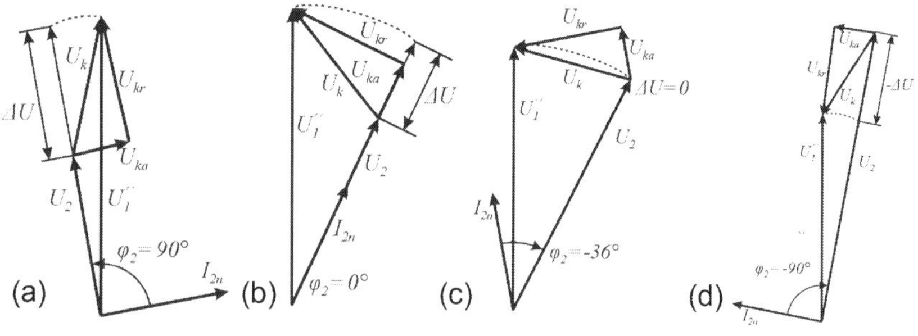

FIGURE 4.33 Effect of the nature of the load on voltage regulation Δ U: (a) inductive load, $\phi_2 = 90°$; (b) resistive load, $\phi_2 = 0°$; (c) resistive-capacitive load, $\phi_2 = -36°$; (d) capacitive load, $\phi_2 = -90°$.

$$\Delta U_{\%} \approx u_{\mathrm{ka}\%} \cos\varphi_2 + u_{\mathrm{kr}\%} \sin\varphi_2. \tag{4.57}$$

Equation (4.57) tells us that for a resistive load ($\cos\phi_2 = 1$), the voltage regulation $\Delta\ U_{\%} \approx u_{\mathrm{ka}\%}$, while for a reactive load ($\cos\phi_2 = 0$) $\Delta\ U_{\%} \approx u_{\mathrm{kr}\%}$.

Problem 4.6

1. The transformer has a resistive load ($\cos\phi_2 = 1.0$).
2. The transformer has a resistive-inductive load ($\cos\phi_2 = 0.6$).

Solution

1. The per-unit voltage regulation in the case of the resistive load is

$$\Delta U_{\%} \approx u_{\mathrm{ka}\%} \cos\varphi_2 + u_{\mathrm{kr}\%} \sin\varphi_2 = 2,2 \cdot 1,0 + 5,0 \cdot 0 = 2,2\%$$

2. The voltage regulation in volts in the case of the resistive load is

$$\Delta U = U_{\mathrm{n}} \Delta U_{\%} \cdot 0,01 = 400 \cdot 2,2 \cdot 0,01 = 8,8\mathrm{V}.$$

3. The per-unit voltage regulation in the case of the resistive-inductive load (if $\cos\phi_2 = 0.6$, then $\sin\phi_2 = 0.8$) is

$$\Delta U_{\%} = 2,2 \cdot 0,6 + 5,0 \cdot 0,8 = 5,32\%.$$

4. The voltage regulation in volts in the case of the resistive-inductive load is

$$\Delta U = 400 \cdot 5{,}32 \cdot 0{,}01 = 21{,}28\text{V}.$$

Using the more accurate equation (4.56), the result in point four would be 21.31 V.

Figure 4.34 shows the curve of the voltage regulation as a function of the phase shift angle ϕ_2 of the secondary current for the transformer considered in the problem above. Positive ϕ_2 values correspond to a resistive-inductive load, while negative ϕ_2 values correspond to a resistive-capacitive load. Positive ΔU values correspond to the transformer's voltage dropping below the no-load voltage, while negative ΔU values correspond to the transformer's voltage rising above the no-load voltage (under the effects of a resistive-capacitive load).

4.10 TRANSFORMER VOLTAGE ADJUSTMENT

The output voltage of a transformer can be adjusted by changing the transformer's transformation ratio $k_{12} = w_1/w_2$. Output voltage can be reduced by increasing the number of turns in the primary winding w_1 or decreasing the number of turns in the secondary winding w_2. Output voltage can be increased only by increasing the number of turns in the secondary winding, because reducing the turns in the primary winding is generally not acceptable due to the increase in magnetisation current that this would bring about. The number of turns can be changed by

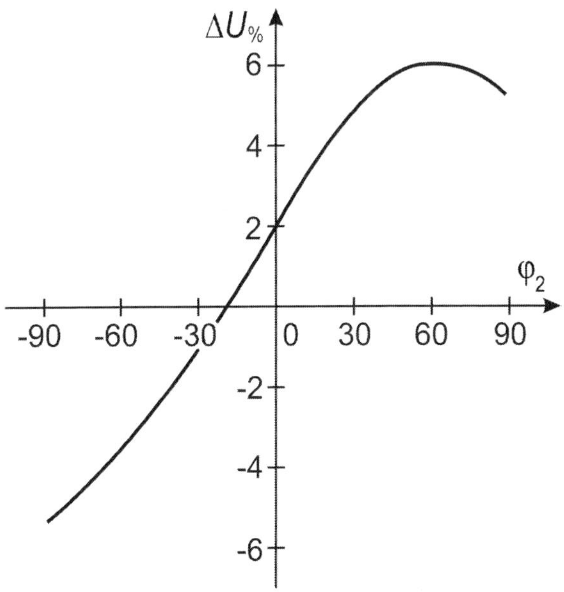

FIGURE 4.34 Relationship between the voltage regulation and the nature of the load for a transformer with the parameters given in Problem 4.6.

means of a tap changer. In power transformers, tap changers are usually placed on the side where the current is lower and the tap changer is easier to construct.

Nearly all transformers in substations are equipped with tap changers for adjusting the voltage. When the load current increases, the voltage across the loads in the grid decreases. This is due to voltage drops on the power lines and transformers. In order to prevent the voltage across the loads from falling below the minimum permissible level, the transformer voltage is increased by means of a tap changer. Conversely, when the load decreases, causing the voltage to increase, the voltage must be lowered. To compensate for daily voltage fluctuations, so-called tap changing under load (TCUL) transformers are used, which are usually operated at a voltage of 110 kV in electrical substations.

Grid transformers which are used to step 6–20 kV medium-voltage grid voltages down to a low-voltage level of 410 V can also be adjusted, but not under load. These transformers are adjusted during installation according to the average voltage level at the installation site. In these transformers, the tap changers may not be operated under load, because the simple mechanical construction of the tap changer does not guarantee the quenching of the arc generated when the circuit is broken. Such transformers can typically be switched between five voltage levels. The options include the rated voltage and the rated voltage $\pm 2 \times 2.5\%$.

TCUL transformers are equipped with two mechanical contact systems (Figure 4.35). One contact system is not switchable under load and is called the selector (contacts SEL1 and SEL2). The second contact system consists of contacts C1 and C2 which are switchable under load. In addition, either a set of resistors for limiting the short-circuit current of one switching step or an inductor known as a reactor RE is required. The switching process shown includes five consecutive switching positions: a, b, c, d, and e. In position a, current enters the centre tap of the reactor, branches into two, and passes through the reactor's winding coil in opposite directions. The branched currents then pass, in parallel, both through the contactors C1 and C2 as well as the selector contacts SEL1 and SEL2. Finally, the current reaches the transformer winding tap X1.

In position b, tap changing is initiated by opening the contactor C2. This interrupts the current in the lower branch. In this de-energised state, the selector contact SEL2 is disconnected from the winding tap X1 and connected to tap X2. The contactor C2 is then closed and two current branches are again established (position c). Now one branch current flows to tap X1 and the other flows to tap X2. This creates a closed loop, which includes the transformer winding between the taps X1 and X2. This part of the winding is a voltage source, and, in this closed loop, the reactor RE behaves as a current-limiting reactance. The different parts of the winding of the reactor RE in this loop are in series, and the reactor's AC resistance is relatively high. For the current in the external transformer circuit, however, the resistance of the reactor RE is much lower because the currents in the equal sides of the windings are opposite in direction and the magnetisation current in the magnetic circuit caused by the load current is close to zero. In some tap changers, the reactor RE is replaced by two resistors. Current only passes through these resistors when the selector contacts SEL1 and SEL2 are connected

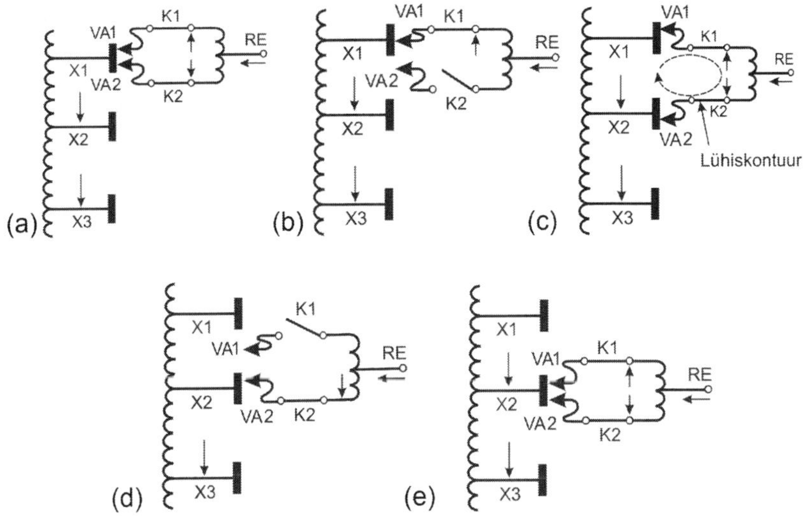

FIGURE 4.35 Adjusting a transformer's voltage under load by means of mechanical switches.

to different transformer taps. At all other times, the resistors are short-circuited by extra contactors (this circuit variant is not shown in the figure).

In position d, the contactor C1 breaks the short-circuit loop. Thereupon, the entire load current of the transformer is shifted over to tap X2. Then, in the de-energised state, the selector contact SEL1 is repositioned from tap X1 to tap X2. Finally, the contactor C1 is closed, at which point the position e is reached. This position is similar to position a, except that the load current is now flowing into tap X2 instead of tap X1. Thereby, the overall number of transformer winding turns connected to the load circuit has changed. And, thus, so have the transformation ratio and output voltage of the transformer. The reactor RE and selector (SEL1, SEL2) are housed in the main oil tank of the transformer, while the contactors C1 and C2 are set in a small separate oil tank to prevent the electric arc generated when the contactors are operated from degrading all of the oil in the transformer.

Tap switching under load by means of mechanical switches is quite a complex process. Complex mechanisms, meanwhile, tend not to be very reliable. Due to this, transformer voltage regulators with thyristor switches are sometimes used (Figure 4.36). The advantage of the thyristor switch is a faster operating speed and the lack of parts that are subject to wear. If, instead of tap X1, we needed to connect tap X2 to the circuit, then we would stop sending control pulses to the thyristor switch V1 (in, e.g., the positive half-period of the grid voltage) and, at the start of the next half-period of the grid voltage (which, in our example, is the negative half-period), we would send a control pulse to the thyristor switch V2. Thus, the switching happens when the grid current passes through zero, no short-circuit loop is formed, and no de-energised state occurs. The tap changer with mechanical contacts considered above required that a short-circuit loop be formed because such tap changers are slow-acting.

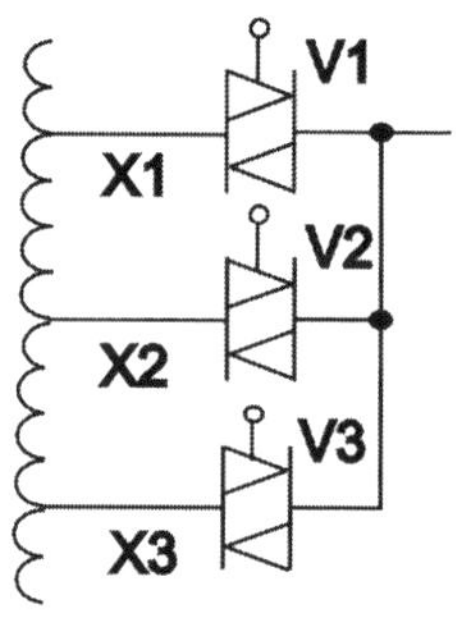

FIGURE 4.36 Use of thyristor switches for switching taps in a transformer winding.

In TCUL transformers, the tap voltage adjustment step is relatively small – 1–2% of the rated voltage – and the number of adjustment steps is around 20. In newer 110 kV transformers, the adjustment range is $\pm 9 \times 1.78\%$ (9 steps up and down, with each step changing the voltage by 1.78%).

4.11 EFFECT OF CHANGES IN LOAD CURRENT ON THE OPERATION OF A TRANSFORMER

Transformers are designed and built to operate at a certain voltage and current called the rated voltage and rated current. The actual voltage and current in an operating transformer almost always differ from the rated values. The voltage will differ relatively little from the rated voltage (generally within ± 10%), while the load current may vary between zero and the rated current. The load on a transformer is characterised by the loading factor β :

$$\beta = I_2 / I_{2n} \approx I_1 / I_{1n}. \tag{4.58}$$

where I_1 and I_2 – the actual primary and secondary current, respectively;

I_{1n} and I_{2n} – the primary and secondary rated current of the transformer, respectively.

The weight and dimensions of a transformer are determined by a third key rating: the rated apparent power S_n. In a single-phase transformer,

$$S_n = U_{1n} I_{1n} = U_{2n} I_{2n}. \tag{4.59}$$

The heat generated in transformer windings depends on the current flowing through the winding, but not on the power factor $\cos\phi$. That is why the indicated rated power of a transformer is not the active power P, but the apparent power S. The efficiency η of a transformer is defined as the ratio of the active power of the secondary side to that of the primary side:

$$h = P_2 / P_1. \tag{4.60}$$

The power lost as heat in a transformer, Δ *P*, can be expressed as the difference between the primary power and the secondary power:

$$\Delta P = P_1 - P_2. \tag{4.61}$$

Transformer heat losses consist of two components. The heat lost in the transformer core is called iron loss. Iron losses in a transformer are the same at no load and at the rated load, because they do not depend on the load current. Iron loss is also called no-load loss P_0. The heat lost in the transformer windings is called copper loss P_{Cu}, and it is proportional to the square of the load current. Copper loss is highest at the maximum operating current, i.e. the rated current. Copper loss at the rated current of the transformer is usually determined through short-circuit testing. That is why it is also called short-circuit loss P_k. Copper losses occurring at a current below the rated current can be calculated using the equation

$$P_{Cu} = \beta^2 P_k, \tag{4.62}$$

where β is the loading factor of the transformer.

Since it is always desirable to minimise losses, it is valuable to know how the efficiency of a transformer changes under load. In view of the above, the power loss of a transformer can be expressed as

$$\Delta P = P_0 + \beta^2 P_k. \tag{4.63}$$

Rewriting the efficiency equation, we get equation (4.60):

$$\eta = \frac{P_2}{P_1} = \frac{P_1 - \Delta P}{P_1} = 1 - \frac{\Delta P}{P_2 + \Delta P} = 1 - \frac{P_0 + \beta^2 P_k}{P_2 + P_0 + \beta^2 P_k}. \tag{4.64}$$

Active power P_2 can be expressed through the power factor $\cos\phi_2$ and the rated apparent power of the secondary side of the transformer S_{2n}:

$$P_2 = \beta P_{2n} = \beta \cos\varphi_2 S_{2n}. \tag{4.65}$$

From equations (4.64) and (4.65), we get the transformer efficiency expressed through the loading factor:

$$\eta = 1 - \frac{P_0 + \beta^2 P_k}{\beta \cos\varphi_2 S_{2n} + P_0 + \beta^2 P_k}. \tag{4.66}$$

In equation (4.66), P_0, P_k, and S_{2n} are constant transformer parameters, while β and $\cos\phi_2$ are variables. The efficiency is zero if either β or $\cos\varphi_2$ equals zero. Efficiency is highest at a load where the constant loss P_0 and variable loss P_{Cu} are equal, i.e. when $P_0 = \beta^2 P_k$. From here, we can find the loading factor β for the maximum efficiency ($\eta = \eta_{max}$):

$$\beta = \sqrt{P_0 / P_k}. \tag{4.67}$$

When designing a transformer, the ratio of no-load loss to short-circuit loss is selectable within certain limits. Normally, transformers are designed in such a way that the no-load is several times lower than the short-circuit loss. For example, if $P_0/P_k = 0.25$, then $\eta = \eta_{max}$ with a loading factor of $\beta = \sqrt{0{,}25} = 0.5$. In other words, efficiency is maximal when the transformer is running at half capacity. Such parameters are often practical, because substation transformers operate under full load only for short periods of time. Most of the time, a substation transformer is underloaded. Figure 4.37 shows the typical relationship between efficiency and the loading factor.

During rated operation, large substation transformers (more than 10 MVA) have losses of 1% or less, while small transformers (10 kVA) have losses of 3–4%. Generally, transformers are more efficient than other energy converters (electrical machines, semiconductor converters).

4.12 WINDING CONFIGURATIONS AND VECTOR GROUPS OF THREE-PHASE TRANSFORMERS PARALLEL OPERATION OF TRANSFORMERS

4.12.1 Labelling of Connection Terminals, Winding Configurations

Three-phase transformers have at least 6 windings, because each phase will have a high-voltage winding and a low-voltage winding. As each winding has a beginning and end, connecting the 6 windings requires 12 connection terminals. In

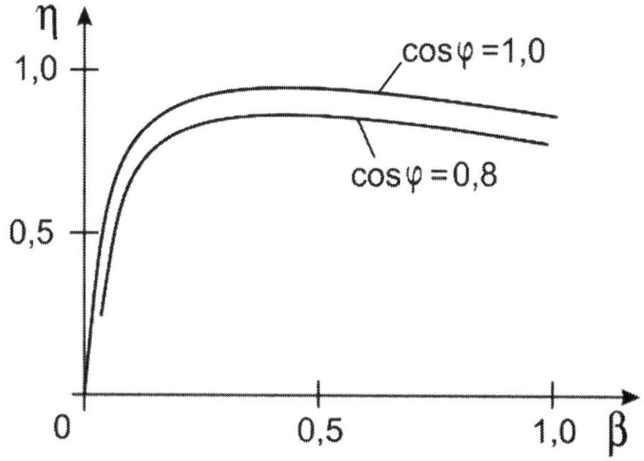

FIGURE 4.37 Relationship between a transformer's efficiency and loading factor for two different power factor values.

order to connect a transformer correctly into an electrical circuit or electrical grid, it is necessary to know for each terminal which winding it belongs to and whether it is the beginning or end of that winding. To prevent errors, terminals are labelled. Table 4.1 shows a German labelling system.

High-voltage winding terminals are denoted by uppercase letters, while low-voltage winding terminals are denoted by lowercase letters. The beginnings of windings are labelled with the first letters of the alphabet, while the ends of windings are labelled with the last letters of the alphabet. The electromotive force in a winding is considered positive if, in the positive half-period of the magnetic flux in the leg, it is directed from the end of the winding towards the beginning, and if the current produced by this electromotive force leaves the terminal at the beginning of the winding.

In English-language literature, the beginnings of windings are designated using a system comprising only the lowercase letters a, b, c and the same lowercase letters plus an apostrophe a', b', c'. Winding ends are not labelled.

These labels are used both on the actual terminals of a transformer as well as in transformer circuit diagrams (Figure 4.38). In the diagram of a three-phase transformer, the windings are drawn as perpendicular to the symbol of the magnetic circuit, which is the thicker straight line (Figure 4.38b); meanwhile, in the diagram of a single-phase transformer, the windings are shown as parallel to the magnetic circuit symbol.

Diagrams are drawn according to the same rules that apply to writing. Energy enters the transformer from the left side (primary winding) and flows to the right.

When connecting a three-phase transformer to an electrical circuit, the amount of connection wires needed can be reduced by interconnecting some of the terminals at the transformer. This is done either using a star connection (primary winding in Figure 4.39) or a delta connection (secondary winding in Figure 4.39). The winding configuration is selected based on practical considerations. For example, at line voltages of 35 kV and more it is appropriate to connect the windings in star and to connect the winding connection point to earth. This results in a condition where the voltage of the line wires relative to the earth is $\sqrt{3}$ times lower than the

TABLE 4.1
Labelling of transformer winding beginnings and ends

Winding	Phase 1	Phase 2	Phase 3
High-voltage winding:			
Beginning	A	B	C
End	X	Y	Z
Low-voltage winding:			
Beginning	a	b	c
End	x	y	z

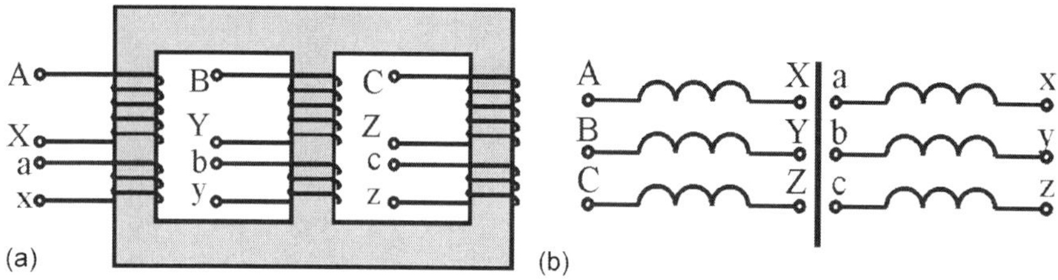

FIGURE 4.38 Spatial arrangement of windings on the magnetic circuit of a three-phase transformer (a) and a depiction of the electrical circuit of the same transformer with the windings not connected (b).

voltage between the line wires. This reduces the cost of the insulation, including overhead power line insulators.

The point of connection of star-connected windings is called the neutral point, denoted by n or 0. Star connection is also used in low-voltage grids, because it enables incandescent lights to be inserted between the phase and neutral at a voltage of 230 V, but more powerful loads (electric motors) to be set at 400 V. At a lower voltage, incandescent lights have better luminous efficacy, and lower voltages are also better for powering all kinds of electronic devices. For electrical grids, on the other hand, a voltage of 400 V is more advantageous, because it means lower currents and, therefore, smaller cross-sectional areas for wires.

a b c

d Y/Δ, or **Y-Δ**, or **Yd**

At the same time, for the reduction of higher current harmonics in an electrical grid, a delta connection is a better choice. For this reason, it is desirable for one of the windings – either the low-voltage winding or the high-voltage winding – to be connected in delta.

For electrical grids and the quality of the mains voltage, an interconnected star connection, or zigzag connection, is an even better option than a delta connection. In a zigzag connection, the winding of each phase consists of two halves which have an equal number of turns and are positioned on different legs of the transformer and connected in series. Moreover, the series connection is not a standard one (normally, the end of the first half would be connected to the beginning of the second half), as the beginning and end of the second half are switched around. Figure 4.40a shows the magnetic circuit of a transformer with a star-connected primary winding and a zigzag-connected secondary winding.

The phase of the magnetic flux in leg A of the transformer (the left-most leg in the figure) is determined by the phase of the phase A voltage on the primary side (phasor diagram in Figure 4.40b). The voltage across the half of the phase A secondary winding that is positioned on leg A (the winding half denoted by $\frac{1}{2}a$) has

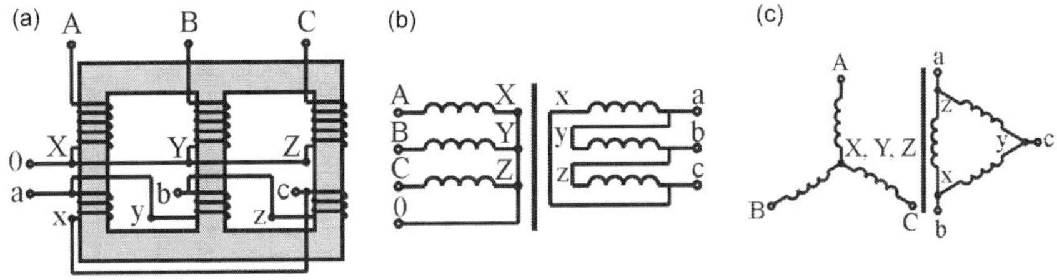

FIGURE 4.39 Spatial arrangement of windings on the magnetic circuit of a three-phase transformer (a), standard depiction of the electrical circuit of the same transformer (b), easy-to-remember electrical circuit diagram (c), and the electrical circuit symbol (d) in the case of star-connected primary windings and delta-connected secondary windings.

the same phase. The second half of the phase A secondary winding is positioned on leg B (this winding half and the corresponding voltage phasor are denoted by $-\frac{1}{2}b$). Since the beginning and end of the winding half $-\frac{1}{2}b$ on leg B are switched around, the direction of the phasor $-\frac{1}{2}b$ in the diagram is the opposite of that of the phasor of the phase B voltage on the primary side. The phasor a of the phase A voltage on the secondary side is the sum of the voltage phasors of the winding halves $\frac{1}{2}a$ and $-\frac{1}{2}b$. The diagram shows that the transformer's primary voltage system (phasors A, B, and C) and secondary voltage system (phasors a, b, and c) are not in phase. The zigzag connection results in a 30° difference between the phase angles of the secondary and primary voltages. Normally, the phase of the secondary voltage (phasor a) is considered to be shifted relative to the phase of the primary voltage (phasor A), which in this case means a shift angle of 330° (the phasors are rotating counterclockwise).

The diagram also shows that, due to the phase shift, the total voltage across the windings $\frac{1}{2}a$ and $-\frac{1}{2}b$ is less than the arithmetic sum of the voltages across these windings (the arithmetic sum is obtained when there is no phase shift). The voltage decreases by $2/\sqrt{3} = 1.15$ times. In order to compensate for this decrease in voltage, the number of turns in the winding needs to be increased by 1.15 times, compared to a Y/Y or Y/Δ connection. This is the downside of a zigzag connection.

One of the advantages of a zigzag connection, however, is that, in the case of an unbalanced load, there is no significant difference in phase voltages on the secondary side, because the load current of each phase is obtained through windings arranged on two different legs of the transformer. Another advantage is that the third harmonic current and other harmonic currents divisible by three do not reach the primary side from the secondary side. While these currents are in

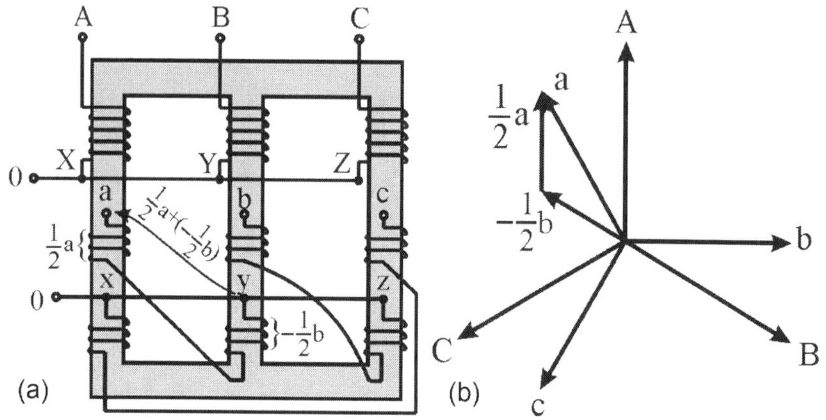

FIGURE 4.40 Arrangement of windings on a magnetic circuit for a star-zigzag connection (a) and the corresponding phasor diagram (b).

phase in all three phase wires A, B, and C, in the secondary winding halves on any given leg of the transformer, these harmonic currents are opposite in phase and cancel out.

Transformer winding configurations are indicated using the following symbols: star connection – Y (or y on the low-voltage side); delta connection – Δ or D (or d on the low-voltage side); zigzag connection – Z (or z on the low-voltage side). In the case of a star connection and a zigzag connection, the common point of all of the three phases (or the neutral point, 0 in Figure 4.40) may be brought out, in which case we will have a four-wire transmission line. If the neutral point has not been brought out, we will have a three-wire line. If the transformer is equipped with a neutral point connection terminal, the symbol of the winding configuration is followed by the neutral symbol n or the subscript 0. For example, 'Dyn' refers to a delta-star connection where the connection point of the windings has been brought out on the secondary side. In the former Soviet Union, this transformer winding configuration was denoted by Δ/Y_0.

4.12.2 Parallel Connection of Transformers. Formation of Equalisation Currents

Parallel connection of transformers is common in electrical grids. In substations, two or three transformers are operated in parallel (Figure 4.41). Normally, having multiple transformers enables energy to continue to be supplied to loads even when one transformer is malfunctioning or undergoing repairs. This significantly increases the security of supply. Second, when demand is low (usually in summer), one transformer may be switched off to reduce transformer losses due to no-load current.

The secondary windings of two transformers may only be joined together if the maximum value and phase angle of the voltage of the corresponding phase (e.g. phase A) of one transformer are equal to those of the same phase of the other transformer. Differences in the maximum value and phase of the voltage lead to the formation of equalisation currents, which are generally unacceptable. Voltage differences can be identified using a voltmeter before the corresponding bars of the secondary side of the transformer are linked together (Figure 4.41). In energised transformers, the voltage measured by the voltmeter must be zero or negligibly low.

Figure 4.42 shows the formation of an equalisation current when the secondary voltage phases of the two transformers are equal but the effective values differ. The switchover to parallel operation is achieved by means of the switch S1.

When the S1 switch is open, only the transformer T1 is connected to the output circuit, to which it supplies its secondary voltage – in this case, 400 V. In the equivalent circuits of the transformers, the short-circuit impedances are, in this case, equal, and their per-unit value is $\overset{*}{z}_{\mathrm{eq}} = 0.05$ (Figure 4.42b). When the switch S1 is closed (Figure 4.42c), an equalisation current I_{eq} is formed in the transformer windings due to the difference in output voltages between the transformers (in this case, $\Delta U = 408 - 400 = 8$ V). The direction of the equalisation current is different in the windings of the two transformers. The transformer T2, whose secondary voltage is higher, emits the current, which then enters the secondary winding of the transformer T1, whose secondary voltage is lower. The differently directed currents produce differently directed voltage drops across the resistances in the equivalent circuit. This causes the output voltage of the transformer T1 to

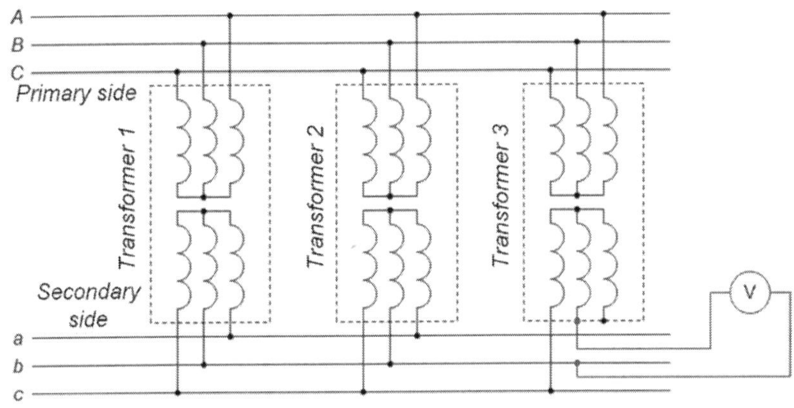

FIGURE 4.41 Switching transformers over to parallel operation and checking for voltage differences.

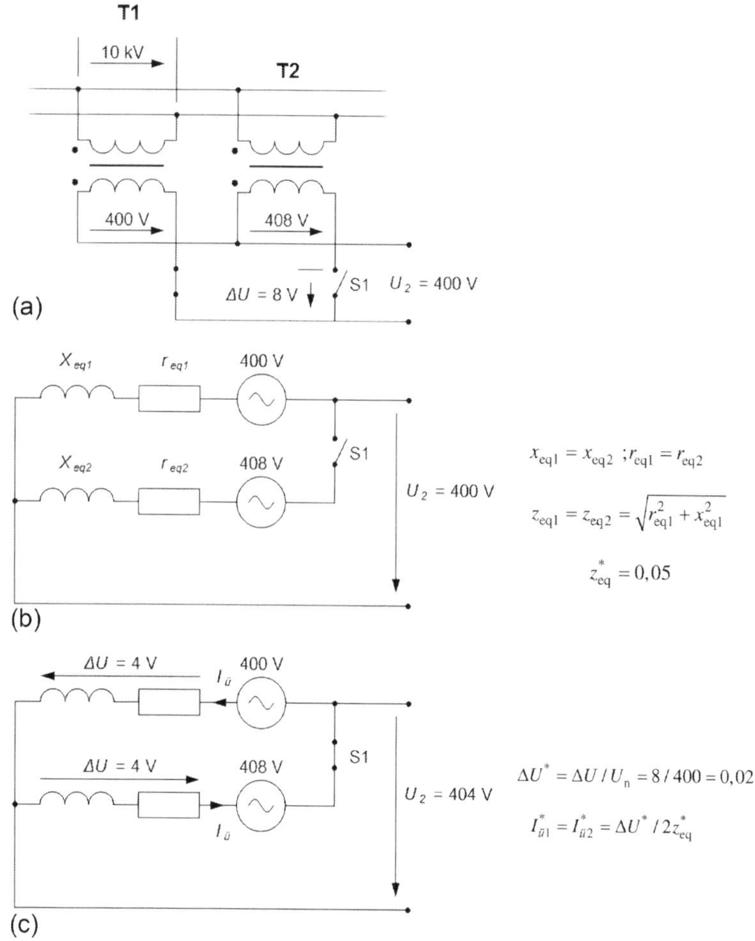

FIGURE 4.42 Formation of an equalisation current when the transformation ratios of two parallel-operated transformers are not perfectly equal: (a) connection diagram of the transformers, (b) equivalent circuit referred to the secondary side when the secondary side of the transformer T2 is disconnected, (c) same equivalent circuit when the transformers are connected in parallel.

increase and that of the transformer T2 to decrease, as a result of which the secondary voltages of the two transformers become equal (resulting, in this case, in a voltage of 404 V). However, the transformers will carry a load current even if the external circuit is not consuming power. In this case, the 2% voltage difference produces an equalisation current equal to 20% of the rated current. This equalisation current causes unnecessary heat losses in both transformers, and, moreover, the transformers can no longer be operated at full load, because they are already loaded to the extent of the equalisation current.

Equalisation currents also occur when the effective values of the secondary voltages of parallel-connected transformers are equal but the phase angles are different. This happens with parallel-connected three-phase transformers that utilise different winding configurations. In our example, in the first transformer, both the primary winding and the secondary winding are connected in star, while the second transformer has a star-connected primary winding and a delta-connected secondary winding (Figure 4.43a). In this case, the phase A voltage of the second transformer leads that of the first transformer by 30 degrees (Figure 4.43b).

If the secondary terminals of the transformers T1 and T2 are not yet connected for parallel operation, the voltage difference between the phase A secondary voltage phasors U_{A2} and U_{A1} are determined by the phasor ΔU (Figure 4.43c), the length of which is

$$\Delta U = 2U_{\text{A}} \sin 15° = 0,518U_{\text{A}},$$

where $U_{\text{A}} = U_{\text{A1}} = U_{\text{A2}}$ are the lengths of the secondary voltage phasors (the lengths of the phasors are proportional to the effective values of the corresponding voltages). When the terminals a, b, and c of the transformers T1 and T2 are wired together (as in Figure 4.43a), closed current loops are formed in each phase. Figure 4.43 shows the simplified equivalent circuit of these current loops. The impedances $\overset{*}{z}_{\text{k1}}$ and $\overset{*}{z}_{\text{k2}}$ are the short-circuit impedances of the transformers T1 and T2 (Figure 4.42b also shows the components x_{eq} and r_{eq} of these impedances).

In this case, $\overset{*}{z}_{\text{k1}} = \overset{*}{z}_{\text{k2}} = 0.056$, corresponding to short-circuit voltages of the transformers in Figure 4.43a. The equalisation current formed in the equivalent circuit under consideration can be found easily by using Ohm's law:

$$\overset{*}{I}_{\ddot{u}} = \frac{\Delta U^*}{\overset{*}{z}_{\text{k1}} + \overset{*}{z}_{\text{k2}}} = \frac{0,518}{0,056 + 0.056} = 4,63.$$

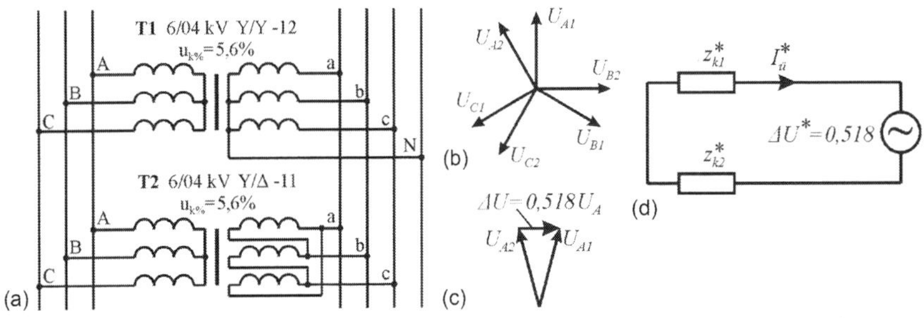

FIGURE 4.43 Formation of an equalisation current when the phase shift angles of the secondary voltages of parallel-connected transformers are not equal: (a) connection diagram; (b) phasor diagrams for the secondary voltages when the terminals a, b, and c have not yet been wired together; (c) voltage difference ΔU in phase A; (d) simplified equivalent circuit for finding the equalisation current.

Since, by definition, the per-unit rated current of a transformer equals one ($I_n^* = 1$), the calculated result indicates a current of 4.63 times the rated current. An equalisation current of this magnitude is unacceptable in transformers, and, in substation transformers, the circuit breaker would quickly switch the transformers off from the primary side if such a current were to arise.

4.12.3 Concept of Vector Groups

In order to avoid potential problems due to differences in the phase angles of the secondary voltages of transformers, the concept of vector groups has been introduced. When two transformers have the same vector group, the secondary voltage phase angles of the transformers are equal.

Vector groups are determined using a simple clock face with clock hands. The line voltage phasor of the high-voltage side is placed in the role of the minute hand and is always set to 12 o'clock on the clock face. Meanwhile, the line voltage phasor of the low-voltage side takes the position of the hour hand (the hour hand is shorter than the minute hand) and is set in such a way that the angle between the two clock hands is equal to the angle between the phasors.

For example, for the transformer shown in Figure 4.44a, the primary voltage phasor U_{AB} would be set to 12 o'clock. The secondary voltage phasor is identical in direction to the primary voltage phasor and would thus also be set to 12 o'clock. The 'time' would therefore be 12 or 0 o'clock. This gives the symbol of the winding configuration and the vector group as Yyn0.

In the transformer in Figure 4.44b, however, the secondary voltage phasor U_{ab} lags behind the primary voltage phasor U_{AB} by 330°, which means that the hour

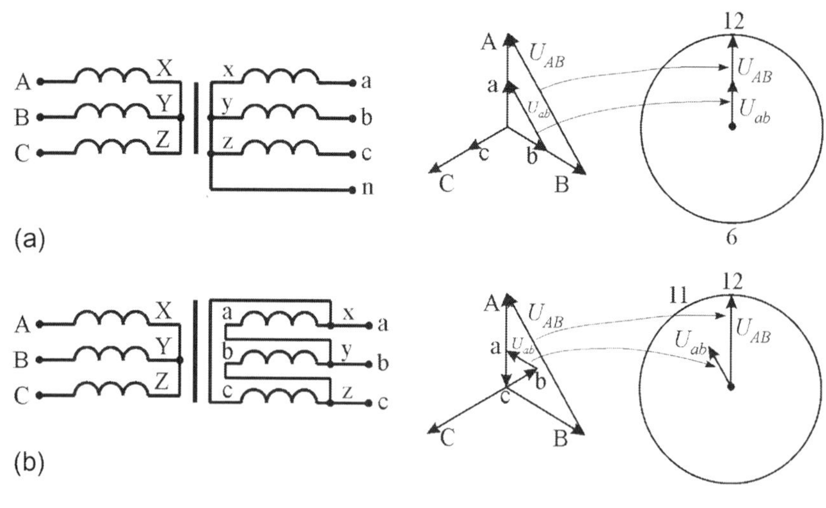

FIGURE 4.44 Determining the vector groups of three-phase transformers from phasor diagrams: (a) winding configuration and vector group Yyn0; (b) winding configuration and vector group Ynd11.

hand must be set to 11 o'clock. In this case, the clock reads the eleventh hour. This clock reading is used as the number of the vector group. Thus, the symbol of the winding configuration and the vector group becomes Ynd11.

Vector group symbols are used for brief transformer descriptions. For example, the main parameters of the transformer T2 shown in Figure 4.43 may be given briefly as 6/0.4 kV Yd11. The description above contains the following information:

- rated line voltage in high-voltage side: 6 kV;
- rated line voltage in low-voltage side: 0.4 kV;
- winding configuration in high-voltage side: star;
- winding configuration in low-voltage side: delta;
- vector group: Yd11.

Different vector groups are also possible. Using various techniques, it is possible to produce phase shifts between the primary voltage U_{AB} and secondary voltage U_{ab} from zero up to 330° in 30° steps. Figure 4.45 shows how to get the vector group Yy6. Here, the winding ends X, Y, and Z are connected on the primary side, while the beginnings a, b, and c are connected on the secondary side. In this case, the ends x, y, and z of the secondary winding are connected to the output terminals a, b, and c on the secondary side of the transformer. The line voltage phasor U_{ab} of the secondary side lags behind the line voltage phasor U_{AB} of the primary side by 180°.

Even-numbered vector groups (0, 2, 4, 6, 8, and 10) can be obtained if the high-voltage and low-voltage windings have the same configuration (Yy, Dd, or Zz). Conversely, odd-numbered vector groups (1, 3, 5, 7, 9, and 11) can be obtained if the high-voltage and low-voltage windings have different configurations (Yd, Dy, Yz, etc.). Having a large number of vector groups is not suitable for electrical grids.

In Europe, the most commonly used vector group types are 0 (or 12) and 11, while in the US, 1 is also used. The Yyn winding configuration does not work well with unbalanced loads, which is why transformers with this configuration are becoming less common in low-voltage grids. In various devices (such as converters), other vector group types than 11, 0, and 1 are used as well.

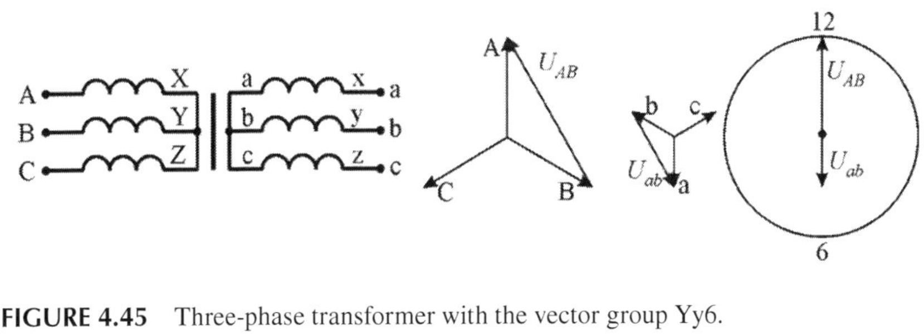

FIGURE 4.45 Three-phase transformer with the vector group Yy6.

4.12.4 Conditions for the Parallel-Operation of Transformers

When connecting transformers in parallel, the following conditions must be met.

1. **The windings to be connected in parallel must have the same phase sequence and the same phase name** (e.g. the beginning terminal of phase A of one transformer may be connected to the beginning terminal of phase A of another transformer). Voltages of different phases have different phase shift angles, and different phases must not be wired together, because this would result in unacceptably high equalisation currents.
2. **The transformers must have the same transformation ratio.** Then, if the primary voltages are equal (which is ensured by the parallel connection of the primary windings), so are the secondary voltages. Given equal secondary voltages, there are no undesirable equalisation currents (equalisation currents were discussed in Section 4.12.2). A small difference in transformation ratios, however, is normally admissible (e.g. ± 0.5% of their average value).
3. **The transformers must have the same vector group.** Given the same vector group, all voltages of the same phase (e.g. phase A) on the secondary side will have the same phase shift angle, which means that no equalisation currents due to phase differences are produced.
4. **The short-circuit voltages of the transformers must be equal ($u_{k1} = u_{k2} = u_{k3}$).**

Meeting the first three conditions is necessary to prevent equalisation currents from arising between the transformers already at no load. The last, fourth condition, however, serves a different purpose. The short-circuit voltages must be equal for the overall load current to be divided between the transformers in proportion to their rated power values. This condition is expressed mathematically in the equality of the loading factors

$$\beta_{\mathrm{load}\,1} = \beta_{\mathrm{load}\,2} = \beta_{\mathrm{load}\,n} = \cdots, \tag{4.68}$$

where the transformer loading factor $\beta_{\mathrm{load}} = I / I_{\mathrm{n}}$. If the loading factors are equal, the load current will be higher for the transformer with the higher power, lower for the transformer with the lower power, and equal for transformers with an equal power. This is the optimal distribution of loads. Now let us prove that the equality of the short-circuit voltages is both necessary and sufficient to ensure the equality of the loading factors. The voltage drops across the equivalent impedances Z_{k1} and Z_{k2} of two parallel-connected transformers are equal (Figure 4.46):

$$\Delta U = Z_{k1} I_1 = Z_{k2} I_2. \tag{4.69}$$

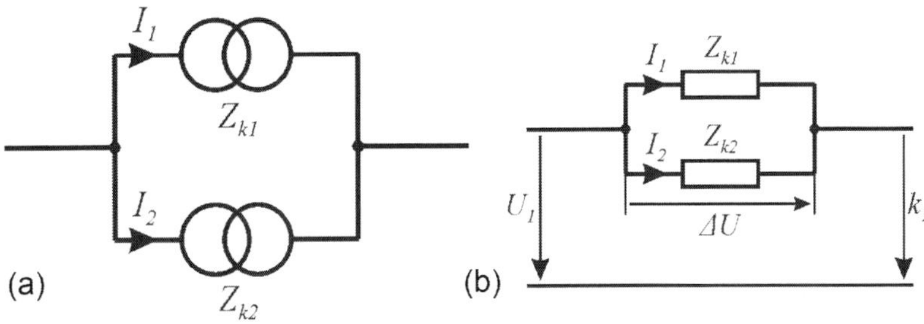

FIGURE 4.46 (a) Two parallel-operated transformers (a) and (b) their equivalent circuit (b).

The equivalent short-circuit impedance Z_k can be expressed as

$$Z_k = \overset{*}{z}_k Z_B = \overset{*}{u}_k \frac{U_n}{I_n}, \tag{4.70}$$

since $\overset{*}{z}_k = \overset{*}{u}_k$ and the base impedance $Z_B = Z_n = U_n / I_n$.

The load current I can expressed via the loading factor β and the rated current I_n

$$I = \beta \cdot I_n \,. \tag{4.71}$$

By replacing Z_k and I in equation (4.69) with the expressions in equations (4.70) and (4.71), we get

$$U_n \overset{*}{u}_{k1} \beta_1 = U_n \overset{*}{u}_{k2} \beta_2. \tag{4.72}$$

Since the rated voltages of parallel-connected transformers are equal,

$$\overset{*}{u}_{k1} \beta_1 = \overset{*}{u}_{k2} \beta_2. \tag{4.73}$$

Assuming that the short-circuit voltages in equation (4.73) are equal, i.e. $\overset{*}{u}_{k1} = \overset{*}{u}_{k2}$, it follows that $\beta_1 = \beta_2$. Therefore, the short-circuit voltages of the transformers being equal ensures that the load currents are divided proportionally.

4.13 UNBALANCED LOAD IN THREE-PHASE TRANSFORMERS

Single-phase loads (e.g. lighting fixtures) may subject a three-phase transformer to unbalanced loading. The effects of an unbalanced sinusoidal load can be analysed using the method of symmetrical components. Every unbalanced three-phase current system (or voltage system) can, in principle, be replaced with three balanced systems which, together, form an unbalanced system that is equivalent to the original system. Such a replacement is shown in Figure 4.47.

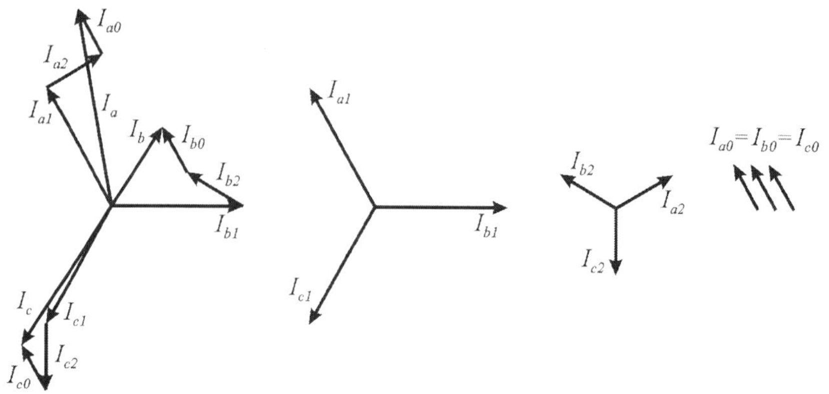

FIGURE 4.47 Expression of three-phase unbalanced current as symmetrical components.

Each phasor I_a, I_b, and I_c of a three-phase current system consists of three component phasors:

$$\underline{I}_a = \underline{I}_{a1} + \underline{I}_{a2} + \underline{I}_{a0};$$

$$\underline{I}_b = \underline{I}_{b1} + \underline{I}_{b2} + \underline{I}_{b0}; \tag{4.74}$$

$$\underline{I}_c = \underline{I}_{c1} + \underline{I}_{c2} + \underline{I}_{c0}.$$

The first set of components is selected in such a way that the components form a positive sequence I_{a1}, I_{b1}, and I_{c1}. The second set of components forms a negative sequence I_{a2}, I_{c2}, and I_{b2}, where the order of the phasors has been changed. The third set of components forms a zero sequence, where the phasors are all in phase with each other. The positive and negative sequence components will pass through a three-phase transformer without any significant differences. However, problems may arise with the zero-sequence currents, which are all in the same phase in all three windings of the transformer. Figure 4.48 shows the zero-sequence currents for a star connection with a neutral conductor and a delta connection.

In the case of a star connection, the current in the neutral conductor is three times higher than that in any single winding. In the case of a delta connection, the zero-sequence currents can flow in a **closed loop** formed of the three windings, if they are induced there by the other, star-connected winding of the transformer. A delta connection prevents zero-sequence currents from occurring in the line conductors, because the current cannot enter simultaneously from the terminals a, b, and c. For the same reason, zero-sequence currents cannot occur in a star-connected system without a neutral conductor.

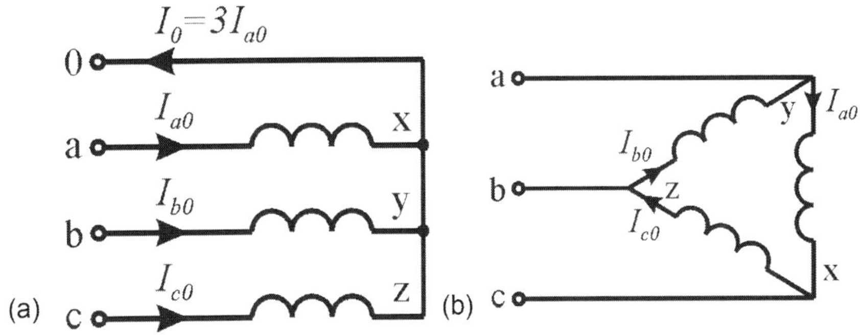

FIGURE 4.48 Zero-sequence currents in transformer windings in the case of a star connection with a neutral conductor (a) and a delta connection (b).

Zero-sequence currents can have a frequency of either 50 Hz, if they are caused by the phase currents being unbalanced, or 150 Hz (third harmonic) and 450 Hz (ninth harmonic), if the phase currents are balanced, but the shape of the current curve is distorted.

If one of the transformer windings is connected in delta or zigzag (e.g. Y/Δ, Y/Z, Δ/Δ, Δ/Y), the zero-sequence component does not pass through the transformer but is cancelled out in the transformer. This occurs without any significant distortions in the voltage.

In the case of a Y_0/y_0 connection, the zero-sequence component passes through the transformer, but, again, without leading to distortions in the voltage. In the case of a Y/y_0 connection, the zero-sequence component cannot pass through the transformer and is also not cancelled out in the transformer. In such a case, the sinusoidal zero-sequence currents will begin to magnetise the magnetic circuit, resulting in a voltage imbalance (Figure 4.49).

If the transformer has three legs, the magnetic flux of the zero-sequence current passes through the air gap and the transformer tank. Since the magnetic resistance across this path is high, the resultant voltage imbalance is not very large. However, in transformers that have side yokes, a Y/y_0 connection can only be used by installing a third, delta-connected winding on the transformer ($Y/y_0/d$). Then the zero-sequence currents can cancel out and there is no voltage imbalance.

If the zero-sequence currents are higher harmonics, a noticeable distortion of the voltage curve will also occur in the case of a Y/y_0 connection. Since the Y/y_0 configuration has drawbacks, the zigzag configuration Y/z_0 is sometimes used instead of it, although this increases the cost of the transformer.

When choosing a winding configuration, it is also necessary to bear in mind that it determines the ratio between the primary and secondary voltages of the transformer. In the case of a D/d or Y/y connection, the ratio of the line voltages on the primary and secondary side at no load is equal to the ratio of the number of turns of wire in the transformer windings

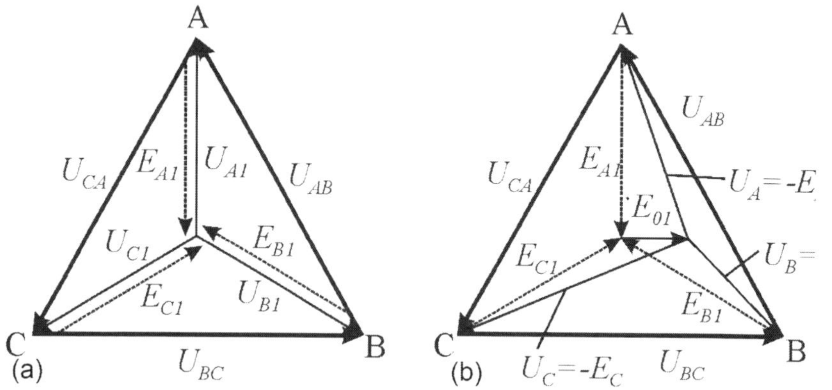

FIGURE 4.49 Distortion of the system of phase voltages in a Y/y_0 winding configuration due to zero-sequence currents.

$$\frac{U_{1l}}{U_{2l}} = \frac{w_1}{w_2}, \tag{4.75}$$

but in the case of a D/y connection,

$$\frac{U_{1l}}{U_{2l}} = \frac{w_1}{\sqrt{3}w_2}. \tag{4.76}$$

4.14 AUTOTRANSFORMERS

In conventional transformers, the primary and secondary windings are electrically, aka galvanically, isolated. In conventional transformers, energy transfer is achieved solely via a magnetic field. When a residential building is supplied with power from a 10 kV grid via a 10/0.4 kV transformer, galvanic isolation prevents what is a high-voltage potential relative to earth from entering the building. However, in some cases, it is more expedient to instead use transformers where the primary and secondary windings are electrically connected. Such transformers are known as autotransformers.

Figure 4.50 shows a transformer in which a 100 V primary voltage U_1 is converted into a 50 V secondary voltage in two different ways. The secondary voltage is obtained both from the galvanically isolated winding w_2 and the tap (between points A and B) on the primary winding w_1. The variable magnetic flux passing through the transformer core induces the same electromotive force in each winding turn. In this example, an emf of 1 V is induced in each turn. Due to this, a voltage of 50 V is obtained both from the secondary winding w_2, which has 50 turns, and the section of the primary winding w_1 which also has 50 turns. In the latter case, however, point A belongs directly to both the primary and secondary sides of the transformer. Therefore, there is no galvanic isolation.

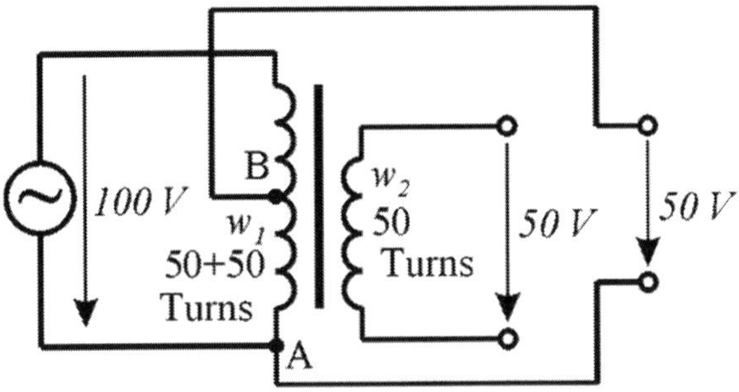

FIGURE 4.50 Two different ways of stepping down AC voltages with a transformer: separately with the secondary winding w_2 and with a tap on the primary winding.

If the primary and secondary sides of a transformer are electrically connected, then the transfer of energy between the primary and secondary sides can occur not only through magnetic coupling but also through electrical coupling. This is illustrated in Figure 4.51. The transformer has two windings: w_1 and w_2. These windings can be seen in two ways: either as a single winding with a tap or two windings connected in series.

The primary voltage, which in this case is the higher voltage U_{HV}, is supplied to the total number of turns in the windings w_1 and w_2. The secondary voltage, which is the lower voltage U_{LV}, is determined by the number of turns in the winding w_1. If we consider the transformer as an ideal transformer, then $U_{LV} = U_1 = E_1$.

In an autotransformer, there are two separate power flows. The apparent power flowing from the high-voltage side to the low-voltage side is the power flowing through the transformer. This can be expressed through the parameters of the high-voltage side.

$$S_{\text{through}} = U_{\text{upper}} \cdot I_{\text{upper}} = U_{\text{upper}} \cdot I_2 \tag{4.77}$$

as well as the parameters of the low-voltage side

$$S_{\text{through}} = U_{\text{lower}} \cdot I_{\text{lower}} = U_{\text{lower}} \left(I_1 + I_2\right). \tag{4.78}$$

The power transferred via the transformer's magnetic field is the transformer's calculated apparent power

$$S_{\text{calc}} = U_1 \cdot I_1 = U_2 \cdot I_2. \tag{4.79}$$

In a conventional transformer, $S_{\text{calc}} = S_{\text{through}}$, but in an autotransformer, these two power values are not equal. Based on Figure 4.51 and equations (4.77) and (4.79),

$$\frac{S_{\text{calc}}}{S_{\text{through}}} = \frac{U_2 I_2}{U_{\text{upper}} I_{\text{upper}}} = \frac{\left(U_{\text{upper}} - U_{\text{upper}}\right) I_{\text{upper}}}{U_{\text{upper}} I_{\text{upper}}} = 1 - \frac{1}{k_{\text{tr}}} = \frac{k_{\text{tr}} - 1}{k_{\text{tr}}}. \tag{4.80}$$

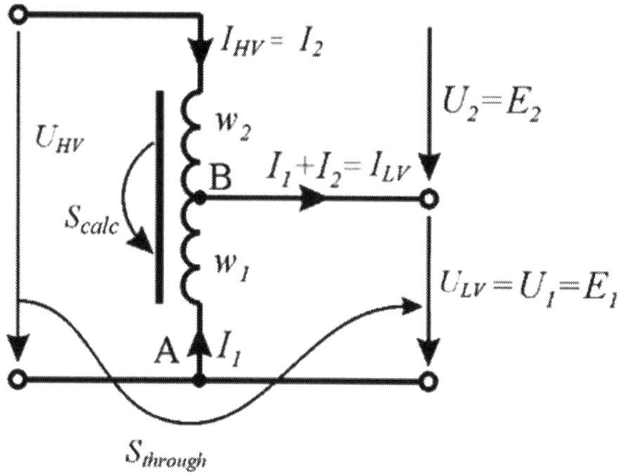

FIGURE 4.51 In an autotransformer, the calculated energy flow through the magnetic field between the windings w_1 and w_2 (S_{calc}) is lower than the energy flow transferred from the high-voltage side to the low-voltage side, i.e. the energy flow through the transformer ($S_{through}$).

TABLE 4.2
Ratio calculated power to throughput power in an autotransformer at different transformation ratios k_{tr}

k_{tr}	$S_{calc} / S_{through}$
1.0	0
1.5	0.33
2.0	0.5
2.5	0.6
10	0.9

Here, the autotransformer's transformation ratio $k_{tr} = U_{HV} / U_{LV} = (w_1 + w_2)/w_1$. Equation (4.80) shows that, if the transferred power is the same, an autotransformer with a lower transformation ratio k_{tr} will have a lower calculated power (and, therefore, a lower weight and smaller dimensions). Table 4.2 lists some corresponding values for the ratio $S_{calc} / S_{through}$.

Since $S_{calc} < S_{through}$, autotransformers are more cost-efficient than conventional transformers. An autotransformer with the same weight and price as a conventional transformer can handle more power. That is why autotransformers are sometimes also called savings transformers. However, if the transformation ratio of the transformer is high, the savings will not be that great. Autotransformers are normally only used if k_{tr} £ 2.5.

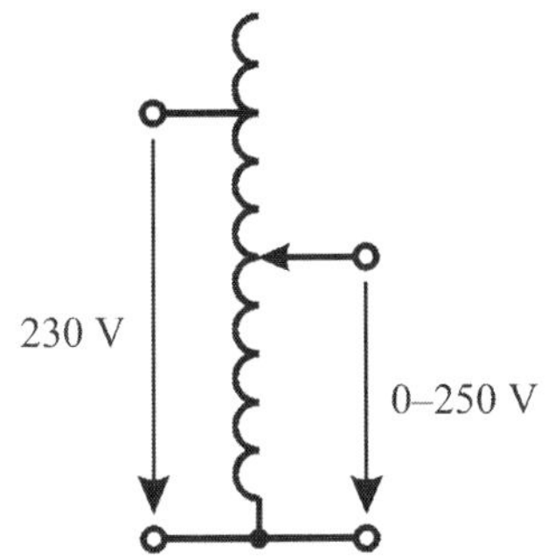

FIGURE 4.52 Laboratory autotransformer.

The autotransformers utilised in energy systems are three-phase transformers. They may be used, for example, to link together a 220 kV grid and a 330 kV grid.

The design of autotransformers is also used as the basis for constructing adjustable-voltage transformers. Such transformers usually have a single-layer winding where the insulation of the winding wires has been removed to expose their surface so that a contact brush can be slid along its length (see Figure 4.52). As the brush moves, the location of the winding tap changes, which in turn changes the number of winding turns and the voltage.

Such autotransformers are used in laboratories.

Problem 4.7

An autotransformer has a primary voltage of 230 V and a secondary voltage of 200 V. The apparent power transferred from the transformer's primary side to the secondary side at rated conditions is 2,000 VA. Find the currents in the windings w_1 and w_2 at rated conditions (Figure 4.51) and the calculated power of the autotransformer. Consider the transformer to be an ideal transformer.

Solution

1. The current in the winding w_2 is

$$I_{\text{upper}} = S_{\text{through}} / U_{\text{upper}} = 2000 / 230 = 8,7\text{A}.$$

2. The calculated power of the autotransformer is

$$S_{\text{calc}} = U_2 \cdot I_2 = \left(U_{\text{upper}} - U_{\text{lower}}\right) \cdot I_{\text{upper}} = \left(230 - 200\right) \cdot 8,7 = 261\text{VA}.$$

3. The current in the winding w_1 is

$$I_1 = S_{\text{calc}} / U_{\text{lower}} = 261 / 200 = 1,3\text{A}.$$

4.15 VOLTAGE AND CURRENT MEASURING TRANSFORMERS

Voltage transformers and current transformers are used to measure high voltages and high currents. Transformers that are used for such purposes are called instrument transformers. Designing electrical measuring instruments for high voltages (e.g. 110 kV) or high currents (e.g. 1,000 A) is not sensible. Such instruments would be very large and expensive, as well as dangerous for the operator in high-voltage applications.

A solution to this problem when measuring voltages is offered by the fact that the secondary voltage of a transformer is proportional to the primary voltage at no load. Current transformers, specifically, operate on the principle that the secondary current of a transformer is proportional to the primary current when the magnetisation current is negligibly low. Measuring instruments may thus be connected to the secondary circuit of a transformer, instead of directly to the measured circuit. The rated value of the secondary voltage (or secondary current) of the transformer can be selected in a way that ensures that the measuring instrument is simple and inexpensive. A rated voltage of 100 V is generally chosen for measuring voltages, while a rated current of 5 A is chosen for measuring currents. Measuring instruments are then connected to the measured electrical circuit via the appropriate transformers, as shown in Figure 4.53.

In the example above, the voltage of the measured circuit is 110,000 V and this is the rated voltage; meanwhile, the voltage across the secondary winding of the voltage transformer is 100 V, which is the rated voltage of the voltmeter, but the voltmeter will read 110 kV. The voltmeter's scale is graduated in kilovolts, and the instrument is designed for use only with an appropriate voltage transformer. In our example, the ratio between the number of turns in the primary and secondary

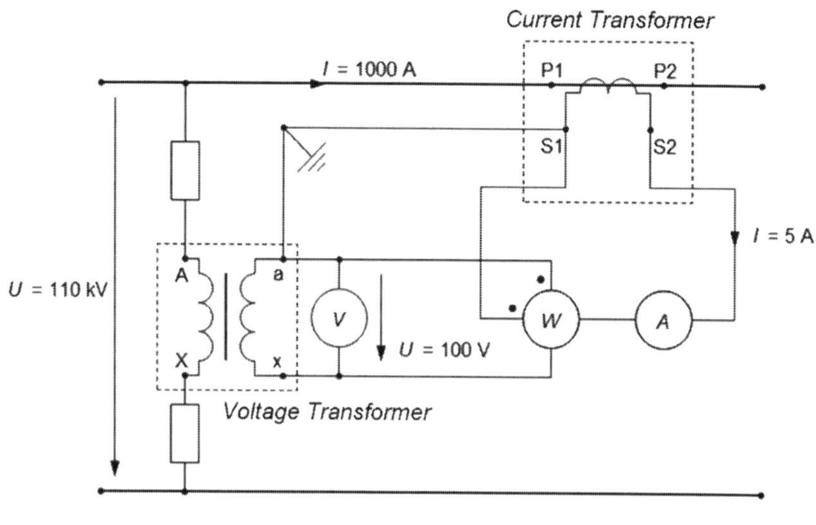

FIGURE 4.53 Connection of measuring instruments into a single-phase circuit via instrument transformers.

windings is 110,000/100 = 1,100. Similarly, if the measured circuit carries a current of 1000 A, the current through the secondary winding of the transformer and the ammeter is 5 A (which is also the rated current), in which case the pointer of the ammeter stands at the end of the ammeter's scale and indicates a current of 1,000 A.

The secondary side of a voltage transformer may be connected to a voltmeter, frequency meter, the potential winding of a wattmeter, or the potential winding of an electricity meter. It is also possible to connect several measuring instruments in parallel to a voltage transformer's secondary winding. The electrical resistance of these instruments is high and they will draw very little power from the voltage transformer. Voltage transformers operate at close to no-load conditions. Due to this, the voltage drops across the windings of such a transformer are very small and the transformer will operate nearly like an ideal transformer. This is necessary for obtaining accurate measurement results. In an ideal transformer, $U_2 = (w_2/w_1) \cdot U_1$. As a voltage transformer is loaded with current, the measurement accuracy decreases. In the event of a short circuit, a high current will be produced in the windings of a voltage transformer, causing them to overheat and break down.

The secondary side of a current transformer may be connected to an ammeter, the current winding of a wattmeter, or the current winding of an electricity meter. Up to two or three instruments may be connected at once, but only in series (Figure 4.53). The current of the entire secondary winding of the current transformer must pass through each instrument. The resistance of measuring instruments connected to the secondary circuit of a current transformer is, by design, extremely small. This ensures that the current transformer operates at close to short-circuit conditions. Due to this, the voltages across the transformer windings are low. A low voltage is necessary to minimise the magnetic flux in the magnetic circuit. When the magnetic flux is low, so is the magnetisation current (in the equivalent circuit of the transformer, this will be the current in the magnetisation branch). If the current in the magnetisation branch is negligible, then $I_2 = (w_1/w_2) \cdot I_1$. This is precisely what is needed for measurement accuracy.

For safety reasons, both in current transformers and voltage transformers, one of the terminals of the secondary winding is earthed (Figure 4.53). If the secondary circuit is broken during the operation of a current transformer (e.g. by disconnecting an ammeter), the current in the secondary circuit becomes zero. As a result, all of the current in the primary winding turns into magnetisation current. The magnetic flux increases severalfold and the transformer core begins to overheat. The increasing of the magnetic flux is accompanied by the increasing of the voltage across the secondary winding, which can become dangerously high. That is why it is prohibited to break the secondary circuit of a current transformer when a current is passing through the primary winding.

4.16 TRANSIENT RESPONSES IN TRANSFORMERS

4.16.1 Energisation of a Transformer

Transient responses occur in transformers and electrical machines when their mode of operation changes abruptly (connection to grid, changing of load, short circuit, etc.). This chapter explores the most common transient responses in transformers.

Consider what happens when a single-phase transformer with an open secondary winding (Figure 4.54) is connected to a grid where the voltage is sinusoidal $u_1 = U_{1m}\sin(\omega t+\psi)$.

4.16.1.1 Unsaturated Transformer

First of all, let us suppose that the transformer core is not saturated at all, and therefore the winding inductance $L_{11} = const$. In such a case, the transient response upon connection is given by the equation:

$$U_{1m}\sin\left(\omega t+\psi\right) = r_1 i_1 + L_{11}\frac{\mathrm{d}i_1}{\mathrm{d}t}. \tag{4.81}$$

From the course on the theoretical foundations of electrical engineering, we know that the transient response current i_1 can be expressed as the sum of two components:

$$i_1 = i_1^{'} + i_1^{''}, \tag{4.82}$$

where the first component

$$i_1^{'} = I_{1m}\sin\left(\omega t+\psi-\varphi_0\right), \tag{4.83}$$

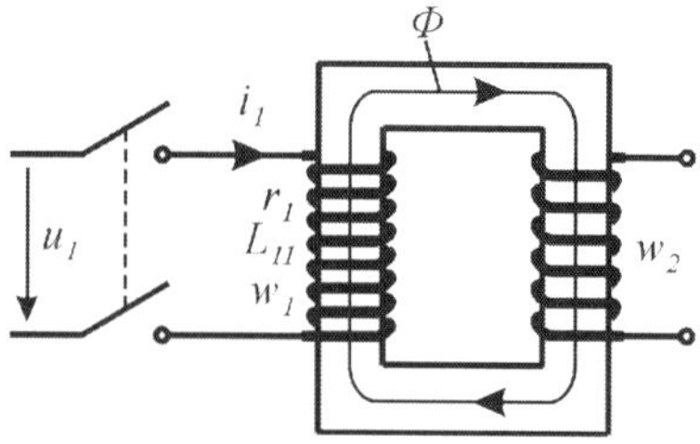

FIGURE 4.54 Diagram for connecting a transformer operating at no load to a grid.

where

$$\left.\begin{aligned} I_{1m} &= \frac{U_{1m}}{\sqrt{r_1^2 + \omega^2 L_{11}^2}} \\ \varphi_0 &= arctan \frac{\omega L_{11}}{r_1} \end{aligned}\right\} \tag{4.84}$$

is a sinusoidal steady-state current, or forced current, which is produced by the voltage u_1 applied to the primary terminals, while the second component

$$i_1^{"} = -I_{1m} \sin\left(\psi - \varphi_0\right) e^{-\frac{t}{T_1}} \tag{4.85}$$

is an aperiodic transient current, which is not maintained by any external source of electromotive force, and thus drops to zero over a constant time period T1 = L11/r1.

That the solution above is correct can be verified by plugging the values of i_1 and its components from equations (4.82)–(4.85) into equation (4.81). $i_1^{'}$ satisfies equation (4.81), while $i_1^{"}$ satisfies the equation

$$0 = r_1 i_1 + L_{11} \frac{di_1}{dt}.$$

Figure 4.55b shows that when $t = 0$, $i_1 = i_1^{'} + i_1^{"} = 0$. Thus, the initial value of the transient current is always equal to, but the inverse of, the initial value of the steady-state current. If, at the time of switching, the initial phase ψ of the voltage is such that $\psi - \phi_0 = 0$, then $i_1^{"} = 0$ and a steady state arises immediately in the circuit (Figure 4.55a). If, however, $\psi - \phi_0 = \pm \pi/2$, then $i_1^{"}$ reaches its maximum possible value, which, at $t = 0$, equals the maximum intensity of the alternating current I_{1m} (Figure 4.55b). The maximum instantaneous value of the current intensity occurs when about half a period has passed from the moment of switching. If the value of T_1 is sufficiently high, $i_{1\ max} \approx 2I_{1m}$.

It follows from the above that the transient current arises when the steady-state current, or forced current, is not passing through zero at the time of switching $t =$ 0 and has some final value $i_{10}^{'}$. The value of the initial transient current $i_{10}^{"} = -i_{10}^{'}$ is such that at $t = 0$, the total current equals zero according to the initial conditions at switching.

The transient response in the AC inductive circuit above also characterises more complex cases. In the transient response that occurs when the operating mode of a circuit is interrupted in any way, a transient aperiodic current is generated, which decays due to energy dissipation in the resistance. The initial value of that aperiodic current is equal to the difference between the instantaneous values

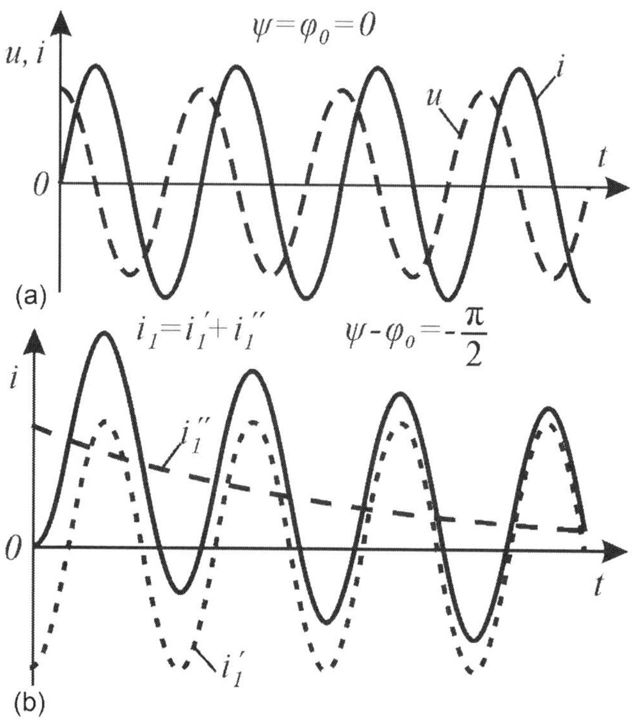

FIGURE 4.55 Inrush current in an unsaturated transformer.

of the steady-state currents in the initial mode (i.e. the preceding mode) and the new mode (i.e. the succeeding mode) at the moment of application of the pulse (e.g. voltage pulse) that causes the initial mode to change. The transient current is therefore an inductance-containing inert circuit's reaction to the pulse attempting to change the operating mode of the circuit. This transient current smooths the transition to the new mode, preventing instantaneous finite changes in current, which are not possible in an inert circuit.

4.16.1.2 Saturated Transformer

In a real transformer, it is imperative to take into account the considerable saturation of the core. Thus, $L_{11} \neq const$, and the last component of equation (4.81) must be replaced with $\mathrm{d}(L_{11}i_1)/\mathrm{d}t$, where

$$L_{11}i_1 = \Psi = w_1 F \tag{4.86}$$

is the flux linkage of the winding.

In this case, it is expedient to substitute variable flux Φ for the current i_1. According to equation (4.86)

$$i_1 = \frac{w_1 F}{L_{11}}, \frac{\mathrm{d}(L_{11} i_1)}{\mathrm{d}t} = w_1 \frac{\mathrm{d}F}{\mathrm{d}t}$$

and so, instead of equation (4.81), we get

$$\frac{U_{1\mathrm{m}}}{w_1} \sin(\omega t + \psi) = \frac{r_1}{L_{11}} F + \frac{\mathrm{d}F}{\mathrm{d}t}. \tag{4.87}$$

Integrating this equation is relatively difficult, because, here, $L_{11} \neq const$. The latter can be determined from the complex non-linear relationship $i = f(\Phi)$, which is a magnetic characteristic of the transformer. However, since the first component on the right side of equation (4.87) is insignificant compared to the second component due to r_1 being small, just as the voltage drop $r_1 i_1$ is small compared to the electromotive force $-\mathrm{d}(L_{11} i_1)/\mathrm{d}t$, L_{11} may be considered to be approximately constant when integrating equation (4.87). Under these conditions, equation (4.87) may be expressed as the sum of two components

$$F = F' + F'', \tag{4.88}$$

where Φ' is the instantaneous value of the steady-state flux and Φ'' is the instantaneous value of the transient flux.

The instantaneous value of the sum of the magnetic fluxes Φ' and Φ'' is given by

$$F = -F_{\mathrm{m}} \cos(\omega t + \psi) + (F_{\mathrm{m}} \cos\psi \pm F_0) e^{-\frac{r_1 t}{L_{11}}}. \tag{4.89}$$

The most expedient moment for switching is when $\psi = \pi/2$ (the voltage is passing through the peak at the time of switching) and $\Phi_0 = 0$. In such a case

$$F = -F_{\mathrm{m}} \cos\left(\omega t + \frac{\pi}{2}\right) = F_{\mathrm{m}} \sin \omega t, \tag{4.90}$$

which means that the normal mode with a variable sinusoidal flux is reached instantaneously and there is no transient response. The most inexpedient moment would be that when $\psi = 0$ (the voltage is passing through zero at the time of switching) and a plus sign must be used in equation (4.89), i.e. when the flux Φ_0 opposes the flux Φ' at $t = 0$. In such a case, based on equation (4.89),

$$F = -F_{\mathrm{m}} \cos \omega t + (F_{\mathrm{m}} + F_0) e^{-\frac{r_1 t}{L_{11}}}. \tag{4.91}$$

Figure 4.56 shows the shape of the curve $\Phi = f(t)$ for this case. The flux Φ reaches its maximum value about half a period after switching on, i.e. when $\omega t \approx \pi$. Thereat

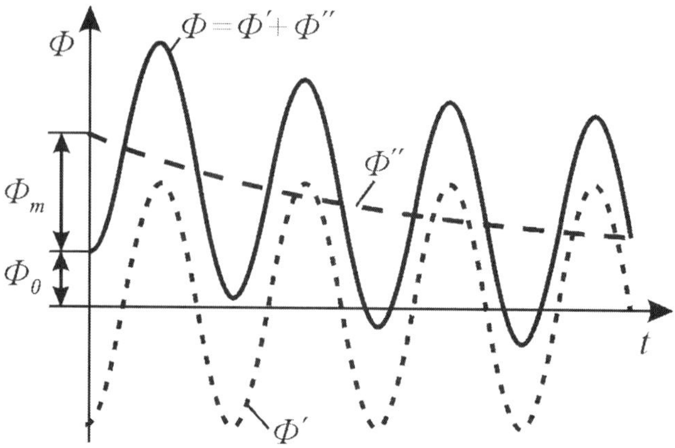

FIGURE 4.56 Changing of the magnetic flux Φ when the transformer is plugged into the grid at an undesirable moment.

$$e^{-\frac{r_1 t}{L_{11}}} = e^{-\frac{r_1 \grave{A}}{\omega L_{11}}} \approx 1\,, \tag{4.92}$$

because normally $r_1 << \omega L_{11}$. Thus, based on equation (4.91),

$$F_{\mathrm{max}} \approx 2F_{\mathrm{m}} + F_0\ . \tag{4.93}$$

Therefore, Φ_{max} is more than two times higher than the normal value of the operating flux, as a result of which the core becomes exceedingly saturated, which in turn leads to relatively high magnetisation currents.

Figure 4.57 shows the oscillogram of the inrush current in a moderately saturated medium-power transformer.

In many cases, the peak inrush currents at switch-on can be 100–150 times higher than the peak steady-state no-load current, and, thus, several times higher than the rated maximum current. Such inrush currents pose challenges for the designing and configuration of protective current relays for transformers, as they may cause the protective relays to trip unexpectedly when the transformer is plugged into the grid. However, at zero saturation, as we saw above, the peak inrush current can only be two times higher than the peak steady-state no-load current.

4.16.2 Abrupt Short-Circuits in a Transformer

4.16.2.1 Short-Circuit Currents

In Section 4.7, we looked at steady-state short circuits on the secondary terminals of a transformer. Under operating conditions, short circuits generally occur

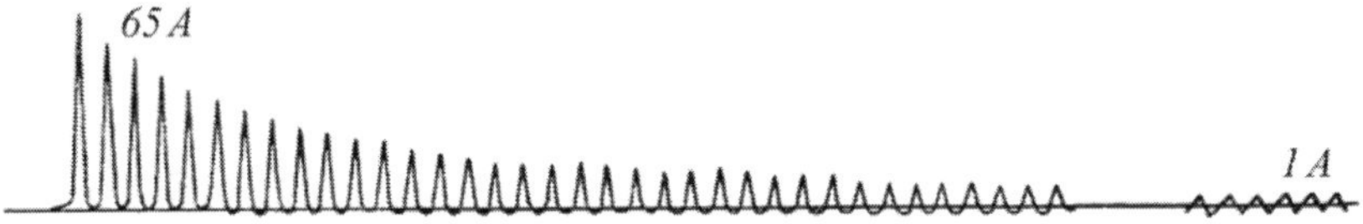

FIGURE 4.57 Oscillogram of inrush current in a saturated transformer.

abruptly due to various failures in the electrical grid (dielectric breakdown or mechanical damage to the insulation, erroneous actions by personnel, etc.). This is accompanied by a sudden transient response in the transformer, giving rise to high currents. These, in turn, involve a risk of damage to the transformer. Let us now explore what happens during an abrupt short circuit on the secondary terminals of a single-phase transformer (Figure 4.58a). Assume that the effective value of the primary voltage U_1 does not change throughout. The features of the process are the same for short circuits on the secondary terminals of a three-phase transformer.

In a sustained short circuit, the magnetisation current is relatively small compared to the total current in the winding. This is also the case for abrupt short circuits. In addition, due to the short-circuit current being high, the voltage drop across the primary winding's resistance r_1 and parasitic inductive reactance x_1 is relatively high. Indeed, normally $z_1 \approx z_2'$, and according to the simplified equivalent circuit of the transformer, in a short circuit, $I_1 z_1 \approx 0{,}5E_1$. Due to this, the electromotive force E_1 and core flux Φ are nearly twice as low as their rated value, and the core is not saturated. For the same reasons (low voltage and magnetic flux), in an abrupt short circuit, the magnetisation current may also be disregarded, and thus calculations may be performed based on the simplified equivalent circuit of the transformer (Figure 4.58b). The parameters of this circuit

$$r_k = r_1 + r_2' \text{ja } L_k = \frac{x_k}{\omega} = \frac{x_1 + x_2'}{\omega}$$

may be considered constant.

Based on the circuit in Figure 4.58b, the process of an abrupt short circuit on the secondary terminals of the transformer is described by the differential equation

$$U_{1m} \sin(\omega t + \psi) = r_k i_{1k} + L_k \frac{di_{1k}}{dt}. \tag{4.94}$$

This equation is completely analogous to equation (4.81), which is only natural, since both equations essentially describe the switching of a reactor with constant parameters to a sinusoidal voltage. In the present case, the short-circuit current i_{1k} also consists of two components: steady-state current i_{1k}' and transient current i_{1k}'':

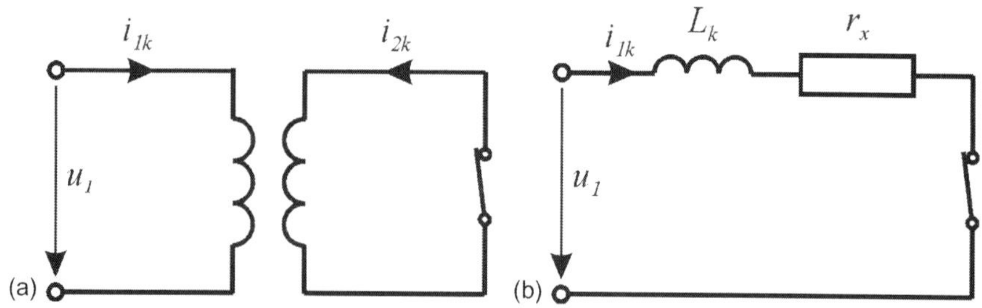

FIGURE 4.58 (a) Trasformer short circuit and (b) its equivalent circuit.

$$i_{1k} = i_{1k}^{'} + i_{1k}^{''}. \tag{4.95}$$

To find the instantaneous values of the steady-state current and the transient current and, thereby, the total current i_{1k}, we can put together the following equation:

$$i_{1k} = I_{1km}\sin\left(\omega t + \psi - \varphi_k\right) - I_{1km}\sin\left(\psi - \varphi_k\right)e^{-\frac{r_k t}{L_k}}. \tag{4.96}$$

The short-circuit current is inductive in nature and $0 < \psi_k < 90°$. The load on the transformer is usually also inductive, and thus $0 < \phi_{load} < 90°$. Under these conditions, the preceding load causes the aperiodic transient current to decrease and, thereby, the peak short-circuit current to also decrease. And conversely: if the preceding load is capacitive in nature, the aperiodic transient current will increase. In practice, capacitive loads almost never occur for transformers, and thus the short circuit at no load expressed by equation (4.96) may be deemed the most undesirable case. This is the case we will be referring to hereafter.

The curve $i_{1k} = f(t)$ will be shaped differently, depending on the size of the angle $\psi - \phi_k$. The most common cases are analogous to those shown in Figure 4.55.

At $\psi - \phi_k = \pm\pi/2$, the aperiodic transient current and peak currents are maximal (see Figure 4.55b). For $\psi - \phi_k = \_\pi/2$, equation (4.96) gives us

$$i_{1k} = -I_{1km}\cos\omega t + I_{1km}e^{-\frac{r_k t}{L_k}}. \tag{4.97}$$

The maximum current, or peak current, $i_{1k\ peak}$ is reached when $\omega t = \pi$, i.e. when a time period of $t = \pi/\omega$ has passed from the start of the short circuit. According to equation (4.97),

$$i_{1kl} = I_{1km}\left(1+e^{-\frac{\pi r_k}{x_k}}\right). \tag{4.98}$$

The multiplier in the parentheses in the equation above is called the peak factor

$$k_l = 1+e^{-\frac{\pi r_k}{x_k}}. \tag{4.99}$$

The peak factor shows by how many times the peak short-circuit current exceeds the maximum value of the steady-state short-circuit current. Depending on the value of r_k/x_k, the value of the factor k_{peak} can range from 1 to 2. In powerful transformers, k_{peak} – 1.7–1.8, while in smaller ones, k_{peak} = 1.2–1.3. For example, in a transformer with a power of 1,000 kVA, the short-circuit voltage and its components are:

$u_{k\%} = 6{,}5\%, u_{ka\%} = 1{,}5\%$ and $u_{kr\%} = 6{,}32\%$.

The corresponding

$$k_l = 1+e^{-\frac{\pi r_k}{x_k}} = 1+e^{-\frac{\pi u_{ka}}{u_{kr}}} = 1+e^{-\frac{1{,}5\pi}{6{,}32}} = 1+e^{-0{,}745} = 1{,}475$$

peak short-circuit current is

$$\frac{100}{u_{k\%}}k_l = \frac{100}{6{,}5}\cdot 1{,}475 = 22{,}7$$

times higher than the rated maximum current.

In transformer short circuits where damage to insulation causes some of the turns in a winding to short-circuit, the current multipliers in the damaged part of the winding will be even greater, because while the voltage or electromotive force across that part of the winding will decrease in proportion to the number of turns, the inductive reactance will decrease in proportion to the square of the number of turns.

Short-circuit currents result in increased heating of the transformer windings and the generation of large electromagnetic forces acting on the windings.

Electrical systems and grids utilise fast-acting protective relays or fuses, which will disconnect the damaged part of the grid and any damaged equipment in a fraction of a second after the start of a short circuit. This prevents external short circuits from heating transformer windings to dangerous temperatures. In the event of an internal short circuit, however, the wires of the short-circuited turns will generally melt, causing the transformer to break down.

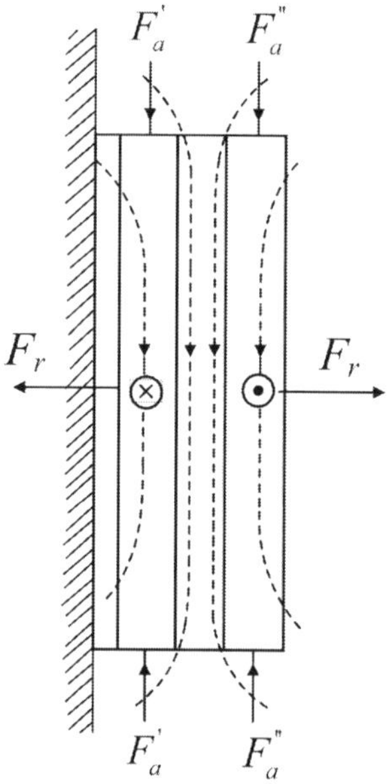

FIGURE 4.59 Nature of magnetic leakage field and electromagnetic forces in a transformer during a short circuit.

Figure 4.59 shows the magnetic leakage field of a transformer with concentric cylindrical windings. From the figure, it is evident that the windings are affected by two kinds of electromagnetic forces: first, radial forces $\boldsymbol{F}_r$, which attempt to compress the inner winding radially and pull the outer winding apart, and, secondly, axial forces $\boldsymbol{F}_a$, which attempt to compress both windings longitudinally.

The electromagnetic forces acting upon the windings are proportional to the current squared and can get quite large during short circuits. Consequently, one of the crucial concerns when designing a transformer is ensuring proper fastening (bracing) of the windings.

5 Induction Machines

Induction machines, just like synchronous machines, are AC machines. Both types of machines contain a similar stator that generates a rotating magnetic field in an air gap. Induction machines are primarily used as motors, but under certain conditions can also function as generators. Three-phase induction motors are the most common type of electric motor. Compared to other types of motors, they have the simplest construction and are relatively inexpensive and reliable.

5.1 DESIGN AND OPERATING PRINCIPLE OF INDUCTION MACHINES

Refer to Figures 1.5 and 1.6 in the first chapter for what an induction motor, or induction motor, looks like. In Figure 1.6b, the rotor has been removed from inside the stator and we can see that it is essentially a cylinder mounted on a shaft. This cylinder is made of steel sheets.

Figure 5.1 shows a crosssection view of another induction motor. The rotor and stator are both visible and can both be seen to contain slots for windings. The slots on the stator can vary widely in shape (e.g. slots 1, 2, and 3 in Figure 5.1b). The stator is fixed to the body of the motor (usually pressed into it).

The exterior of the rotor also has slots in it for housing the winding. The possible shapes of these slots are also shown in Figure 5.1b (slots 4, 5, and 6). The rotor – just like the stator – is constructed from steel sheets to reduce eddy current losses. The rotor winding may be designed either as a **short-circuited winding** (Figure 5.2a) or, similarly to the stator winding, as a three-phase winding composed of winding coils insulated from the rotor (Figure 5.2b). In the former case, the motor would be what is known as a **squirrel-cage induction motor**, while in the latter case it would be a **slip ring-rotor induction motor**. Squirrel-cage induction motors are the most common type of induction motor. Their short-circuited winding consists of a number of bars connected by shorting rings. The winding is made from aluminium by die casting. During the casting process, all uninsulated slots are filled at the same time and shorting rings are formed along with the adjoining fan blades needed for ventilation. These types of windings are also known as squirrel-cage windings.

In a slip-ring rotor, the windings of the three phases are usually connected in star, and the beginning of each winding is connected to one of three slip rings, which are mounted on the shaft and insulated from each other. Via brushes that touch the slip rings, the rotor winding is connected to an external circuit containing a speed adjustment circuit. Recently, slip ring-rotor induction motors have often come to be supplanted by squirrel-cage motors with a frequency converter, which also enables the speed to be adjusted.

DOI: 10.1201/9781003404088-5

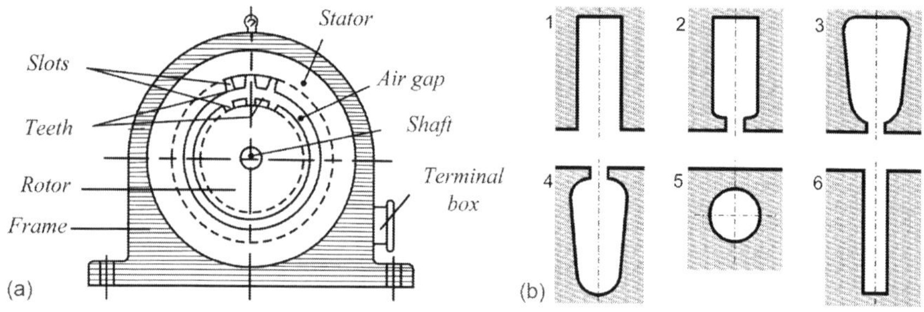

FIGURE 5.1 Schematic crosssection view of an induction motor (a) and possible slot shapes (b).

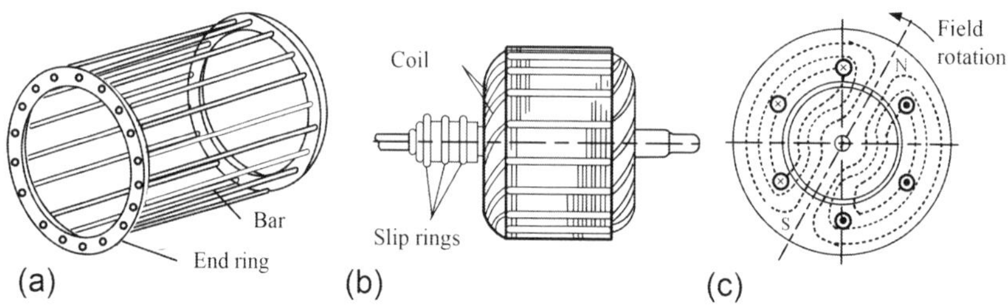

FIGURE 5.2 Squirrel-cage winding (a), slip-ring rotor (b), and two-pole magnetic flux field (c) for an induction motor.

To reach a better understanding of the operating principle of induction motors, it is useful to start from two extreme operating modes, which are **standstill** and **ideal no-load operation**. When the rotor of an induction motor is halted and prevented from rotating by braking, the motor operates as a transformer. When the stator winding is connected to the grid, a current is generated. The AC field of the stator winding passes through the air gap and induces a voltage in the rotor winding. If the rotor is fitted with a squirrel-cage winding, the situation that arises is analogous to the short-circuit operation of a three-phase transformer (Figure 5.3). In the rotor winding, a voltage and current with the frequency of the supply voltage are generated.

However, if the shaft is not halted and an external power source is used to make the rotor of the induction motor rotate at the same speed and in the same direction as the magnetic field of the stator winding, then no voltage is induced in the rotor winding. This is because the field of the stator winding is then stationary relative to the rotor, and a stationary field cannot induce a voltage in a set of wires. With no voltage, no current can arise in the rotor winding (Figure 5.4).

The aforementioned rotational speed of a rotor is called the **synchronous speed** (by analogy with synchronous machines). At the synchronous speed, an induction machine operates in absolute no-load mode, as there is also no electromagnetic torque due to the absence of current in the rotor winding. Meanwhile,

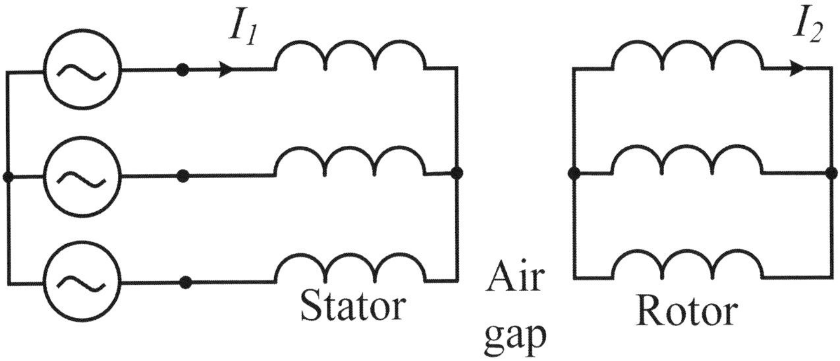

FIGURE 5.3 An induction motor with a halted squirrel-cage rotor functions like a short-circuited transformer.

the stator winding carries a current corresponding to a transformer's no-load current. **Thus, an induction motor with a halted rotor is equivalent to a short-circuited transformer, while an induction motor at ideal no-load is equivalent to a transformer in no-load mode.** However, the no-load current of an induction motor is several times higher than that of a transformer, because, unlike transformers, induction motors have an air gap. To reduce the no-load current, it is desirable for the air gap in an induction motor to be as small as possible (around 0.5 mm).

When the external power source shown in Figure 5.4 is switched off, a new condition arises, as shown in Figure 5.5. The rotor speed begins to decrease gradually due to friction and air resistance. This causes the wires in the rotor winding to begin 'slipping' behind the stator field rotating at synchronous speed, which induces a voltage in the wires. Since the resistance in the squirrel-cage winding is small, even a low voltage will produce a relatively high current. According to the right-hand rule, the current in the wires opposite the north pole in the figure will flow towards the plane of the figure, while the current in the wires opposite the south pole will flow out from the plane of the figure. The current in the rotor produces

- an induced force and torque in the rotor wires, and
- a magnetic field for the rotor winding (with the poles positioned at the points N and S on the rotor in Figure 5.5).

The direction of the induced force is such that the resulting torque follows the direction of rotation of the stator field (the direction of the force can be determined using the left-hand rule; the force is not indicated in the figure). The produced torque is what makes the induction machine rotate. Thus, **the field of the stator winding induces both an electromotive force and a mechanical force on the rotor wires of the induction motor** (if the current loop of the rotor windings is closed).

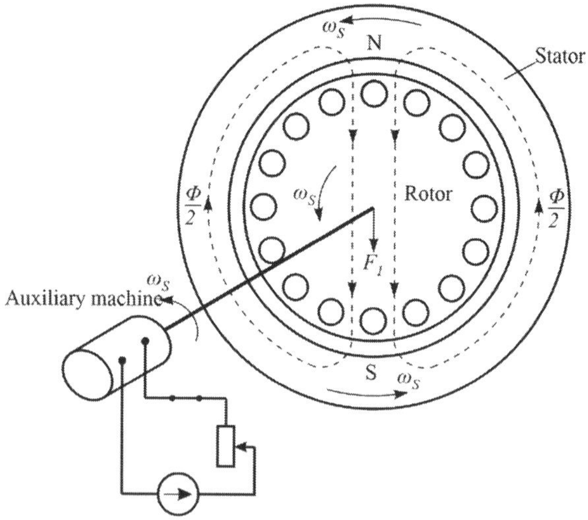

FIGURE 5.4 Induction machine maintained at synchronous speed, i.e. at ideal no-load, by an external power source.

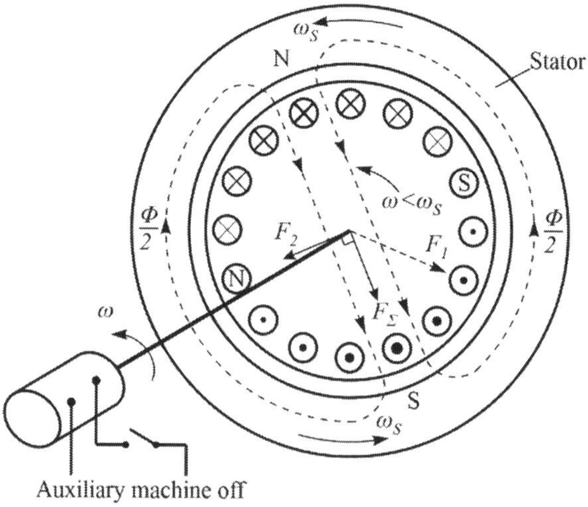

FIGURE 5.5 Lightly loaded induction motor.

When the induction motor is only lightly loaded, the **rotor's angular velocity** ω is nearly equal to the synchronous speed ω_S (e.g. $\omega = 0.999\,\omega_S$). In the rotor winding, an electromotive force with a very low angular frequency ω_2 is induced. In a two-pole motor, this angular frequency is equal to the **speed of the rotor relative to** the rotating stator field. If the stator field is rotating at the synchronous angular velocity ω_S and the angular velocity of the rotor relative to the stationary stator is ω, then the angular velocity of the rotor relative to the stator field, ω_2, is given by

$$\omega_2 = \omega_S - \omega \tag{5.1}$$

or

$$n_2 = n_S - n, \tag{5.2}$$

if the rotational speed is measured in revolutions per minute. The speed of the rotor relative to the stator field, ω_2 or n_2, is also called the **slip speed**. The per-unit slip speed with respect to the speed of the stator field is called the **slip** s:

$$s = \frac{n_2}{n_S} = \frac{\omega_2}{\omega_S} = \frac{n_S - n}{n_S} = \frac{\omega_S - \omega}{\omega_S}. \tag{5.3}$$

With a two-pole 50-Hz power supply, the synchronous speed n_S = 3,000 rpm.

If speed of the motor is $n = 0.999\,n_S$, the slip speed $n_2 = n_S - 0.999\,n_S = 0.001\,n_S = 0.001 \cdot 3{,}000 = 3$ rpm. Thus, the rotor slowly slips behind the rotating field to only three revolutions per minute. Meanwhile, the rotor speed relative to the stationary stator is $n = n_S - n_2 = 3{,}000 - 3 = 2{,}997$ rpm. In such a case, the slip is $s = n_2/n_S = 3/3{,}000 = 0.001$, or 0.1%. The slip equation (5.3) may also be written as

$$n_2 = sn_S \text{ ehk } \omega_2 = s\omega_S. \tag{5.4}$$

The synchronous speed ω_S, which expresses the rotational speed of the stator field in radians per second, depends on the number of pole pairs p and the angular frequency of the supply voltage, $\omega_1 = 2\pi f_1$, where f_1 is the frequency of the grid voltage.

$$\omega_S = \frac{\omega_1}{p} = \frac{2\pi f_1}{p}. \tag{5.5}$$

By analogy,

$$n_S = \frac{n_1}{p} = \frac{60 f_1}{p}. \tag{5.6}$$

The table below shows a series of synchronous speeds n_S and rated speeds n_n for induction motors at different numbers of pole pairs p at a rated slip of 4% and at f_1 = 50 Hz. At a 60 Hz grid frequency, the rotational speeds will be higher; for example, at $p = 1$, the synchronous speed n_S = 3,600 rpm.

Speed, rpm	***p*** 1	2	3	4	6
n_S	3,000	1,500	1,000	750	500
n_n	2,880	1,440	960	720	480

Changing the number of pole pairs does not change the ratio between the stator current frequency f_1 and the rotor current frequency f_2. Irrespective of the number of pole pairs,

$$f_2 = sf_1. \quad (5.7)$$

In a lightly loaded induction motor, if, for example, $s = 0.1\%$, the rotor current frequency will be very low: $f_2 = 0.001 \cdot 50 = 0.05$ Hz. At this frequency, the rotor winding's inductive reactance $x_2 = 2\pi f_2 L_2$ is also very low. Due to this, the rotor current and voltage are practically in phase. The direction of the magnetic field generated by the rotor current (indicated by the magnetomotive force vector F_2) will then differ from the direction of the magnetic field resulting from the stator and rotor (vector Φ, which produces an electromotive force and a wave of current) by about 90° (Figure 5.5).

Increasing the load torque drawn from the induction motor causes the following processes to occur:

1. The rotor speed decreases – this causes both the slip and the rotor current frequency to increase.
2. As the frequency increases, the induced electromotive force increases and, therefore, so does the rotor current.
3. As the rotor current increases, the electromagnetic torque of the machine increases, balancing out the increase in the load torque.

However, as the frequency increases, so does the rotor winding's inductive reactance x_2, which slows the increase of the rotor current. From a certain slip value onwards (approximately 25%), the torque of the motor can no longer increase. Thus, the speed of an induction motor depends on the load. Nevertheless, near the rated load, the change in speed is usually only a few percent and the speed itself is relatively close to the rotational speed of the stator field. **A rotating induction motor is essentially a transformer whose primary winding and secondary winding are in motion relative to each other. Due to this, the primary frequency and secondary frequency are unequal. This 'transformer' is constructed in such a way that the secondary current combined with the magnetic field of the primary current produces the necessary torque.**

Problem 5.1

There is a three-phase induction motor with the following operating ratings:

$P_n = 7.5$ kW; $U_n = 3 \cdot 400$ V; efficiency $\eta_n = 0.85$; power factor $\cos\phi = 0.89$; rated rotational speed $n_n = 1{,}440$ rpm; frequency $f = 50$ Hz.

Find the rated slip s_n, the frequency of the rotor current at rated conditions f_{2n}, the motor's rated torque T_n, and the current drawn from the grid at rated conditions I_{1n}.

Solution

1. The rotational speed of the magnetic field in the induction motor is

$$n_S = \frac{60f}{p} = \frac{60 \cdot 50}{p} = \frac{3000}{p}.$$

   If $p = 1$, then $n_S = 3{,}000$; if $p = 2$, then $n_S = 1{,}500$, etc.

2. Since, at rated conditions, the rotational speed n_n of the induction motor differs from the rotational speed of the magnetic field by only a few percent, then n_S must equal 1,500 rpm.
3. The slip at rated conditions is

$$s_n = \frac{n_S - n_n}{n_S} = \frac{1500 - 1440}{1500} = 0{,}04.$$

4. The frequency of the rotor current at rated conditions is

$$f_{2n} = f \cdot s_n = 50 \cdot 0{,}04 = 2\text{Hz}.$$

5. The motor's rated torque is

$$P_n = T_n \cdot \omega_n \; ; T_n = \frac{P_n}{\omega_n} \; ; \omega_n = \frac{\pi \cdot n_n}{30};$$

$$T_n = \frac{30P_n}{\pi \cdot n_n} = \frac{30 \cdot 7500}{\pi \cdot 1440} = 49{,}74\text{Nm}.$$

6. The current drawn from the grid is

$$P_n = \sqrt{3}U \cdot I_{1n} \cos\varphi_n \cdot \eta_n;$$

$$I_{1n} = \frac{P_n}{\sqrt{3} \cdot U \cos\varphi_n \cdot \eta_n} = \frac{7500}{\sqrt{3} \cdot 400 \cdot 0{,}89 \cdot 0{,}85} = 14{,}31\text{A}.$$

5.2 ELECTRICAL BALANCE EQUATION FOR THE STATOR OF AN INDUCTION MACHINE

The rotating magnetic field generated by the three-phase current in a stator induces a counter-electromotive force in each turn of the stator winding

$$e' = -\frac{dy}{dt}. \tag{5.8}$$

where the flux linkage of the rotating field with sinusoidal distribution

$$y = F_p \sin \omega t.$$

The effective value of the electromotive force (5.8) based on the transformer electromotive force equation

$$E' = 4,44 f_1 F_p, \tag{5.9}$$

where Φ_r is not maximum flux, but the constant value of the rotating magnetic flux.

The stator winding has a resistance and a parasitic inductive reactance. Across these resistances, the stator current produces a voltage drop. This voltage drop and the transformer emf must be balanced out by the grid voltage. The stator electrical balance equation thus takes the following form:

$$\underline{U}_1 = -\underline{E}_1 + \underline{I}_1 \left(r_1 + jx_1 \right). \tag{5.10}$$

It is not fundamentally different from the electrical balance equation for a transformer's primary winding. After all, in an induction motor, too, energy is converted by means of a magnetic field. The only difference is that the internal voltage drop is greater, because here, due to the air gap, the leakage flux is much greater than in a transformer.

5.3 ELECTROMOTIVE FORCE AND CURRENT IN THE ROTOR OF AN INDUCTION MACHINE

The rotating magnetic flux is coupled to the rotor winding and induces in it an electromotive force E_2, which produces the rotor current I_2. When the rotor is not rotating ($n_2 = 0$), then, based on the transformer electromotive force equation, the electromotive force induced in one phase of the rotor winding is

$$E_{2n} = 4,44 f_2 w_2 k_{w2} F_p = 4,44 f_1 w_2 k_{w2} F_p,$$

where w_2 is the number of turns in one phase and k_{w2} is the winding factor. For a standing rotor, the same as for a transformer, the relationship $f_2 = f_1$ holds for the frequency of the emf induced in the rotor winding. In a squirrel-cage rotor, the number of phases is considered to be equal to the number of bars (slots): $m_2 = Z_2$. Then $w_2 = 1/2$ and $k_{w2} = 1$, because dividing a squirrel-cage winding and shortening the step is impossible.

When the rotor of the induction motor is rotating, then the frequency $f_2 < f_1$, and, based on equation (5.7),

$$f_2 = \frac{pn}{60}. \tag{5.11}$$

where the relative speed of the rotating magnetic flux and rotor is $n = n_1 - n_2$. Plugging this into equation (5.11), we get:

$$f_2 = \frac{p(n_1 - n_2)}{60} \cdot \frac{n_1}{n_1} = \frac{pn_1}{60} \cdot \frac{n_1 - n_2}{n_1} = f_1 s$$

and the emf induced in one phase of the rotating rotor

$$E_2 = 4{,}44 f_1 s w_2 k_{w2} F_p = E_{2n} s. \tag{5.12}$$

Impedance of rotor winding phase

$$z_2 = \sqrt{r_2^2 + (2\pi f_2 L_2)^2},$$

where r_2 is the resistance of one phase and L_2 is the parasitic inductance. Since $2\pi f_2 = 2\pi f_1 s = \omega s$, then the rotor phase current

$$I_2 = \frac{E_2}{\sqrt{r_2^2 + (s\omega L_2)^2}} \tag{5.13}$$

or, replacing E_2 per the relationship in (5.12),

$$I_2 = \frac{E_{2n} s}{\sqrt{r_2^2 + (s\omega L_2)^2}} = \frac{E_{2n}}{\sqrt{\left(\frac{r_2}{s}\right)^2 + (\omega L_2)^2}}.$$

This tells us that the rotor current depends on the slip and increases as the rotational speed decreases.

The multiphase current of the rotor also produces a rotating magnetic field, the speed of which relative to the rotor is given by

$$n_{Sr} = \frac{60 f_2}{p},$$

while the rotor rotates in the same direction at the speed n_2. The speed of the rotor's magnetic field relative to the stator is given as the sum of n_S and n_2, based on which we can rearrange the equation to get

$$n_{Sr} + n_2 = \frac{60 f_1}{p} + n_2 = n_1 \frac{n_1 - n_2}{n_1} + n_2 = n_1. \tag{5.14}$$

Thus, the magnetic field of the rotor of the induction motor rotates synchronously with the magnetic field of the stator, with the rotor rotating at the speed $n_2 < n_1$. That the magnetic fields of the stator and rotor rotate synchronously at any slip value is inherent to the transfer of power via a rotating magnetic field in an induction motor.

5.4 EQUIVALENT CIRCUIT AND PHASOR DIAGRAM OF AN INDUCTION MOTOR

In order to enable the phasors of the stator and rotor parameters to be depicted in a single diagram based on an equivalent circuit of the induction motor, the rotor winding's parameters and electromotive force must also be referred to the stator winding. As in transformers, the referral must be performed without causing the powers and phase shift angles to change.

The system of equations describing the electrical aspects of an induction motor is as follows:

$$\left.\begin{aligned} \underline{U}_1 &= -\underline{E}_1 + \underline{I}_1 \left(r_1 + \mathrm{j} x_1 \right) \\ 0 &= \underline{E}_2 - \underline{I}_2 \frac{r_2}{s} - \underline{I}_2 \mathrm{j} x_2 \\ \underline{I}_1 &= \underline{I}_2^{'} + \underline{I}_{10} \end{aligned}\right\}.$$

The corresponding equivalent circuit for these equations is shown in Figure 5.6.

To draw the phasor diagram, we need to replace the values of the rotor parameters with the referred values, for which we must split the quantity $r_2^{'}/s$ into two components as follows:

$$\frac{r_2^{'}}{s} = \frac{r_2^{'}}{s} - \frac{r_2^{'} s}{s} + r_2^{'} = r_2^{'} + r_2^{'} \frac{1-s}{s}. \tag{5.15}$$

This gives us the system of equations corresponding to the phasor diagram (Figure 5.7):

$$\left.\begin{aligned} \underline{U}_1 &= -\underline{E}_1 + \underline{I}_1 \left(r_1 + \mathrm{j} x_1 \right) \\ -\underline{E}_2^{'} &= \underline{I}_2^{'} \left(r_2^{'} + \mathrm{j} x_2^{'} \right) + \underline{I}_2^{'} r_2^{'} \frac{1-s}{s} \\ \underline{I}_1 &= \underline{I}_{10} + \underline{I}_2^{'} \end{aligned}\right\}, \tag{5.16}$$

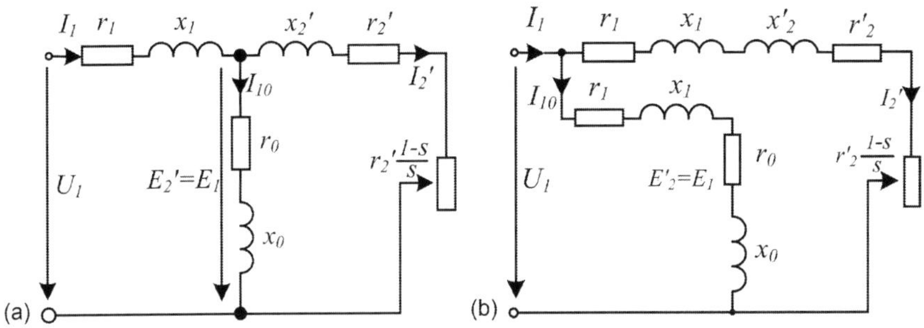

FIGURE 5.6 T-shaped (a) and Γ-shaped (b) equivalent circuit for an induction motor.

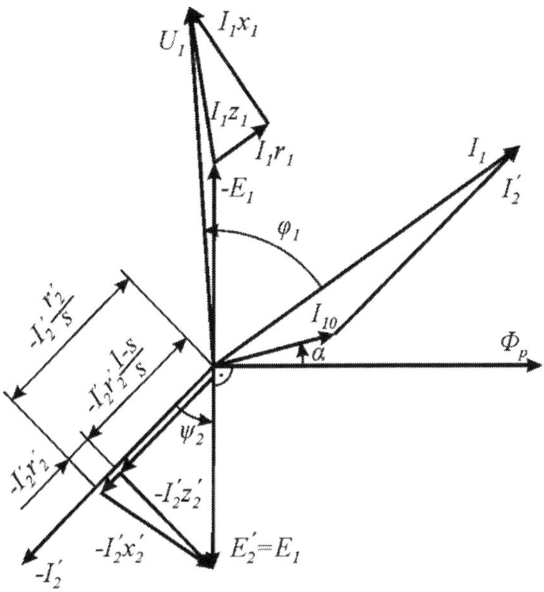

FIGURE 5.7 Phasor diagram of an induction motor.

where the angle of shift between the emf E_2 and rotor current I_2 can be found via the parameters from the relationship

$$y_2 = \arctan\frac{x_2^{'} s}{r_2^{1}}.$$

The angle ϕ_1 arises between the voltage phasor U_1 and the stator current I_1 and can be calculated on the basis of the power factor:

$$\varphi_1 = \arccos\frac{P_1}{S_1},$$

where P_1 is the active power consumed by the induction motor and S_1 is the total apparent power. The phasor diagram is drawn in a similar way as the phasor diagram for a loaded transformer. The only difference is that, here, the rotor winding is short-circuited and therefore there is no voltage U_2. In addition, in Figure 5.7, the resistive voltage drop is split into two components based on equation (5.15), of which $I_2^{'}r_2^{'}$ is associated with the copper losses in the rotor (heat loss in rotor winding) and $I_2^{'}r_2^{'}\dfrac{1-s}{s}$ with the conversion of energy into mechanical energy.

In order to obtain a T-shaped equivalent circuit, we must first replace the stator's counter-electromotive force in the first equation of the system (5.16) with the voltage drop produced by the parameters of an equivalent circuit corresponding to ideal no-load conditions:

$$-\underline{E}_1 = \underline{I}_{10}\left(r_0 + \mathrm{j}x_0\right),$$

where r_0 is the resistance of the equivalent circuit, in which the no-load current I_{10} produces a power loss equal to the iron loss, and the inductive reactance x_0 is associated with the reactive energy of the rotating magnetic flux Φ_r. Then, knowing that $E_2^{'} = E_1$, we must plug the value of $E_2^{'}$ into the first equation. This gives us the system of equations corresponding to the equivalent circuit in Figure 5.6:

$$\left.\begin{aligned} \underline{U}_1 &= \underline{I}_1\left(r_1 + \mathrm{j}x_1\right) + \underline{I}_{10}\left(r_0 + \mathrm{j}x_0\right) \\ \underline{U}_1 &= \underline{I}_1\left(r_1 + \mathrm{j}x_1\right) + \underline{I}_2^{'}\left(r_2^{'} + \mathrm{j}x_2^{'}\right) + \underline{I}_2^{'}r_2^{'}\frac{1-s}{s} \\ \underline{I}_1 &= \underline{I}_{10}^{'} + \underline{I}_2^{'} \end{aligned}\right\}.$$

The variable resistance $r_2^{'}\dfrac{1-s}{s}$ can be seen as a resistive load that is dependent on the rotor's rotational speed (slip). For example, when the rotor is not moving, i.e. $n_2 = 0$ ($s = 1$), then

$$r_2^{'}\frac{1-s}{s} = 0,$$

which corresponds to a short-circuit state in the induction motor. When $s = 0$, however, then

$$r_2^{'}\frac{1-s}{s} = \infty,$$

which correspond to a no-load state in the motor – the rotor circuit is broken in the equivalent circuit, i.e. $I_2^{'} = 0$.

5.5 POWER BALANCE OF AN INDUCTION MOTOR

When an induction motor is loaded, it draws power (expressed through phase quantities) from the supply grid:

$$P_1 = m_1 U_1 I_1 \cos\varphi, \tag{5.17}$$

some of which is dissipated in the stator winding as copper losses

$$P_{\text{Cu1}} = m_1 I_1^2 r_1. \tag{5.18}$$

The remaining power is converted into the power of the rotating magnetic field, some of which is dissipated in the stator as iron losses P_{Fe1}, which are composed of eddy current losses and hysteresis losses.

Through the air gap, the rotating magnetic field supplies electromagnetic power, or air gap power, to the rotor:

$$P_{\text{em}} = P_1 - P_{\text{Cu1}} - P_{\text{Fe1}}, \tag{5.19}$$

which can be expressed through the parameters of the induction motor as

$$P_{\text{em}} = m_2 I_2^2 \frac{r_2}{s} = m_1 I_2'^2 \frac{r_2'}{s} = \frac{U_1^2}{\sqrt{\left(r_1 + \frac{r_2'}{s}\right)^2 + \left(x_1 + x_2'\right)^2}} \cdot \frac{m_1 r_2'}{s}. \tag{5.20}$$

Part of the electromagnetic power is dissipated in the rotor winding as copper losses

$$P_{\text{Cu2}} = m_2 I_2^2 r_2 = m_1 I_2'^2 r_2', \tag{5.21}$$

while the rest is converted into the motor's mechanical power

$$P_{\text{mech}} = P_{\text{em}} - P_{\text{Cu2}}, \tag{5.22}$$

which, in accordance with the rearrangement in (5.15) and based on the above relationships, can be written as

$$P_{\text{mech}} = m_1 I_2'^2 \frac{r_2'}{s} - m_1 I_2'^2 r_2' = m_1 I_2'^2 r_2' \frac{1-s}{s}. \tag{5.23}$$

The iron losses in the rotor are disregarded here due to how small they are, since, under load conditions, the frequency of the rotor current is relatively low, namely $f_2 \approx 0.5$–3 Hz. Meanwhile, the frequency of the stator current is 50 Hz.

If we also take into account the mechanical losses $P_{\text{mech loss}}$, which consist primarily of the losses in the machine's bearings and ventilation losses, as well as various additional losses P_{add} which are also subtracted from the mechanical power, we get the useful mechanical power on the shaft of the induction motor:

$$P_2 = P_{\text{mech}} - P_{\text{mech loss}} - P_{\text{add}}. \tag{5.24}$$

Additional losses represent approximately 1% of the consumed power P_1.

The relationships (5.17)–(5.24) are well illustrated by the power diagram shown in Figure 5.8, which corresponds to the system of equations

$$\left.\begin{aligned} P_{\text{em}} &= P_1 - P_{\text{Cu1}} - P_{\text{Fe1}} \\ \text{P}_{\text{em}} &= P_2 + P_{\text{mech loss}} + P_{\text{Cu2}} + P_{\text{add}} \end{aligned}\right\}.$$

The efficiency of the induction motor is

$$\eta = \frac{P_2}{P_1}$$

or expressed through total losses,

$$\sum P_{\text{loss}} = P_{\text{Cu1}} + P_{\text{Fe1}} + P_{\text{Cu2}} + P_{\text{mech loss}} + P_{\text{add}}$$

knowing that $\sum P_{\text{loss}} = P_1 - P_2$,

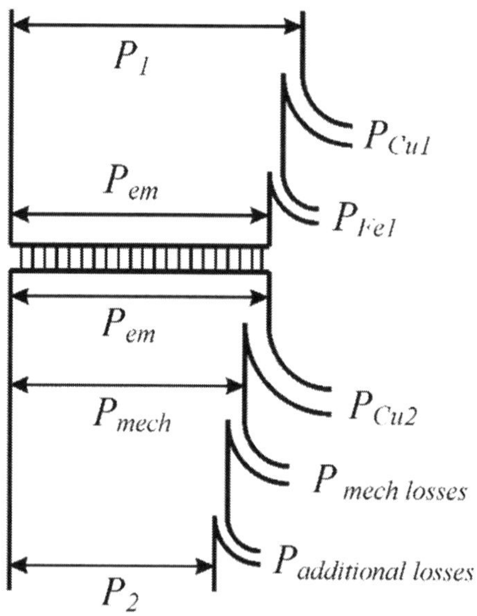

FIGURE 5.8 Power diagram of an induction motor.

$$\eta = \frac{P_2}{P_2 + \sum P_{loss}}.$$

The efficiency of induction motors at rated load ranges from 70% to 95% where the upper end of this range applies to more powerful motors. The rated (aka 'nominal') power P_{2n} of an induction motor, which will also be marked on the data plate of the machine, is the useful power on the shaft at the rated rotational speed n_{2n}. If the rated voltage U_{1n} is applied to the stator winding, the rated current I_{1n} is also generated in the stator winding. In addition to these quantities, the data plate will also show the rated power factor $\cos\phi_{n}$, rated efficiency η_{n}, and the type of connection of the stator winding.

5.6 TORQUE OF AN INDUCTION MOTOR

The interaction between the rotor current and the rotating magnetic field produces an electromagnetic torque which is directly proportional to the electromagnetic power and inversely proportional to the angular velocity of the rotating field:

$$T = \frac{P_{em}}{\omega_1}, \tag{5.25}$$

If the power is measured in watts and the angular velocity in radians per second, the unit for the torque will be N m.

Based on the parameters of the equivalent circuit, the electromagnetic torque is given as

$$T = \frac{m_1 p U_1^2 \frac{R_2'}{s}}{2\pi f_1 \left[\left(r_1 + \frac{R_2'}{s} \right)^2 + \left(x_1 + x_2' \right)^2 \right]} = f(s). \tag{5.26}$$

Here, we can see that the motor's electromagnetic torque is proportional to the voltage squared. The relationship between the torque and the slip (5.26) or the rotor's rotational speed n_2 is called the mechanical characteristic of the induction motor. In Figure 5.9, this relationship is shown as $s = f(T)$ and $n_2 = f(T)$, where the characteristic consists of a number of characteristic points.

At the moment when the motor is started, the rotor is not rotating, $n_2 = 0$ and, based on equation (5.3), $s = 1$. The developed starting torque T_{st} can be found using equation (5.26):

$$T_k = \frac{m_1 p U_1^2 R_2'}{2\pi f_1 \left[\left(r_1 + R_2' \right)^2 + \left(x_1 + x_2' \right)^2 \right]}.$$

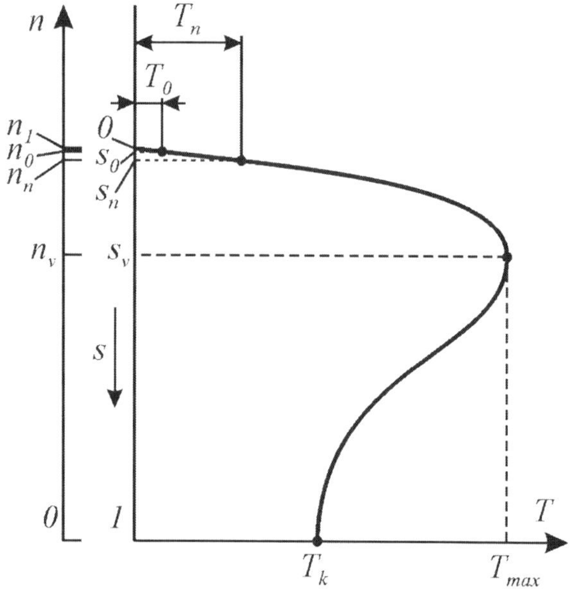

FIGURE 5.9 Mechanical characteristics of an induction motor.

This torque causes the rotor to start rotating, resulting in the rotational speed increasing and the slip decreasing. When the slip reaches a certain value s_{br} known as the **breakdown slip**, the torque is at its maximum value T_{max}. The rotational speed at the breakdown slip is n_{br}. The breakdown slip can be found from the first derivative of the function (5.26) with respect to slip when set equal to zero. It turns out that the breakdown slip depends only on the parameters of the motor:

$$s_v = \frac{R_2'}{\sqrt{r_1^2 + \left(x_1 + x_2'\right)^2}}. \tag{5.27}$$

Plugging this into equation (5.26) gives us the maximum electromagnetic torque

$$T_{max} = \frac{m_1 p U_1^2}{4\pi f_1 \left[r_1 + \sqrt{r_1^2 + \left(x_1 + x_2'\right)^2}\right]}. \tag{5.28}$$

The value of the maximum torque, or breakdown torque, does not depend on the resistance of the rotor circuit.

If the rotational speed continues to increase and, hence, the slip continues to decrease, the torque will decrease. Proceeding thus along the characteristic curve

(Figure 5.9), we eventually reach the rated (aka 'nominal') torque point T_n, which corresponds to the rated power P_n, where the rotor is at the rated rotational speed n_n and the rated slip s_n. When the load on the motor drops to zero, the rotational speed rises to the no-load rotational speed n_0, which corresponds to the slip value s_0. At this point, the motor will only develop a no-load torque T_0, which is associated with no-load losses.

In order to bring the torque developed by the motor to zero (at which point the slip, too, will equal zero and the rotor's rotational speed n_2 will be equal to the rotational speed n_1 of the magnetic field of the stator), the no-load torque would need to be overcome by means of an auxiliary motor, using its energy to cover the no-load losses.

The region of normal operation on the mechanical characteristic curve is at $T_0 < T \leq T_n$. At $T > T_n$, the motor is overloaded. When the resistive torque increases until it exceeds the maximum torque, the rotational speed of the rotor quickly drops to zero. The ratio between the maximum torque and the rated torque is called the motor's **overload potential**

$$l_T = \frac{T_{\text{max}}}{T_{\text{n}}}, \tag{5.29}$$

which ranges from 1.7 to 3.0 in the most commonly used motors at the rated voltage.

5.7 START-UP OF AN INDUCTION MOTOR

The quantities that are relevant to the start-up process of electric motors include the starting current, starting torque, starting time, and starting efficiency.

In squirrel-cage induction motors, the starting current $I_{1\text{start}}$ at the moment of starting ($s = 1$) is 4.0–7.0 times higher than the rated current $I_{1\text{n}}$. The exact ratio depends on the rated power and rated speed of the motor. This ratio $\lambda_I = I_{1\text{start}}/I_{1\text{n}}$ is called **the starting current multiplier**.

The ratio of the starting torque T_{st} ($s = 1$) to the rated torque T_{n} is called the **starting torque multiplier** $\lambda_{\text{stT}} = T_{\text{st}}/T_{\text{n}}$, which, under established standards, must not be less than 0.7–1.8 in squirrel-cage induction motors.

The start-up method must be chosen in accordance with the following requirements:

1. At start-up, the motor must develop a torque that exceeds the resistive torque of the appliance to such a degree that the rotor can start to rotate and its rotational speed can increase to the rated speed.
2. The starting current must be limited to a degree where it does not poses a hazard to the motor nor interfere with the normal operation of the supply grid.
3. The start-up circuit must be as simple as possible, ensuring maximum reliability and minimum cost.

Squirrel-cage induction motors may be started in a number of ways. These include direct connection to the grid, starting at a reduced voltage, and starting via a frequency converter. Voltage reduction can be achieved by connecting an inductive reactance, an autotransformer, or a thyristor-containing circuit known as a soft starter. Where a frequency converter is used for starting, the frequency of the voltage applied to the motor during start-up will be lower than the grid frequency and the voltage itself will be lower than the grid voltage. After start-up, the same frequency user can be used to adjust the rotational speed. Where voltage reduction circuits are used, these are disconnected from the electrical circuit after start-up, following which the rotating motor is connected directly to the grid.

For the duration of the start-up process of a slip ring-rotor induction motor, resistors are usually inserted into the motor's rotor circuit to reduce the starting current as well as increase the starting torque.

If the starting torque of the motor is sufficiently high and the starting current neither poses a hazard to the motor nor interferes with the operation of the grid, then the motor is **started by connecting it directly to the grid**. In such a case, the high current that arises after starting subsides relatively quickly. The motor enters a mode of operation determined by the load torque of the appliance where the motor current is equal to or less than the rated current. Due to its simplicity, start-up by direct connection to the grid is the most common method. With this method, the switching circuit consists only of a three-pole switch and electrical safety devices (Figure 5.10a).

The starting time depends primarily on the masses that are being made to rotate. Since the starting current I_{1start} is 4–7 times the rated current, and the copper loss increases in proportion to the square of the current, the maximum starting time is limited by the temperature tolerance of the winding insulation. Direct connection to the grid is probably not a viable method for appliances where the movable masses are very large. On the other hand, the inrush current that occurs during start-up leads to a voltage drop in the supply grid. This short-term change in voltage interferes with the operation of other loads, especially lighting fixtures that are supplied with power from the same grid. Acceptable voltage fluctuation levels are 10–15% of the rated voltage – another limiting condition for starting squirrel-cage induction motors by direct connection to the grid.

In industrial enterprises, the capacity of the substations and distribution grids is usually high enough to almost always enable squirrel-cage induction motors to be started by connecting them directly to the grid. With particularly powerful distribution devices, it is even possible to use this method for starting induction motors whose rated power is 1,000–2,000 kilowatts.

Sometimes, however, the substation will not be sufficiently powerful, in which case starting a motor in this way may interfere with the operation of other loads. One way to overcome this limitation **is to use a squirrel-cage induction motor with improved starting performance**. In such motors, the starting current multiplier is lower (λ_I = 3.0–5.0) and the starting torque multiplier is higher (λ_{stT} = 1.0–3.0). A greater starting torque and the consequent decrease in starting current can be achieved by increasing the resistance of the rotor's squirrel-cage winding.

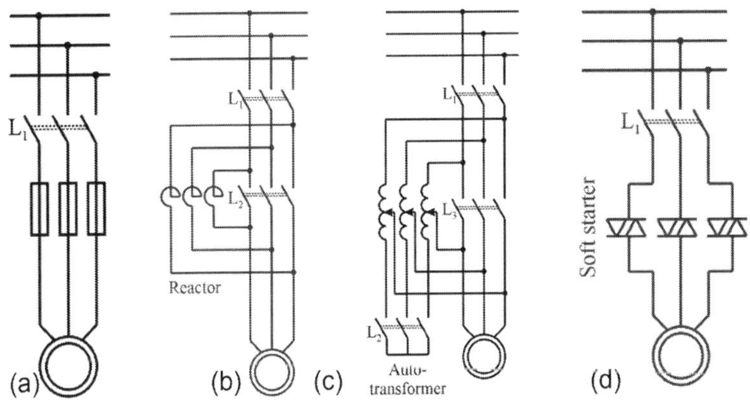

FIGURE 5.10 Possible start-up circuits for a squirrel-cage induction motor, (a) direct online starter, (b) reactor starer, (c) autotransformer starter, (d) soft startter.

This also results in a greater rated slip for the motor, yielding a 'gentler' mechanical characteristic. The winding resistance can be increased by utilising bars that have a higher electrical resistivity or cutting deep slots or double slots in the rotor.

Figure 5.11 shows a comparison of the mechanical characteristics of a conventional squirrel-cage induction motor (1), an induction motor with a rotor winding with increased electrical resistivity (2), an induction motor with a rotor with deep slots (3), and an induction motor with a rotor with double slots (4). The starting of induction motors with improved starting performance is usually done by means of direct connection to the grid.

The downside of using rotors with increased electrical resistivity, however, is lower efficiency. Increasing the resistance of the rotor winding always reduces the efficiency of the machine. When using deep slots or double slots, on the other hand, the special design of the slots provides a higher resistance for the rotor winding only during start-up, when the slip is high. During operation, the resistance of the rotor winding is lower, resulting in greater efficiency. The resistance of the rotor winding is reduced because during operation the frequency of rotor current is lower and the lower-frequency current also enters the deeper parts of the slots and the thicker bars deeper down.

Another way to start induction motors in a weak supply grid is **to start them at a reduced voltage**. At a lower voltage, the current drawn from the grid will also be lower, as will the voltage drop across the grid. However, since the motor's torque will be proportional to the square of the voltage across the stator winding, using a reduced voltage will also result in a reduced starting torque. This is one of the downsides of this starting method. Starting at a reduced voltage is only possible when the motor is in no-load mode or with a reduced load torque.

Figures 5.10b, c, and d show different voltage reduction circuits. In Figure 5.10b, starting is achieved by means of coils known as reactors. The resistance of the reactors reduces the voltage and the current. In Figure 5.10c, starting is achieved by means of an autotransformer, which reduces the voltage applied to

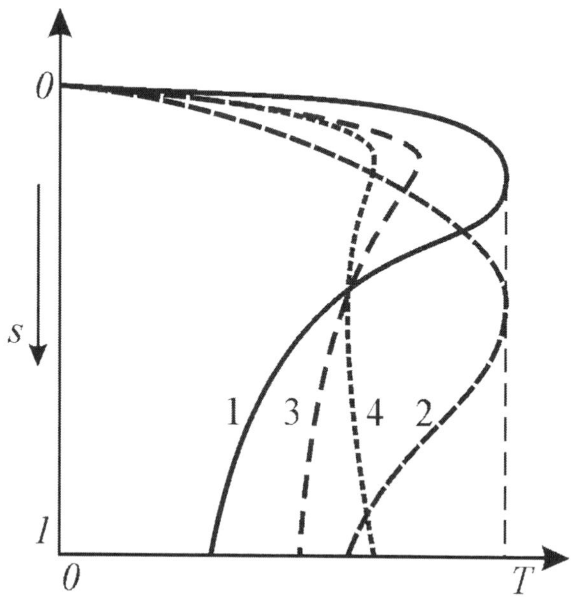

FIGURE 5.11 Mechanical characteristics of squirrel-cage induction motors with improved starting performance.

the motor at start-up. At the same time, the autotransformer also reduces the current drawn from the grid. Figure 5.10d depicts a starting circuit with a thyristor regulator, which is inserted in front of the motor. Such regulators are also called **soft starters**. The thyristors are used to reduce the duration of the half-periods of the grid voltage applied to the motor, thereby reducing the average and effective value of the voltage.

Slip ring-rotor induction motors were used before the introduction of thyristors and power transistors, and, to a lesser extent, still are in cases where:

- the resistive torque of the appliance at start-up exceeds the starting torque of the motor, preventing the motor from starting. Once the motor is rotating, the resistive torque decreases, falling below the rated torque of the motor;
- the starting current needs to be limited in the supply grid, and using a squirrel-cage induction motor that is started at a reduced voltage is not practicable due to the reduced starting torque;
- the movable masses in the appliance that need to be rotated by the motor are so large and the starting time is, as a result, so long that the increase in temperature due to copper losses could compromise the insulation of the windings.

Today, in most such cases, squirrel-cage motors combined with a frequency converter are preferred to slip ring-rotor motors. However, slip ring-rotor motors are still used in older equipment.

Slip ring-rotor induction motors are started using a starting rheostat as shown in Figure 5.12a. The starting rheostat SR is a three-phase resistor connected to the rotor circuit and is only designed for short-term operation. For the moment of starting ($n_2 = 0$), the resistance r_l of the starting rheostat may be set at a value where the referred resistance $R_2' = r_2' + r_{l1}'$ of one phase of the rotor circuit brings the breakdown slip, in accordance with equation (5.27), to a value of $s_{br} = 1$ (Figure 5.12b), which corresponds to the torque $T_{st} = T_{max}$. At the same time, the current is reduced in the rotor circuit (Figure 5.12c) as well as the supply grid. As the rotational speed of the rotor increases, the resistance of the starting rheostat is gradually reduced – there occurs a transition from the rheostat characteristic $l1$ to a new rheostat characteristic l and from there to $r_l = 0$, at which point the rotor circuit is short-circuited. Then $R_2' = r_2'$, which means that we have transitioned to the natural mechanical characteristic n. By the end of the starting time t_{st}, the current will also have reached the rated value I_n (Figure 5.12c).

The number of steps in the starting rheostat and the resistance at each step are selected based on the torque or current conditions. For example, if the limiting condition is the current in the supply grid, the maximum permissible current I_{max} (Figure 5.12c) and, according to the optimal starting time, $I_{min} > I_n$ are ensured, based on which the required resistances r_l are calculated.

5.8 CHANGING THE ROTATIONAL SPEED OF INDUCTION MOTORS

The rotational speed of the rotor of an induction motor can be changed based on the equation

$$n = \frac{60 f_1}{p}\left(1 - s\right).$$

This equation indicates that the rotational speed changes when the supply current frequency f_1, number of pole pairs p, or slip s changes.

Slip can be changed by connecting an additional resistance to the rotor circuit. This is only possible for slip ring-rotor induction motors. Moreover, the rotational speed will decrease as the additional resistance is increased only when the motor is loaded. At no load, the additional resistance will not affect the rotational speed. When an additional resistance is utilised, the mechanical characteristics will look similar to the starting rheostat characteristics shown in Figure 5.12b. However, when changing the speed, the additional resistances must be designed for continuous operation – a starting rheostat cannot be used for this purpose. Increasing the slip increases the rotor current, which in turn increases the copper losses in the rotor. In addition, there will be further energy losses in the additional resistors,

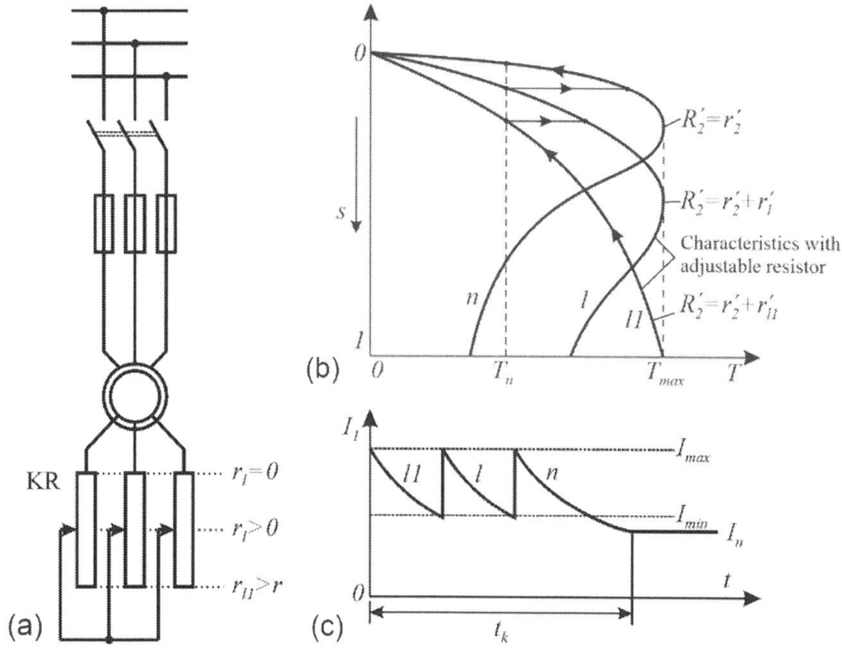

FIGURE 5.12 Start-up circuit (a), starting characteristics (b), and changes in the stator current at start-up (c) for a slip ring-rotor induction motor.

as a result of which the efficiency of the motor will decrease. In terms of energy consumption, changing the rotational speed in this way is uneconomical. Due the above factors, the speed of slip ring-rotor induction motors used to be adjusted only downwards from the rated speed, within the range of n_n–0.7 n_n. Today, the method of adjusting the speed by changing the slip has mostly fallen out of use.

In squirrel-cage induction motors, the rotational speed can also be incrementally by changing the number of pole pairs p. The number of pole pairs can be changed by switching taps in the stator winding. Two-speed motors are the most common type of multi-speed motor; in these, the switchover to a different synchronous speed may occur at a constant power or at a constant torque.

Switching coil groups over from star connection to star connection according to the scheme shown in Figure 5.13a enables the number of pole pairs to be increased two-fold, halving the rotational speed. Since the current does not change in this switching process, neither does the power, and, consequently, the torque increases twofold. The mechanical characteristics corresponding to the different numbers of pole pairs are shown in Figure 5.13b.

When coil groups are switched from a star connection to a double star connection (Figure 5.14a), the number of pole pairs decreases twofold, causing the rotational speed to increase twofold. The supply current as well as the power will

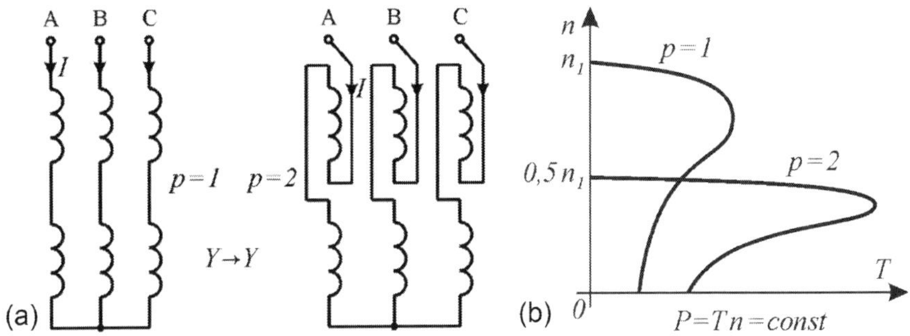

FIGURE 5.13 Changing the speed of an induction motor at a constant power, (a) star (Y) connection with one pole pair and (b) star (Y) connection with two pole pairs.

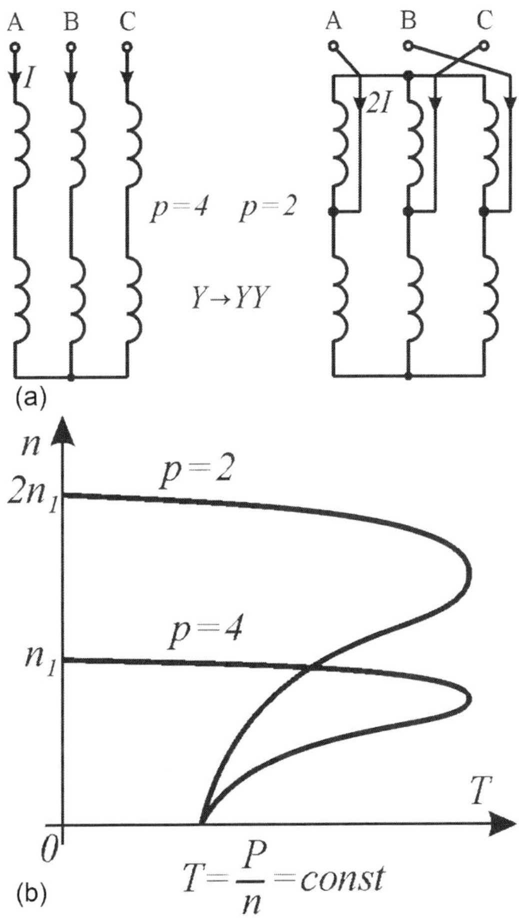

FIGURE 5.14 Changing the speed of an induction motor at a constant torque. (a) star (Y) connection with one four pairs and (b) double star (YY) connection with two pole pairs.

then also increase twofold. The torque, therefore, does not change. The mechanical characteristics corresponding to the different numbers of pole pairs are shown in Figure 5.14b.

Adjusting the rotational speed by changing the frequency of the stator voltage
Changing the stator voltage frequency f_1 enables the rotational speed of a squirrel-cage induction motor to be changed seamlessly within a relatively wide range. Seamless frequency changing can be achieved by means of a **frequency converter**. Today, frequency converters are the most common means of controlling the rotational speed of an induction motor.

Normally, the frequency converter is supplied with power from either a single-phase or three-phase AC grid. The first stage in the frequency converter is a **rectifier**, where the 50 Hz or 60 Hz AC voltage is rectified (Figure 5.15).

The second stage in the frequency converter is an **inverter**, where the DC voltage is converted to a variable-frequency three-phase AC voltage. The inverter output is usually connected to a three-phase induction motor with a rated frequency equal to the grid frequency and a rated voltage equal to the rated voltage of the grid. The shaft of the induction motor is coupled to an appliance, whose operation is controlled by the frequency converter. The motor can generally operate both in motor mode and in generator mode.

The connection between the rectifier and the inverter is called the **DC link**. The DC link usually also includes a high-capacity capacitor (capacitor C in Figure 5.15) for smoothing out the pulsating rectified voltage. This occurs in motor mode. If, however, the motor is operated in generator mode (e.g. during the slowing of the appliance), energy flows, via the motor, from the appliance to the inverter, which will then operate as a rectifier. From the inverter operating in rectifier mode, current is then transmitted to the DC link, where it begins to increase the voltage across the smoothing capacitor. In order to remove the energy from the smoothing capacitor, the first-stage rectifier must then switch over to inverter mode and transmit the energy received from the appliance to the grid. This, however, means that the first stage of the frequency converter needs to be a two-way AC/DC converter. Such converters are expensive and are used relatively rarely.

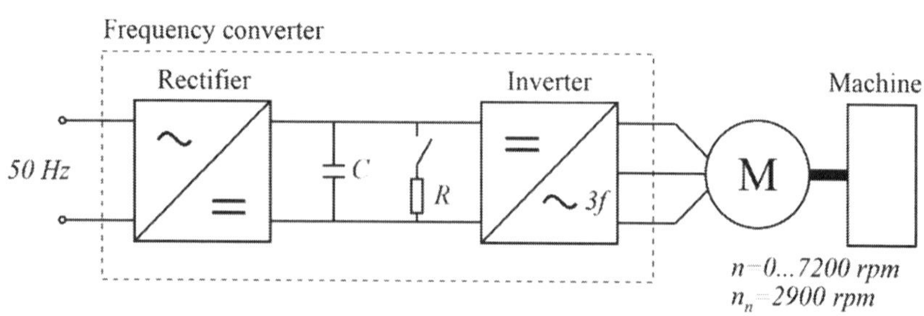

FIGURE 5.15 Simplified circuit of a frequency converter.

Instead, the first stage will often still consist of an inexpensive diode rectifier, which cannot return energy to the grid.

In order to prevent the voltage across the capacitor in the DC link from getting too high when using a diode rectifier, a resistor is employed (resistor *R* in Figure 5.15), which is connected in parallel (usually by means of a relay contact) to the capacitor when the voltage reaches an excessive level. In this resistor, the energy generated during the slowing of the appliance is converted into heat. This solution is practical in cases where the appliance is not returning all that much energy.

The frequency converter can step the frequency both down (from 50 Hz to only a few hertz) and up from the grid frequency. The upper limit is usually 2.5 times the rated frequency, i.e. 125 Hz. At this upper frequency limit, motors will be running at about 2.5 times the rated speed. Such speeds are within the tolerable range for motors.

At the rated frequency, the rated voltage is applied to the motor and a rotating magnetic flux Φ_r will be produced in the air gap of the machine, the magnitude of which can be found from the electromotive force equation for induction machines

$$F_p = \frac{U_1}{f_1} \cdot 4,44 w_1 k_{w1}, \tag{5.30}$$

where U_1 – stator voltage;

f_1 – frequency of stator voltage;

w_1 – number of turns in the stator winding phase;

k_{w1} – winding factor of the stator winding.

The increasing of the magnetic flux Φ_p during speed adjustment is undesirable, as this would cause the magnetic circuit to become saturated and the no-load current to increase significantly. In order to prevent the magnetic flux Φ_r from increasing when the frequency f_1 decreases, the ratio U_1/f_1 needs to be kept constant. This can be done by decreasing the voltage by the same amount as the frequency.

For the above reason, below the rated frequency, the magnetic flux is kept constant, as a result of which the maximum torque of the motor will also remain constant. This can be seen in the mechanical characteristic curves for the different frequencies shown in Figure 5.16.

In the figure, the torque and speed values are shown as per-unit quantities with respect to the rated torque and rated synchronous speed. A positive torque corresponds to motor mode, while a negative one corresponds to generator mode. If, during adjustment, the torque is kept constant and the speed is reduced, the power decreases with the speed.

When the speed is raised above the rated speed, the frequency is raised above the rated frequency (above 50 Hz). Equation (5.30) shows that an increase in the frequency f_1 is accompanied by a decrease in the magnetic flux Φ_p, and thus there is no risk of saturation of the magnetic circuit. Due to this, the voltage is usually not increased when the frequency is raised above the rated frequency.

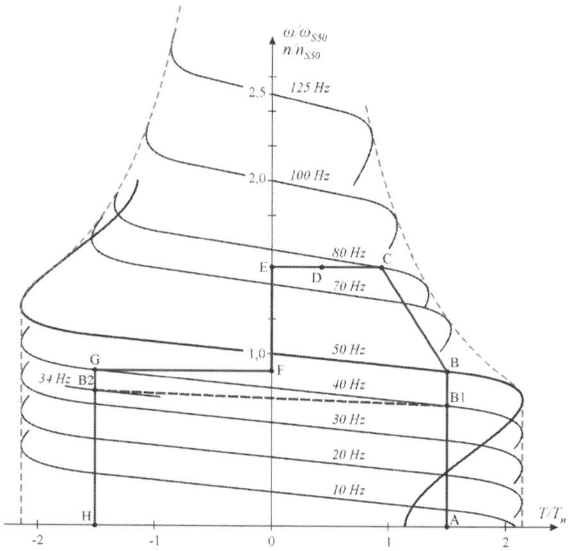

FIGURE 5.16 Mechanical characteristics of an induction machine at different adjusted frequencies.

However, the decrease in Φ_r will result in a decrease in the maximum torque of the machine. This is usually accepted, and, at higher speeds, the mechanical characteristics will look like those shown in Figure 5.16. At speeds above the rated speed, the power will not decrease, because the speed will increase in proportion to the decrease in torque, which means that their product remains constant.

Commonly used types of frequency converters include those of a type where, when the frequency is raised above the rated frequency, the voltage is raised above the rated voltage, so that the torque does not decrease. In such a case, the power, too, will increase when the frequency increases, and at speeds greater than the rated speed, the power of the motor can be kept continuously above the rated power. However, raising the voltage above the rated voltage is challenging in regards to the design of the frequency converter, and will make the converter expensive.

If a DC power source (such as an electric car battery) is used, then no rectifier is required in the first stage of the frequency converter (Figure 5.17).

Replacing the rectifier with a battery does not fundamentally change the frequency adjustment process for the induction motor. Let us explore the operation of a frequency converter in different modes in a running electric vehicle. If the electric vehicle needs to get from point A to point B and there are no unexpected obstacles on the road, the operation of the powertrain of the vehicle may be divided into five stages, which are shown in Figure 5.18.

The speed of the vehicle (which is directly proportional to the rotational speed of the motor) and the torque and power of the motor are shown as time-dependent quantities. The figure shows five stages: start-up and acceleration at a constant

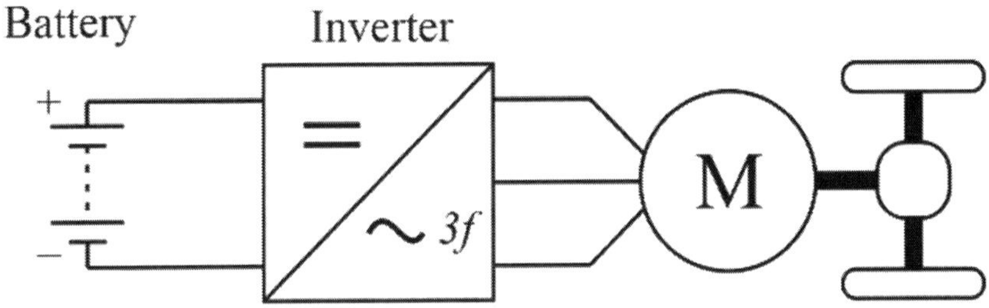

FIGURE 5.17 Powertrain of an electric car with an induction motor.

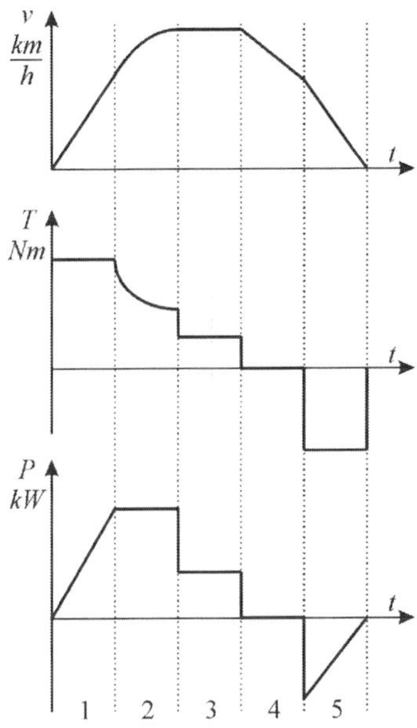

FIGURE 5.18 Typical change in electric vehicle speed, motor torque, and power during a journey from point A to point B on a uniform horizontal road.

torque, acceleration at a constant power, driving at a steady speed, travelling with the engine off, and braking until the speed drops to zero. Let us consider these stages together with the induction motor mechanical characteristics shown in Figure 5.16.

Stage 1. Travelling starts from standstill. A constant torque is applied to the traction motor via a converter. The constant torque produces a constant tractive force. The tractive force causes the vehicle to accelerate. The acceleration is directly proportional to the applied force and inversely proportional to the mass of the vehicle. During acceleration, the speed of the vehicle and the rotational speed of the motor in the powertrain increase. As the rotational speed of the induction motor increases, so do the frequency and voltage of the current supplied to the motor by the converter. During the first stage, the frequency rises to the rated frequency. In Figure 5.16, the first stage occurs along the line segment AB, where the torque is 1.5 times the rated torque and the frequency rises to 50 Hz.

Stage 2. During the second stage, the frequency of the frequency converter rises above the rated frequency and, simultaneously, the torque starts decreasing, because the voltage is no longer boosted. The power of the motor remains constant. In Figure 5.16, this stage occurs along the line segment BC. The frequency rises to 80 Hz.

Stage 3. During the third stage, the speed of the vehicle no longer changes. This is possible if the torque of the motor balances out the counter-torque produced by the motion resistance. On a uniform horizontal road, the motion resistance torque is constant. The torque developed by the motor drops exponentially during the transition from the second stage to the third stage (transition from point C to point D in Figure 5.16), because the vehicle stops accelerating. The third stage occurs in point D in Figure 5.16, where torque, speed, and power are all constant.

Stage 4. The powertrain is switched off. The torque of the motor drops to zero (line segment DE in Figure 5.16). The vehicle continues travelling under the effect of inertia, and the speed drops (line segment EF).

The frequency is selected by the program in the control circuit of the frequency converter according to the rotational speed of the motor, the converter frequency, and the position of the pedals. Most frequency converters are equipped with a programmable control circuit and can be used in motors with different speed and power ratings (within certain limits) utilised for performing different tasks.

Stage 5 consists in braking. Braking is used to stop at the intended destination. The direction of the torque during braking is the opposite of that during starting. In Figure 5.16, the braking torque arises along the line segment FG, where, at point G, the torque is negative. The necessary braking torque occurs when the frequency converter shifts the frequency to 40 Hz. During braking, the traction motor is in generator mode. The energy from the generator passes through the frequency converter and enters the battery (Figure 5.17).

During braking, the torque is kept constant and the frequency of the frequency converter is lowered. In Figure 5.16, this occurs along the line segment GH.

When driving in real life, unexpected obstacles may be encountered on the road, in which case the vehicle will need to brake. For example, when picking up speed (stage 1), point B1 may have been reached on the mechanical characteristic curve, when, suddenly, it may be necessary to switch over from the acceleration torque to a braking torque. To do this, the frequency may be reduced from 40 Hz to 34 Hz, putting as at point B2.

Of course, drivers will never actually adjust the frequency directly. Instead, they operate the vehicle by means of gas and brake pedals. The frequency is selected by the program in the control circuit of the frequency converter according to the rotational speed of the motor, the converter frequency, and the position of the pedals. Most frequency converters are equipped with a programmable control circuit and can be used in motors with different speed and power ratings (within certain limits) that are employed in different applications.

6 Synchronous Machines

In the first chapter, we learned about the design and operating principle of synchronous motors, the rotating magnetic fields produced by stators, and the magnetic poles on rotors. We also found out that rotors rotate synchronously with the field of the stator. Now let us take a broader look at synchronous machines. Among other things, this chapter will touch on the operation of synchronous machines as a generator and as a motor, the adjustment of the power factor of synchronous machines, and the torque of synchronous machines.

6.1 PURPOSE OF SYNCHRONOUS MACHINES

In thermal and hydroelectric power plants, electrical energy is mainly produced using synchronous generators. Energy systems consist of a large number of power plants, which is why all of the generators in such a system must have a voltage with a constant frequency (50 Hz). The frequency of every generator depends on the generator's rotational speed n (rpm) and the number of pole pairs p:

$$f = \frac{pn}{60}. \tag{6.1}$$

In synchronous generators, the rotation of the rotor is driven by a **prime mover**, which is usually a steam or water turbine. According to equation (6.1), to obtain a standard frequency of 50 Hz, the condition $pn = 3{,}000$ must be met. Thus, the rotational speed of the rotor of the generator must be 3,000 rpm in the case of one pair of poles, 1,500 rpm in the case of two pairs of poles, etc. The voltage supplied must be in phase across all generators plugged into a common energy system. They must rotate synchronously.

Different turbines have different optimal rotational speeds, which is why different numbers of pole pairs and, therefore, different designs are sometimes used for synchronous generators. Based on the power source, synchronous generators are divided into **turbo generators** and **hydroelectric generators**. In both types of generators, the stator winding, or armature winding, resembles the stator winding of an induction motor – it is a three-phase distributed winding. The rotor, or inductor, of a synchronous machine is an electromagnet whose excitation winding is supplied with direct current, usually from a DC generator, also called an **exciter**, mounted on the same shaft as the generator. The exciter's power equals 1–2% of the rated power of the synchronous generator.

Nowadays, the direct current needed for the excitation of a synchronous generator is produced by means of a rectifier positioned on the rotor of the machine, and no brushes, which have a low reliability, are needed. Figure 6.1 shows a brushless excitation circuit. The rectifier receives power from a small three-phase generator

 DOI: 10.1201/9781003404088-6

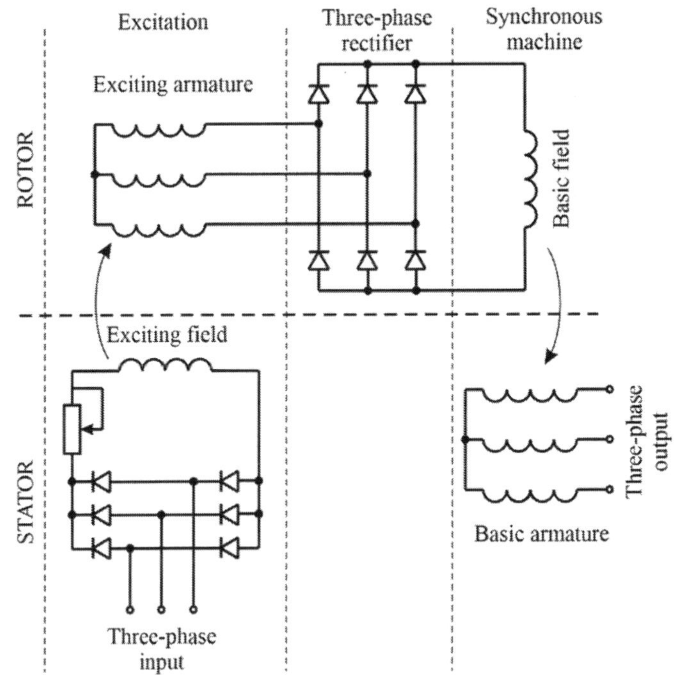

FIGURE 6.1 Brushless excitation circuit where the exciter, rectifier, and generator are mounted on the same shaft.

(excitation generator), which is a synchronous machine of inverse design. The three-phase winding of the excitation generator is mounted on the same shaft as the main generator and is similar to the rotor winding of an induction machine with a slip-ring rotor. The terminals of this winding are connected to the rectifier, which is also mounted on this shaft. The rectified current from the excitation generator is transmitted to the windings at the poles of the main generator. The poles of the excitation generator are stationary and are supplied with power from a small stationary rectifier which receives voltage from an external power supply.

The intensity of the excitation current from the excitation generator can be adjusted by means of a rheostat. This adjustment changes the voltage of the excitation generator and, with it, the excitation current and electromotive force of the large generator.

Water turbines operate at 50–750 rpm, which requires that the synchronous generator have, respectively, 60–4 pole pairs. Where the number of poles is relatively high, it is expedient to use a salient pole design for the rotor, as depicted in Figure 1.12a, where the number of pole pairs is $p = 2$. Synchronous generators that have salient poles are also called hydroelectric generators.

In steam turbines, the rotational speed is usually 3,000 rpm, which means that the synchronous generator must have a single pair of poles. For the average rotor diameter, at such rotational speeds, the linear velocity of the rotor surface can be

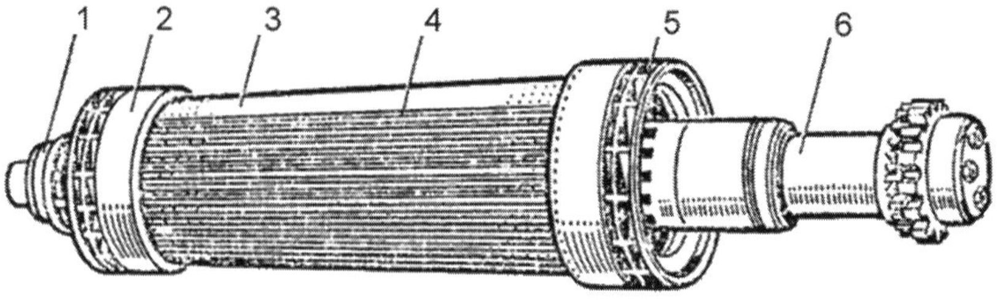

FIGURE 6.2 Rotor with non-salient poles: 1 – slip rings; 2 – retaining ring; 3 – massive magnetic circuit of rotor; 4 – non-magnetic wedge covering a rotor slot; 5 – centrifugal fan; 6 – shaft end, which is connected to a turbine.

up to 170 m/s. To ensure mechanical strength, salient poles cannot be used in this case.

Instead, the rotor here is a steel cylinder (Figure 6.2) with non-salient poles (a cross-sectional view of the rotor is presented in Figure 1.12b). The excitation winding is positioned in the slots on the rotor. Synchronous generators that have non-salient poles are also called turbo generators.

Generators where the prime mover is a diesel engine with a speed of 500–1,500 rpm are called diesel generators, which are built with salient poles (p = 6–2).

The operation of synchronous generators is based on the phenomenon of electromagnetic induction. The magnetic flux, which is produced in the rotor poles by the excitation current and is coupled to the three-phase stator winding, induces a three-phase electromotive force in the stator winding during the rotation of the rotor. In order to induce a sinusoidal voltage, the magnetic flux density B_r in the rotor must vary along the circumference of the rotor according to the law of sines (Figure 6.3).

In this case, the flux density is approximately proportional to the magnetisation excitation. In order to obtain a sinusoidal distribution for the flux density, the magnetisation excitation F (which is equal to the product of the number of turns and the current) must be distributed along the circumference according to the law of cosines. This can be achieved, for example, by changing the number of turns in the coils set in the slots (from three turns to ten turns in Figure 6.3). Another option would be to change the slot pitch, as shown in Figure 1.12b. In a rotor with salient poles, the sinusoidal distribution can be achieved by making the air gap smaller at the centre of the constructional pole end (which is called the **pole shoe**) and larger at the edges (Figure 1.12a). In a larger air gap, the magnetic resistance is greater, and the air gap is penetrated by fewer lines of force. In reality, nevertheless, the flux density distribution is not perfectly sinusoidal, which produces a small distortion in the generator's voltage curve.

A magnetic field with a sinusoidal distribution intersects with the conductors in the stator winding when rotating. This induces a **sinusoidal electromotive force** in all winding wires set in the slots in the stator. The electromotive forces in winding wires positioned in different slots are not in phase. Figure 6.3 shows the

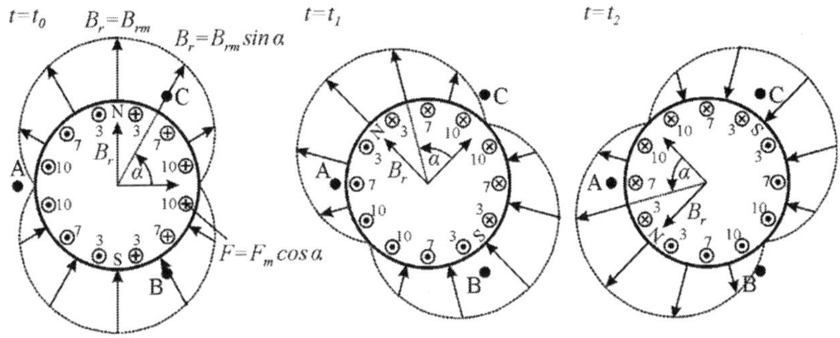

FIGURE 6.3 The required flux density distribution along the surface of the rotor is sinusoidal, for which the magnetisation excitation must be distributed according to the law of cosines. The sine wave of the magnetic field rotating along with the rotor is described by the flux density space vector B_r, the direction of which coincides with that of the amplitude of the sine wave of the magnetic field leaving the north pole.

locations of windings in three different slots (points A, B, and C). These points are offset by 1/3 of a revolution, i.e. 120°. Figure 6.4 shows both the change in flux density over time and the electromotive force e_A, e_B, and e_C induced in each of these points. Note that:

- in the case of a rotating sinusoidal field, the flux density and electromotive force are in phase;
- with two poles, the phase shift in time **measured in electrical degrees** between the electromotive forces of conductors in different slots matches the displacement of the respective slots in **degrees of rotation**.

In Figure 6.4, the flux density and electromotive force are in phase because the electromotive force induced by the moving field is given by the equation $E = Blv$ (see Section 1.10.1). Let it be reiterated that here we have an unchanging field moving in space. This should not be confused with situations where the magnetic flux changes over time. The rotor's magnetic flux does not change over time. A time-varying magnetic flux and the voltage induced by it are shifted in phase by 90° (Faraday's law $e = d\Phi/dt$), but this is not the case here. If we observe a point on the machine's stator (e.g. point A in Figure 6.3), then the magnetic flux density at that point will be different at different points in time, but Faraday's law cannot be applied, because the magnetic flux density is not dependent on time, but on the position of the rotor relative to that point. Magnetic flux can change over time only if magnetic field energy is introduced into or drawn from the space. However, the excitation current is not changed and the magnetic field energy in the rotor is constant.

The sine wave of a rotating magnetic flux is described using the **flux density space vector** B_r, the direction of which coincides with that of the amplitude of the

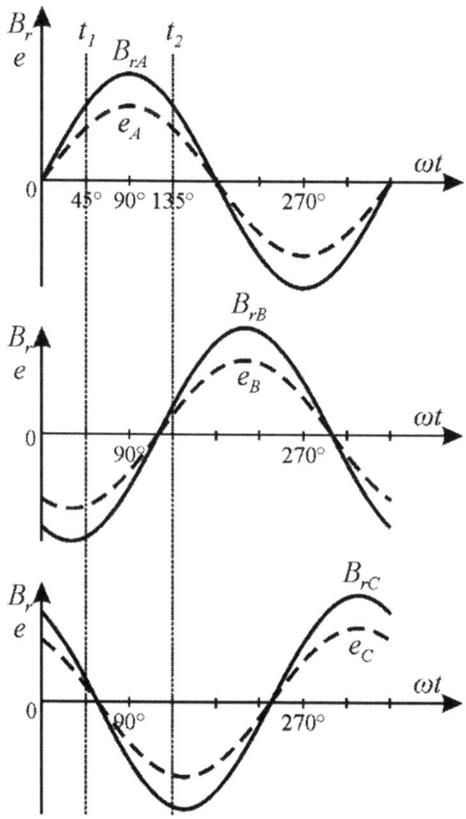

FIGURE 6.4 Changes in flux density at points A, B, and C on a stator and changes over time in the electromotive forces induced in the stator winding parts at those points due to the rotation of a rotor with a sinusoidal flux density distribution.

sine wave of the magnetic field leaving the north pole and the modulus of which is equal to this amplitude.

If a three-phase load is connected to the terminals of the stator winding, then the induced emf will generate a current in it. Thus, the mechanical energy of the prime mover (turbine) is converted into electrical energy by way of electromagnetic induction.

Synchronous machines can also be used as motors. Synchronous motors have a rotor with salient poles. Their rated capacity ranges from 40 to 750 kW and their rotational speed ranges from 125 to 1,000 rpm. What sets them apart from generators is the presence of a start-up winding. The start-up winding (Figure 6.5) is housed in the slots of a pole shoe and takes the form of a squirrel-cage winding, which consists of a number of bars and short-circuiting connections. At start-up, a current is induced in this squirrel-cage winding, producing a torque in a similar way as in an induction machine. When the speed gets close to the synchronous

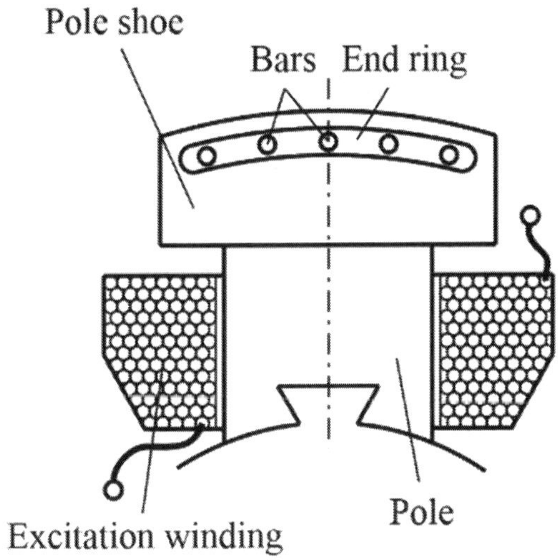

FIGURE 6.5 Start-up winding of a synchronous motor.

speed, current is supplied to the excitation coils and the machine starts to function as a synchronous motor.

Synchronous motors are used in applications where a constant rotational speed and a relatively low starting frequency are needed. The start-up process of a synchronous motor is more complex than that of an induction motor. Where having a constant rotational speed is not an essential condition for the operation of the piece of equipment and where the switching frequency is low, a synchronous motor is also a better option since an over-excited (having a high excitation current) synchronous motor also draws capacitive reactive energy – in addition to active energy – from the supply grid, thereby also compensating for inductive reactive energy in the supply grid. That is why synchronous motors are the motors of choice for driving large fans, centrifugal pumps, compressors, etc.

In energy systems, the power factor can be improved using **synchronous compensators**, which are over-excited synchronous motors run at no load, which consume virtually only capacitive reactive energy. The consumption of capacitive reactive energy can be viewed as the production of inductive reactive energy. The latter is needed by inductive loads, including induction motors. Since there is no load on the shaft of a synchronous compensator, the mechanical structure of the machine may be weaker than that of an analogous synchronous motor.

The behaviour of a synchronous machine when loaded is explained in the mechanical model shown in Figure 6.6, where the ring S and cylinder R are rotating in the same direction at a constant angular velocity $\omega_S = \omega_R$ and are connected by a spring (flexible joint). The ring S represents the magnetic field of a stator, and the cylinder R represents that of a rotor. Meanwhile, the spring represents the interaction of the two fields.

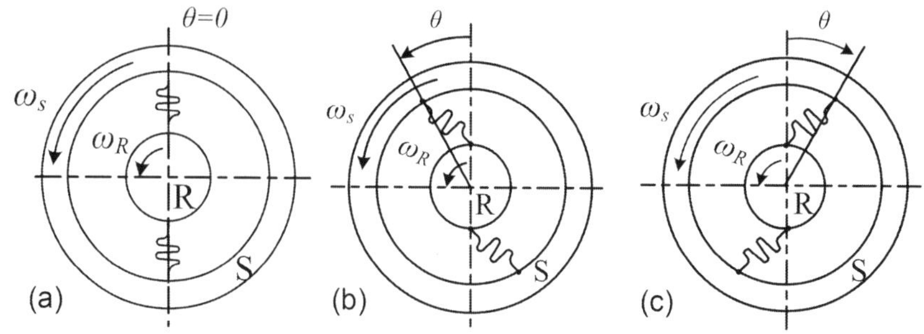

FIGURE 6.6 Explanation of the formation of the load angle of a synchronous machine (a) No load case (b) synchronous motor case and (c) synchronous generator case.

In position a, where $\theta = 0$, there is no transfer of energy via the spring from the ring to the cylinder or vice versa. The situation is similar to the operation of a synchronous machine at no load. In position b, the ring S leads the cylinder R by a certain angle θ , while $\omega_{\mathrm{S}} = \omega_{\mathrm{R}}$. Energy is transferred via the spring from the ring to the cylinder. The situation is similar to what happens in a synchronous motor, where energy is transferred from the stator to the rotor via a magnetic field. In position c, the cylinder R leads the ring S by the angle θ . Here, too, $\omega_{\mathrm{S}} = \omega_{\mathrm{R}}$, and energy is transferred from the cylinder to the ring. The situation is similar to what happens in a synchronous generator, where the mechanical energy of the rotor is converted into electrical energy in the stator winding via a magnetic field (flexible joint).

6.2 VOLTAGE BALANCE IN A SYNCHRONOUS GENERATOR

Every turn of wire in the stator winding is coupled to the magnetic flux of the rotor, which, at no load, varies within $\pm\Phi$. Over the AC half-period, the mean value of the electromotive force induced in one turn is

$$E_{\mathrm{k}}^{'} = \frac{2}{T}\int_{0}^{T/2} e\mathrm{d}t, \tag{6.2}$$

from which, by making the substitution $e\mathrm{d}t = -\mathrm{d}F$, we get

$$E_{\mathrm{k}}^{'} = \frac{2}{T}\int_{+F}^{-F} \mathrm{d}F = \frac{4}{T}F = 4fF. \tag{6.3}$$

For a sinusoidal electromotive force, the effective value is 1.11 times the mean value ($k_{\mathrm{k}} = 1.11$). Thus, the effective value of the electromotive force in one turn is

$$E' = E_{\mathrm{k}}^{'}k_{\mathrm{k}} = 4,44fF \tag{6.4}$$

and the electromotive force induced in one phase of the stator winding, with w turns connected in series, is

$$E = 4,44 fwk_w F. \tag{6.5}$$

Here, the winding factor k_w describes the decrease of the electromotive force E_k' in the turn of winding due to differences in the arrangement of the turns of winding in space.

The magnetic flux Φ of the pole is generated by the excitation current I_e, i.e. $\Phi = f(I_e)$. The relationship $E = f(I_e)$ is called the no-load characteristic of the synchronous generator (see Figure 6.9a), which is also the magnetisation curve of the machine.

When the generator is loaded, the current in the stator is not equal to zero, and then the three-phase stator current I also produces a rotating magnetic field that moves synchronously with the magnetic field of the rotor. The machine's magnetic field that results from the magnetisation excitations of the rotor and the stator differs from the no-load magnetic field. The effect of the magnetic field of the stator (armature) on that of the rotor is called the **armature reaction**.

The phase electromotive force E and phase output voltage U are equal only if there is no current in the stator. The stator current causes a difference to arise between E and U, as it distorts the magnetic field in the air gap (armature reaction) and produces a voltage drop across both the inductive reactance and resistance of the stator winding. The effect of the armature reaction is usually the greatest. The occurrence of the armature reaction is explained in Figure 6.7.

A two-pole rotor is rotating inside a three-phase stator. In Figure 6.7a, no load is connected to the stator. The rotor's magnetic field $\boldsymbol{B}_r$ generates an electromotive force $\boldsymbol{E}_A$, the peak value of which coincides with the direction of $\boldsymbol{B}_r$. The voltage is directed outwards from the plane of the drawing in the upper winding wires and inwards in the lower wires. If there is no load on the generator, there is no armature current, and $\boldsymbol{E}_A$ is equal to the output voltage $\boldsymbol{U}$.

Now let us assume that the generator is connected to a resistive-inductive load and that the phase of the current is lagging behind that of the voltage. This is shown in Figure 6.7b. The current flowing in the stator windings produces a magnetic field $\boldsymbol{B}_s$. The magnetic field $\boldsymbol{B}_s$ in turn generates an electromotive force $\boldsymbol{E}_{stat}$.

Thus, two electromotive forces are generated in the phase winding – the electromotive force of the rotor $\boldsymbol{E}_A$ and the electromotive force of the armature reaction $\boldsymbol{E}_{stat}$. Their sum is the output voltage, $\boldsymbol{U} = \boldsymbol{E}_A + \boldsymbol{E}_{stat}$. The total magnetic field $\boldsymbol{B}\Sigma$ is the sum of the magnetic fields of the stator and the rotor: $\boldsymbol{B}\Sigma = \boldsymbol{B}_s + \boldsymbol{B}_r$.

In an equivalent circuit for one phase of a synchronous machine, the electromotive force of the armature reaction may be considered as an imaginary inductive reactance X, the magnitude of which is such that when the phase current $\boldsymbol{I}_A$ flows through this reactance, a voltage drop that is equal to the electromotive force of the armature reaction is produced across it: $\boldsymbol{E}_{stat} = -jX \cdot \boldsymbol{I}_A$.

In addition to the armature reaction, it is also necessary to take into account the resistance and inductive reactance of the stator winding. If the stator has an inductive reactance of X_A and a resistance of R_A, then the output voltage U is given by

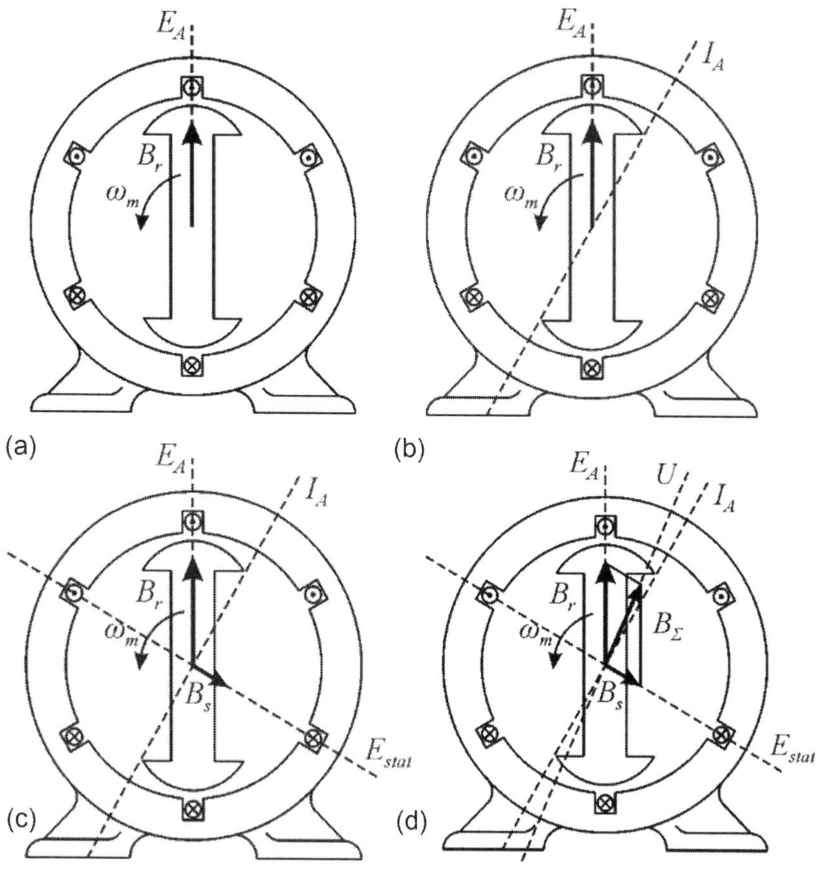

FIGURE 6.7 (a) the rotor's rotating magnetic field B_R generates an emf E_A; (b) a current with a lagging phase angle is generated in the resistive-inductive load connected to the terminals of the generator; (c) the stator current produces its own magnetic field B_s, which generates its own emf E_{stat} in the stator winding of the machine; (d) the field B_s combines with the field B_r, producing the total field $B\Sigma$, while the emf E_{stat} and the emf E_A combine, too, producing the output voltage U.

$$U = E_A - jXI_A - jX_AI_A - R_AI_A. \tag{6.6}$$

Usually the armature reaction's imaginary inductive reactance X and the armature winding's inductive reactance X_A are added together; their sum is called the machine's synchronous reactance $X_S = X + X_A$.

In such a case, the output voltage equation takes the following form:

$$U = E_A - jX_SI_A - R_AI_A. \tag{6.7}$$

Based on this equation it is possible to draw up the equivalent circuit for one phase of the induction machine, as shown in Figure 6.8. The excitation circuit

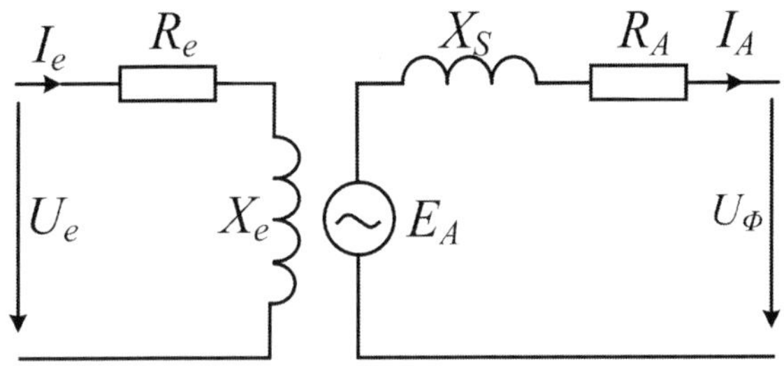

FIGURE 6.8 Equivalent circuit for one phase of a synchronous generator.

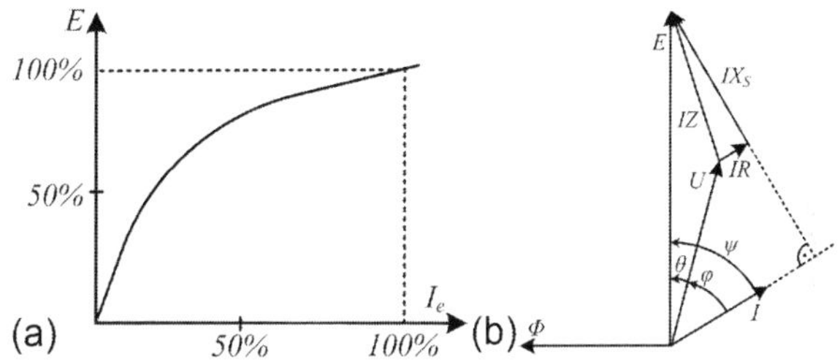

FIGURE 6.9 No-load characteristic (a) and phasor diagram (b) of synchronous generator.

is represented by the excitation coils' inductive reactance X_e and resistance R_e, which consists of the resistance of the winding and the resistance of the excitation current regulator. The excitation circuit is supplied with DC voltage. When the excitation current is adjusted, the machine's electromotive force E_A changes. This source of electromotive force is connected in series to a synchronous inductive reactance, or synchronous reactance, X_S and the armature winding's resistance R_A.

Figure 6.9a shows how the synchronous machine's no-load voltage depends on the excitation current, while Figure 6.9b shows the phasor diagram for the machine. The angle θ between the electromotive force E and the voltage U is the load angle. At no load, this angle is zero and E and U are equal. The angle φ between the voltage U and current I is the phase shift angle of the load current. The phasor of the resistive component IR of the voltage drop produced in the machine under load is parallel to the phasor of the current I, while the inductive component IX_S is perpendicular to the phasor of the current I.

6.3 TORQUE OF A SYNCHRONOUS MACHINE

The electromagnetic power of a synchronous machine is given by

$$P_{em} = 3EU\frac{\sin\theta}{X_S}. \tag{6.8}$$

The electromagnetic torque

$$T_{em} = \frac{P_{em}}{W}, \tag{6.9}$$

where the angular velocity of the rotation of the rotor

$$W = \frac{2\pi f}{p} = \frac{\omega}{p}, \tag{6.10}$$

which, when plugged into equation (6.9) and taking into account the electromagnetic power equation (6.8), gives us:

$$T_{em} = \frac{3p}{\omega}\frac{EU}{X_S}\sin\theta. \tag{6.11}$$

The relationship $T_{em} = f(\theta)$ is called the torque-angle characteristic (Figure 6.10b) of the synchronous machine, where the angle $+\theta$ corresponds to the operation of the synchronous machine in generator mode and the angle $-\theta$ to motor mode. Within the range $-90° < \theta < +90°$, the operation of the synchronous machine is stable, with the maximum electromagnetic torque occurring at the limits of stability:

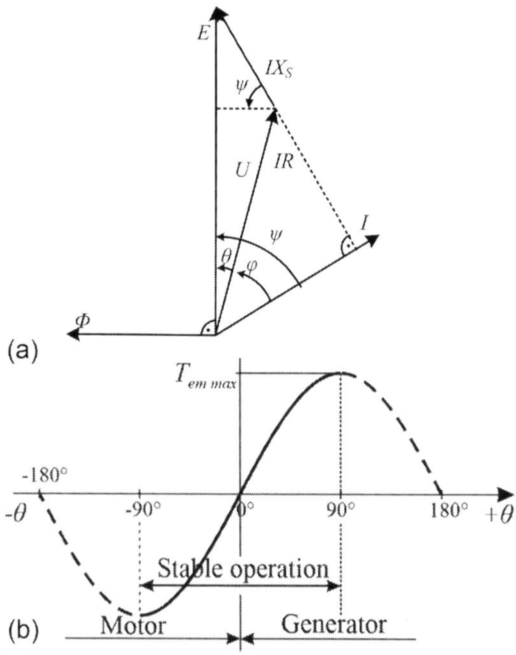

FIGURE 6.10 Simplified phasor diagram (a) and torque-angle characteristic (b) for a synchronous machine.

$$T_{\text{em max}} = \pm \frac{3p}{\omega} \frac{EU}{X_S}. \tag{6.12}$$

Problem 6.1

There is a synchronous motor with a rated power of 500 kW, efficiency of $\eta = 0.95$, rated load angle of $\theta_n = 24°$, and rotational speed of 1,000 rpm. Find the power developed by the motor, the power drawn from the grid, and the load angle when the load torque is 3,500 N m.

Solution

1. The power developed by the motor is

$$P_{\text{mech}} = \omega T = (\pi / 30) nT = (\pi / 30) \cdot 1000 \cdot 3500$$

$$= 366519\,\text{W} = 366,52\,\text{kW}.$$

2. The power drawn from the grid is

$$P_{\text{el}} = P_{\text{mech}} / h = 366,92 / 0,95 = 385,81\text{kW}.$$

3. The rated torque of the motor is

$$T_n = P_n / \omega_n = 500000 / (\pi / 30) \cdot 1000 = 4774,65\text{Nm}.$$

4. The load angle at a load torque of 3,500 Nm is

$$\theta = \arcsin(T \sin\theta_n / T_n) = \arcsin(3500 \sin 24° / 4774,65) = 17,35°,$$

because according to (6.12), the torque is directly proportional to the sine of the load angle.

6.4 TYPES AND FEATURES OF SYNCHRONOUS MOTORS

Conventional synchronous motors feature a three-phase stator winding and electromagnetic poles for which the current can be adjusted (**electrical excitation**). These types of motors generally have a high power rating.

Synchronous machine rotors can also utilise permanent magnets in place of electromagnets. This is seen in **permanent-magnet excited synchronous machines**. It is also possible to have an entirely magnet-free rotor where the magnetic resistance, or reluctance, of the air gap varies across the circumference of the rotor. Machines that utilise this design are known as **synchronous reluctance machines**.

6.4.1 Conventional Electrically Excited Synchronous Motor

This type of motor is started by means of a start-up winding housed in the pole shoes (see Figure 6.5), which allows for so-called induction starting. Figure 6.11a shows a basic induction starting circuit.

The line switch L is used to connect the stator winding to a three-phase voltage system. At this point, the excitation winding is connected to the start-up resistor R_{st} and the changeover switch CS is in position A. The rotating magnetic field of the stator induces a current in the start-up winding, producing a torque in the same manner as in an induction motor. The rotational speed of the rotor starts to increase according to the mechanical characteristic in Figure 6.11b. When the speed of the rotor gets close to the synchronous speed, the changeover switch CS is operated to switch on the excitation current I_e (position B), pulling the motor into synchronism. A sudden transition to mechanical characteristic 2 of the synchronous motor occurs. Since the start-up winding is now rotating synchronously with the magnetic field of the stator (the slip s with respect to the stator field equals 0), no electromotive force is induced in it, and the current in the start-up winding is zero. However, should the resistive torque of the appliance exceed the torque T_{max}, the motor will come out of synchronism and return to the induction characteristic 1.

The starting resistance R_k is selected to be 10–15 times the resistance of the excitation winding, because leaving the excitation winding open or short-circuiting it for the duration of the start-up process is out of the question. The number of turns in the excitation winding is relatively high, due to which the rotating field of the stator induces a considerable electromotive force in the excitation winding at start-up (when the slip changes between $1 \geq s > 0$), which can cause a dielectric breakdown in the insulation in the case of an open excitation winding. Short-circuiting the ends of the excitation winding would prevent dielectric breakdowns. However, this would also cause a relatively high current to arise in the excitation winding, which, together with the rotating field, would give rise to a parasitic torque (curve 3 in Figure 6.11b). Since the excitation winding is not a three-phase winding, the parasitic torque would change direction when the rotating field is at half speed, due to which the resulting mechanical characteristic 4 (Figure 6.11b) would enable only half the rated speed to be achieved at start-up. This speed, however, is insufficient to pull the machine into synchronism. In order to satisfy both conditions at the same time, we can choose an intermediate approach: to connect the excitation winding to the resistor R_k at start-up. Then the start-up winding is neither open nor short-circuited.

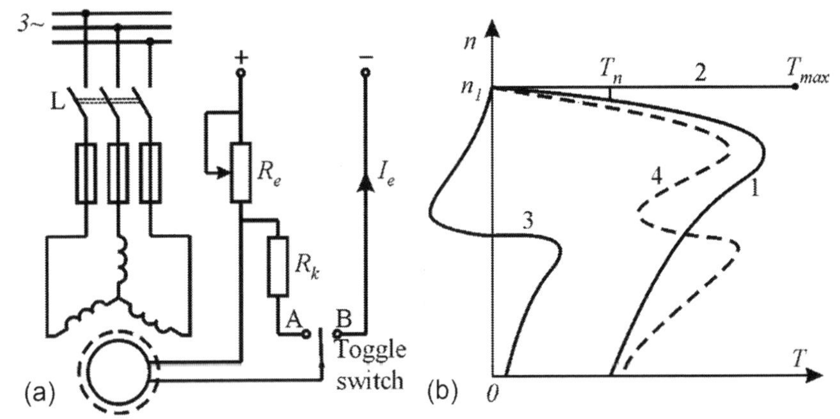

FIGURE 6.11 Starting circuit (a) and mechanical characteristics (b) of a synchronous motor.

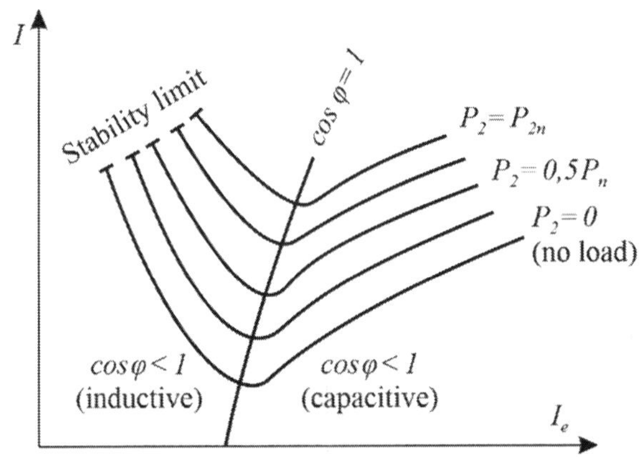

FIGURE 6.12 V-shaped curves for the synchronous motor.

If connecting the stator winding directly to the grid is impossible for the same reasons as described in Section 5.7 for induction machines, it will be necessary to use the same starting methods as for squirrel-cage induction motors.

To change the synchronous motor's excitation current I_e, the excitation rheostat R_e is provided in the excitation circuit (Figure 6.11a).

When the excitation current changes, so does the current I in the stator winding. The relationship $I = f(I_e)$, obtained under the conditions $U = U_n$ = const. and P_2 = const., changes according to the V-curves shown in Figure 6.12.

A lower useful power P_2 is accompanied by a lower stator current. At low excitation currents, $\cos\phi < 1$ and the reactive power is inductive in nature. The phase of the stator current lags behind the phase of the voltage the same way as in an induction motor. However, when the excitation current increases, the stator current decreases through the decreasing of the reactive component of the current, and at a certain excitation current the current will be in phase with the voltage and minimal in intensity. When the excitation current is increased further, the stator current, too, increases, with the motor now drawing capacitive reactive power from the grid – an over-excited synchronous motor acts to compensate for cosϕ. If the power on the shaft $P_2 = 0$, we have a synchronous compensator, which, when over-excited, mainly draws capacitive reactive power from the grid. The stator current leads the voltage by the angle ϕ .

6.4.2 Permanent-Magnet Synchronous Motors

Replacing the electromagnetic poles on the rotor of a synchronous machine with permanent magnets provides for a simpler design. In such a case, no excitation current needs to be supplied to the rotating rotor.

The permanent magnets may be mounted on the surface of the rotor (Figure 6.13) or embedded in recesses cut into the rotor for this purpose (Figure 6.14).

Surface-mounted permanent magnets are fixed in place using an adhesive or a non-magnetic band (such as a stainless-steel band). With surface mounting, less permanent magnetic material is needed than with embedded magnets, and the rotor is also cheaper to manufacture. Rotors with surface-mounted permanent magnets are the most common.

Rotors with embedded permanent magnets are more expensive but have better properties. Using embedded magnets facilitates achieving higher field strengths in the air gap (1 tesla). This gives the machine a higher torque and power. It also provides for greater reliability, as the risk of breakage or demagnetisation of the magnets is lower. Moreover, the distribution of the field across the surface of the pole is better than in the case of surface-mounted magnets. This reduces torque ripple and noise.

The advantage of using permanent-magnet excitation over electrical excitation is a simpler design and the absence of losses related to the excitation current. Using permanent magnets eliminates the need to supply excitation current to the rotating rotor, which is quite challenging. Additionally, it eliminates the need for an excitation power source. Due to their simpler design, permanent-magnet excited machines are also used in very low-power applications where electrically excited machines are impracticable.

FIGURE 6.13 Synchronous machine rotor with surface-mounted permanent magnets.

FIGURE 6.14 Synchronous machine rotor with embedded permanent magnets.

The downside of permanent-magnet excitation is that the excitation field cannot be adjusted. This, unlike with electrical excitation, makes it impossible to adjust the power factor of the machine; and, in the case of a generator, it also precludes the possibility of adjusting the output voltage. Since the excitation field cannot be switched off, it is generally not possible to start permanent-magnet excited synchronous motors by connecting them directly to the grid. The rotating magnetic field of the stator is not able to instantly grab the high-inertia rotor, and the load angle exceeds 180 degrees. As a result, instead of a starting torque, a braking torque is produced (see Figure 6.10b) and the rotor begins to vibrate back and forth at maximum torque. While with electrical excitation, start-up can be achieved with a short-circuited rotor winding, this is unfortunately not the case with permanent-magnet excitation. Starting a permanent-magnet excited synchronous motor requires using a frequency converter. If the rotational speed does not need to be adjusted, this is not a disadvantage. If it does, however, then it is.

A comparison of permanent-magnet synchronous motors and induction motors would reveal that the former have a higher energy conversion efficiency. The rotor windings of induction motors, unlike permanent-magnet synchronous motors, are subject to relatively high losses. Nevertheless, this advantage of permanent-magnet synchronous motors only becomes apparent if the rotational speed of both types of motors needs to be adjusted and they are assisted by a frequency converter. If a single steady speed is required, the efficiency of a powertrain with an induction motor is higher because, in such a case, the induction motor will not need a frequency converter. A permanent-magnet synchronous motor, in contrast, would still need a frequency converter; otherwise, it could not be connected to the grid. And the losses of this frequency converter would reduce the overall efficiency.

6.4.3 Synchronous Reluctance Motors

In the first chapter, we looked at reluctance stepper motors (Figure 1.15), where the stator has salient poles fitted with findings and where the magnetic field of these poles is switched on and off by controlling the current supplied to the windings. Now we will take a look at reluctance motors, where a three-phase winding is used to produce a rotating magnetic field in the stator (and where the stator is

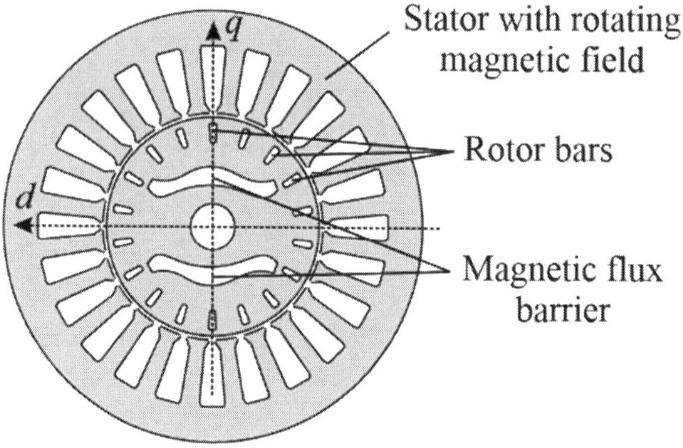

FIGURE 6.15 Cross-sectional view of a synchronous reluctance motor with a single pair of poles.

similar to that of induction and synchronous machines). What sets these motors apart, however, is that their rotor is designed in such a way that its magnetic resistance, or reluctance (which acts against the magnetic flux passing through the rotor), is different in different directions.

Namely, it is lower in the direction of the d-axis (Figure 6.15) and higher in the direction of the q-axis, which is perpendicular to the d-axis. The greater the difference between these reluctances, the higher the torque that the motor is able to develop. In order for the resistance to magnetic flux to be greater in the direction of the q-axis, two slits, called magnetic flux barriers, are cut into each sheet of the bundle of steel sheeting that makes up the rotor. The magnetic flux thus has to pass through two wide air gaps when moving in the direction of the q-axis, making the magnetic resistance greater in this direction than in the direction of the d-axis, where there are no barriers.

In addition to magnetic flux barriers, the rotor also has slots for the bars of a squirrel-cage winding. This winding is similar to the squirrel-cage winding of an induction motor. At start-up, the rotating magnetic field induces an electromotive force in the bars on the rotor, a current is generated, and the starting torque of the motor is developed. This corresponds to an induction start-up. As the speed of the motor increases, the slip and induction torque decrease and, eventually, the rotor is pulled into synchronism due to reluctance torque. The rotor begins to move synchronously with the rotating magnetic field. Reluctance torque occurs when a certain angle, called the load angle, is formed between the direction of the rotating magnetic flux and the direction of the d-axis of the rotating rotor.

In synchronous rotation, the bars of the squirrel-cage winding remain in a stationary field. No voltage is induced and no current is generated in them. This also means that there are no losses in the rotor winding, which is why synchronous reluctance motors have a higher efficiency than induction motors. The downside

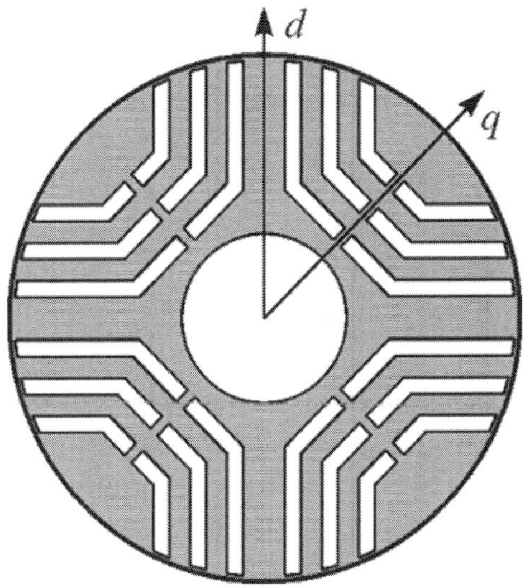

FIGURE 6.16 Rotor of a synchronous reluctance motor with two pole pairs.

of synchronous reluctance motors compared to induction motors, however, is a lower power factor and greater weight and size.

At higher numbers of pole pairs, the rotor of a synchronous reluctance motor is designed in such a way as to produce additional pairs of d- and q-axes. Figure 6.16 shows a rotor with two pole pairs. Here we see three parallel magnetic flux barriers. However, this rotor does not have a squirrel-cage winding, and a machine such as this one, just like synchronous machines, can only be started with a frequency converter.

6.4.4 Adjusting the Rotational Speed of Synchronous Motors

In a synchronous motor, the rotor always moves synchronously with the rotating magnetic field. As such, the rotational speed of the motor can be changed only by changing the rotational speed of the stator field. The stator field's rotational speed n_S is dependent on the number of pole pairs p in the motor and the supply voltage frequency f.

$$n_S = 60f / p. \tag{6.13}$$

Changing the rotational speed of a synchronous motor requires changing the frequency of the supply voltage. This can be achieved by means of a frequency converter with a similar power component as that utilised in frequency converters for induction motors (rectifier, DC link, and inverter, as shown in Figure 5.15). Figure 6.17 shows the mechanical characteristics for frequency adjustment in a synchronous motor.

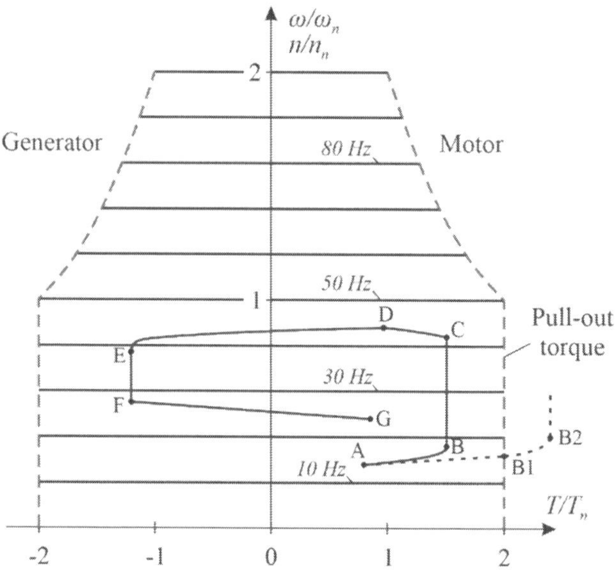

FIGURE 6.17 Mechanical characteristics of a synchronous motor at different adjusted frequencies.

The motor's speed is shown with respect to the rated speed n_n, while the torque is shown with respect to the rated torque T_n. The motor's speed can be adjusted to be higher or lower than the rated speed. Since the stator of a synchronous motor is similar to that of an induction motor, when reducing the frequency with a frequency converter, it is also necessary to reduce the voltage supplied to the motor from the frequency converter (otherwise the stator's magnetic circuit will become saturated).

The rotational speed of a synchronous motor, unlike that of an induction motor, is not dependent on the load torque (the characteristics shown in Figure 6.17 are straight horizontal lines). That is, until the breakdown torque is reached, where the motor's operation becomes unstable. Reaching the breakdown torque is generally unacceptable in synchronous machines because the unstable operation manifests in strong vibrations and dangerously high currents.

Lowering the speed below the rated speed ($\omega / \omega_n < 1$) does not cause the breakdown torque or the slightly lower maximum permissible load torque to change. However, the maximum power decreases in proportion to the decrease in speed, because $P = \omega T$. At speeds greater than the rated speed, however, further increases in speed are accompanied by the decrease in the maximum torque (ωT = const), while the maximum power remains unchanged.

In steady-state operation (point A in Figure 6.17), the torque T_M developed by the motor is equal to the load torque of the appliance T_{load}. In point A, the motor's per-unit speed is 0.28. Given a rated speed of, for example, 3,000 rpm, the speed in this point would be 840 rpm, or 87.96 rad/s. The motor's per-unit torque in this point is 0.8. The load angle then is arcsin 0.4 = 23.58°, because the per-unit breakdown torque is 2 and the relationship between the load angle and the load torque is given by equation (6.11).

Let us see what happens if we increase the angular velocity of the stator's rotating field by 0.1 rad/s and assume that the rotor's speed does not change. Then, after one second, the angle between the space vector of the stator field and the rotor pole will have increased by 0.1 radians, or 5.7 degrees. This angle is the load angle. Thus, if the speed of the stator's rotating field increases by a small amount, the load angle θ starts to increase slowly. This load angle could not change if there did not exist a small difference between the rotational speed of the stator field and that of the rotor for a short while. Therefore, during transient responses to a changeover, the rotor of a synchronous machine may not rotate in perfect synchronism with the stator field.

If we were to increase the angular velocity ω_B of the stator field by means of a frequency converter in such a way that $d\omega_B/dt$ = const, the load angle θ would start to increase in an accelerating manner. According to equation (6.11), when the load angle increases, so does the motor's torque T_M, which becomes higher than the load torque T_{load}. A dynamic torque T_{dyn} arises, which starts to accelerate the rotor,

$$T_{dyn} = T_M - T_{load}. \tag{6.14}$$

The higher this dynamic torque, the faster the increase in the rotor's angular velocity ω_R:

$$d\omega_R / dt = T_{dyn} / J. \tag{6.15}$$

where J is the moment of inertia of the rotating parts of the motor and the appliance (see Section 1.3).

Eventually, the increase in rotor speed becomes equal to the increase in the rotational speed of the stator field:

$$d\omega_R / dt = d\omega_B / dt.$$

This is the case in point B in Figure 6.17. When this happens, the load angle stops increasing (as does the motor's torque). Thereafter, the increase in the speed of the rotor occurs synchronously with that of the stator field. The motor's torque remains constant and exceeds the load torque by the magnitude of the dynamic torque. The motor's speed will increase for as long as the frequency converter is made to increase the frequency. In Figure 6.17, this occurs up to point C. When point C is reached, the rate of increasing the frequency is gradually reduced, and in point D the frequency becomes constant. Then the acceleration of the rotation stops, the dynamic torque drops to zero, and the torque of the motor again becomes equal to the load torque. All of this, however, occurs at a higher per-unit rotational speed of 0.88.

The higher the chosen acceleration dω_B/dt of the rotating field, the higher the resulting acceleration dω_R/dt of the rotor. Note that, according to equation (6.15), this must also result in a higher dynamic torque. Here, it is necessary to ensure that the motor's torque $T_M = T_{load} + T_{dyn}$ does not exceed the breakdown torque T_{max}. The motor's torque must be lower than the breakdown torque, i.e.

$$T_{load} + J\left(d\omega_R / dt\right) < T_{max}. \tag{6.16}$$

Rearranging (6.16) and taking into account that, in this case, the condition dω_R/dt = dω_B/dt is also met, we get the maximum acceleration of the rotation of the stator field, which must not be exceeded:

$$d\omega_B / dt < \left(T_{max} - T_{load}\right) / J. \tag{6.17}$$

In order for the operation of the motor to remain stable when the speed is increased (so that there is no breakdown), the frequency of the frequency converter must not be increased too rapidly. The higher the load torque and the moment of inertia of the rotating parts, the slower the frequency must be increased.

Points B1 and B2 in Figure 6.17 illustrate what happens if the frequency is increased too fast. In order for the rotor to receive the same acceleration as the stator field, it is necessary to apply a torque that is higher than the breakdown torque (point B2). Reaching point B2, however, is impossible, because the motor stops operating in a stable manner when point B1 is reached, which never should be.

Reducing the speed, too, must be done sufficiently slowly as not to reach the maximum torque for generator mode. Figure 6.17 shows how the load point changes along the mechanical characteristic curves when the rotational speed is decreased. When moving from point D to point E, the speed of the stator field decreases by a greater amount than the rotor speed and, as a result, the load angle decreases. Between point E and point F, braking occurs at a constant torque. The magnitude of the braking torque depends on how rapidly the frequency converter lowers the frequency. Between point F and point G, the lowering of the frequency is gradually stopped.

Frequency converters are equipped with a control program that selects the suitable acceleration or deceleration of the rotational speed according to the parameters of the connected motor and the instantaneous value of the torque. Some frequency converters are equipped with programs that can be used to control both synchronous and induction motors. In such converters, the type of motor is defined for the converter during set-up.

6.5 OPERATION OF A SYNCHRONOUS GENERATOR ALONE

The operation of a synchronous generator under load depends on the power factor of the load and, to a large extent, on whether the generator operates alone or in parallel with other synchronous generators. Let us consider a generator that is operated alone to supply power to a load (Figure 6.18).

Based on equations (6.1) and (6.5) and the relationship between the rotational speed and angular velocity $\omega = (\pi/30)\cdot n$, we can get the electromotive force for one phase of the synchronous as

$$E = 0.707 p\omega w k_w F. \tag{6.18}$$

In all generators, the number of pole pairs p, the number of turns in the phase winding w, and the winding factor k_w are constant values. If we say that $0.707pwk_w = K$, equation (6.18) becomes

$$E = K\omega F. \tag{6.19}$$

The rotational speed of the prime mover can be adjusted. This also changes the generator's angular velocity ω. Changes in the angular velocity, in turn, cause a change in the frequency of the generator's output voltage (according to equation (6.1)). According to equation (6.19), moreover, the electromotive force changes as well. The electromotive force can also be changed by changing the flux Φ in the poles.

Problem 6.2

The shaft of an autonomous synchronous generator is rotating at a speed of 1,500 rpm and the generator's terminal voltage is 400 V, 50 Hz. The rotational speed is raised to 1,800 rpm and then the excitation flux is reduced by 20%. Find the resulting voltage and frequency of the generator, knowing that the generator was operating at no load in both cases.

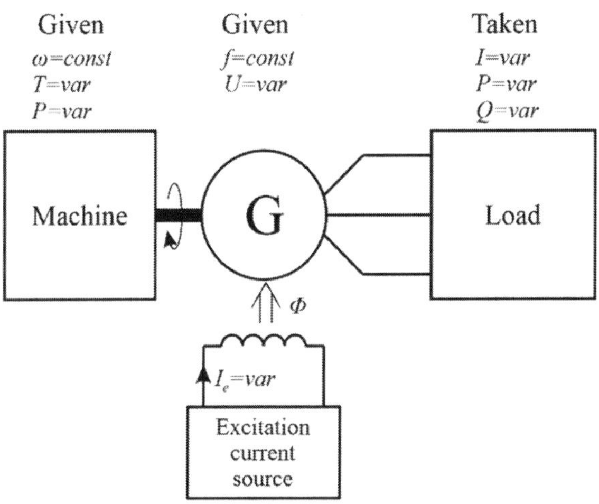

FIGURE 6.18 Lone generator supplying power to a load.

Solution

At no load, the voltage U is equal to the electromotive force E.

1. After increasing the rotational speed, the voltage is

$$U_2 = U_1 \left(n_2 / n_1\right) = 400 \cdot \left(1800 / 1500\right) = 480\text{V},$$

because, according to (6.19), the electromotive force is directly proportional to the angular velocity ω , as is the rotational speed n.

2. After the reduction of the excitation flux, the voltage is

$$U_3 = U_2 \left(1 - 0{,}2\right) = 480 \cdot 0{,}8 = 384\text{V},$$

because the voltage decreases in proportion to the excitation flux.

3. The resulting frequency of the generator is

$$f_2 = f_1 \left(n_2 / n_1\right) = 50 \cdot \left(1800 / 1500\right) = 60\text{Hz},$$

because, according to (6.1), the frequency is directly proportional to the rotational speed.

Loads connected to a generator generally require an invariable frequency and voltage. At the same time, loads draw current that is changing unpredictably. The generator current causes voltage drops within the generator, and the generator's terminal voltage becomes less than the electromotive force (see the equivalent circuit in Figure 6.8 and the phasor diagram in Figure 6.9).

A generator receives mechanical power from a prime mover and converts it into electrical power. In a first-order approximation, the electrical power of a generator and the mechanical power received from the prime mover are determined by the active component of the consumed current.

When the prime mover (such as that of a diesel generator) is loaded, its rotational speed decreases, as does the frequency of the generator. In order to prevent a generator's frequency from dropping, prime movers are equipped with automatic rotational speed regulators.

In order to prevent a generator's voltage from dropping when the generator is loaded, generators are equipped with automatic voltage regulators which increase the excitation current when the voltage drops. Figure 6.19 shows the necessary changes in excitation current for three cases: resistive-inductive load, purely resistive load, and resistive-capacitive load. The required change in excitation current

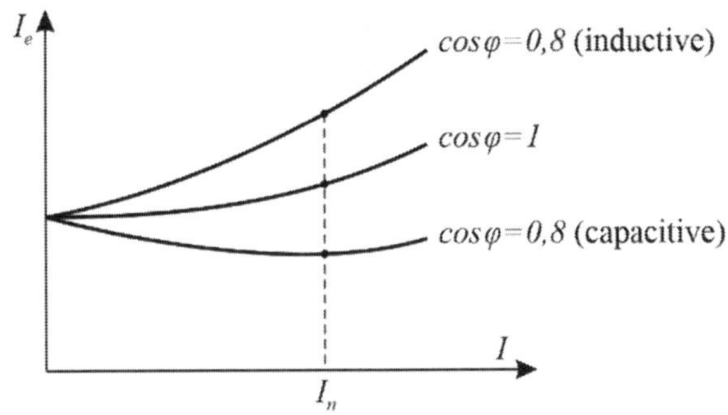

FIGURE 6.19 Necessary changes in excitation current for stabilising the voltage at different load power factors.

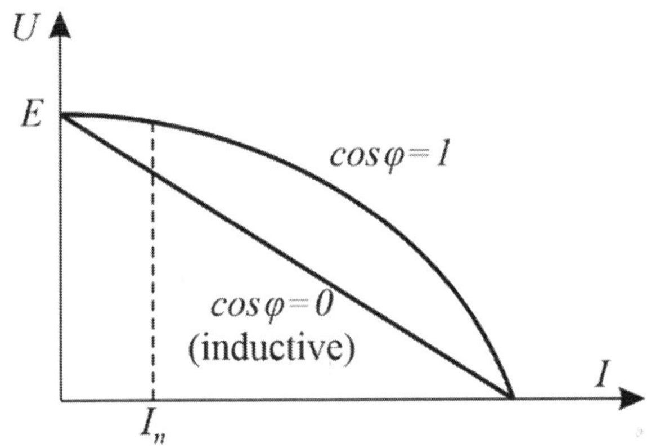

FIGURE 6.20 Decrease in the terminal voltage of a synchronous generator when the load current increases and the excitation current is not adjusted.

is greatest for inductive loads. If there is no automatic regulator, the voltage drops as shown in Figure 6.20.

6.6 PARALLEL OPERATION OF SYNCHRONOUS GENERATORS

Synchronous generators that are operated under a local load independently of other generators are rare and have a low power. In electrical grids, several generators are operated in parallel.

The parallel operation of multiple synchronous generators has a number of advantages.

1. Multiple generators are able to power a higher load than a single generator.
2. Using multiple generators increases the reliability of the system, as when one generator breaks down, power is not lost entirely.

3. Parallel operation of multiple generators enables one or more generators to be switched off for maintenance.

6.6.1 Preconditions for Switching to Parallel Operation

Figure 6.21 shows a synchronous generator G1, which is supplying power to a load, and another generator G2, which needs to be switched over to parallel operation with generator G1 by closing the switch S1. Which conditions must be met before the switch is closed and the two generators are connected?

If the switch is closed at a random point in time, serious damage can occur to the generators, deriving the load of power. At such a random point in time, the instantaneous values of the voltages to the right and left of the switch S1 are different, which means that when the switch is closed, an equalisation current is generated. This current can be very high. To prevent this equalisation current from occurring, the voltage across the lamps connected in parallel to the contacts of the switch (Figure 6.21a) should be as close to zero as possible. For this purpose, the following conditions must be met for switching to parallel operation:

- the effective values of the voltages of the generators must be equal;
- the generators must use the same phase sequence;
- there must be no phase shift between the phase A voltages of the generators;
- the frequency of the added generator must be slightly higher than that of the generator that is already in operation.

These conditions for switching to parallel operation need some clarification. The reason for the first condition is obvious: the voltages of the two generators must be identical, otherwise equalisation currents will occur. The voltages can be equalised by adjusting the excitation current of the added generator.

The second condition ensures that the sequence of the peaks of the phase voltages is the same for both generators. If the phase sequences are different (see Figure 6.21b), one pair of voltages may be in phase (phase A), but the other two pairs will be shifted in phase by 120°. Connecting such generators together would result in high currents in phases B and C, which would damage both machines. To change the phase sequence, two phases need to be interchanged in one of the machines. The phase sequence can be verified using a special-purpose tool. In the case shown in Figure 6.21, an incorrect phase sequence can also be identified based on the incandescent lamps connected to the contacts of the switch. In the case of an identical phase sequence, all three lamps will go out at the same time, if the voltages of both systems are in phase. In the case of different phase sequences, however, the lamp of one phase (e.g. A) will go out, while those of the other two phases will continue to emit light at full voltage.

The instantaneous values of the voltages of different generators will always be perfectly identical if there is no phase shift, as required by the third condition. Phase shifts can be changed as follows. In accordance with the fourth condition,

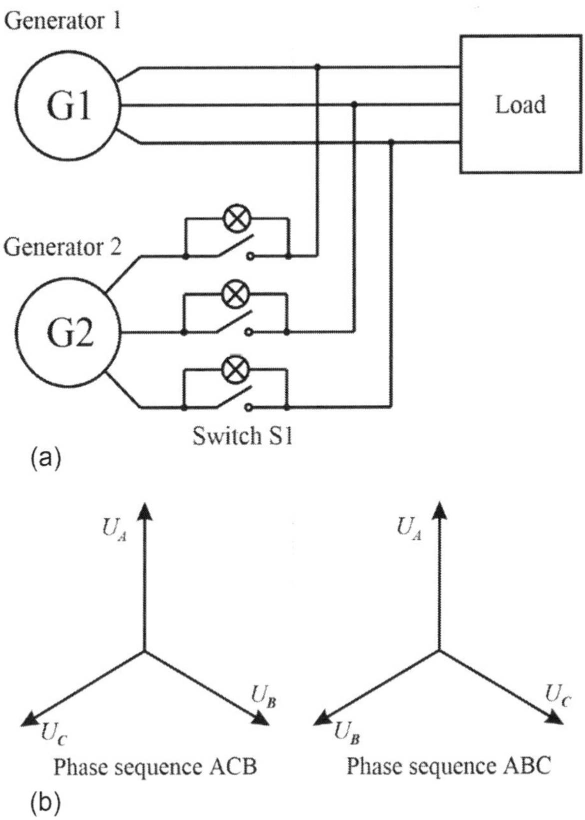

FIGURE 6.21 Connection of a new generator to a running system (a) and two possible phase sequences for a three-phase system (b).

the frequency of the added generator is set to be slightly higher, so that, after it is connected, this generator would supply power to the grid, instead of consuming power as a motor. This frequency adjustment is achieved by slightly adjusting the rotational speed of the primer mover. At near-identical frequencies, the phase angle between the voltages of the two systems will change very slowly. This phase change is monitored, and when the phase angles become equal, the switch connecting the two systems is closed.

How can you tell that two systems are exactly in phase? One simple option would be to monitor the three lamps (Figure 6.21). When all three lamps have gone out, the voltages across them will be small or equal to zero and the systems will be in phase. Unfortunately, this simple method is not very accurate.

A better option would be to use a synchronoscope (Figure 6.22). A synchronoscope is an instrument for measuring the difference between the phase angles of the A-phases of two systems. Since the frequencies of the two systems will differ slightly, the phase angle reading of this instrument will slowly change. If the generator being added is rotating faster than the running system (preferably),

FIGURE 6.22 Synchronoscope.

then the phase angle will increase and the pointer of the synchronoscope will rotate clockwise. If the generator being added is slower, the pointer will rotate counterclockwise. When the pointer of the synchronoscope is in a vertical position, the voltages are in phase, and the switch can be closed to connect the two systems together.

Note that synchronoscopes only monitor one phase and do not provide any information about the phase sequence.

For generators used in large power systems, the entire process of connecting a new generator for parallel operation is controlled automatically by a computer. For smaller generators, this is done manually.

6.6.2 Adjusting the Active and Reactive Power of a Synchronous Generator in a Grid

All generators are driven by a prime mover, which supplies mechanical power to the generator. The most common prime mover is a steam turbine, but it can also be a diesel engine, a gas turbine, or even a wind turbine.

Regardless of their exact type, all prime movers have one thing in common: if the power drawn from them increases, their rotational speed decreases. This decrease in speed is usually non-linear, which is not suitable for a synchronous generator. For this reason, prime movers are usually equipped with a regulating mechanism to make the speed decrease linearly when the power increases.

This regulating mechanism is configured in such a way as to produce a slightly downward-sloping characteristic when the load increases. The downward slope of this characteristic is such that when transitioning from no load to full load, the speed drops by 2–4% (Figure 6.23). By adjusting the regulating mechanism, this characteristic can be shifted up and down. This shifting enables the torque and power transmitted to the generator via the shaft of the prime mover to be adjusted when the generator is connected to the grid (regulator positions 1 and 2). A prime mover cannot change the rotational speed n_S of a generator that is connected to a grid. A generator's rotational speed n_S is determined by the grid frequency f_{grid}, which is a constant value.

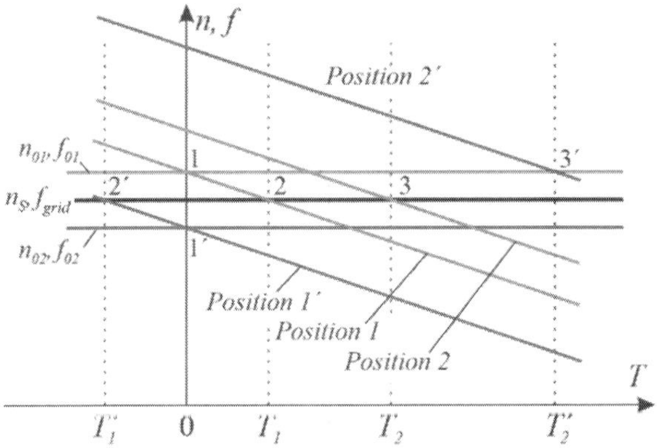

FIGURE 6.23 Relationship between a prime mover's rotational speed and torque at different regulator positions, and the rotational speed (frequency) of the generator before and after connection to the grid.

Before the generator is connected to the grid, the regulating mechanism establishes the no-load speed n_0 and frequency f_0 of the generator.

At first, the regulator is in position 1, the load torque of the generator is equal to zero, and the frequency is slightly higher than the grid frequency (point 1 in Figure 6.23, where the frequency is f_{01}). When the pre-defined values are reached in accordance with the described synchronisation procedure, the generator is inserted into the grid and its frequency becomes equal to the grid frequency f_{grid}. The generator begins to rotate synchronously with the rotating field produced by the grid frequency, and its rotational speed drops slightly. At this lower rotational speed n_S, the prime mover develops the torque T_1 (point 2 in Figure 6.23) and the generator begins to supply power corresponding to this torque to the grid. In order to increase the power supplied to the grid, it is necessary to shift from the regulator position 1 to position 2, the characteristic for which shown in Figure 6.23 intersects with the line corresponding to the synchronous speed n_S in point 3. At this point, the prime mover develops the torque T_2, and the power supplied to the grid has increased.

The torque applied to the shaft of the generator by the prime mover must not be greater than the generator's breakdown torque. Otherwise, the generator will fall out of synchronism and a breakdown will occur (regulator position 2′ and operating point 3′).

If, before connection to the grid, the regulating mechanism is used to select a no-load speed n_{02} and frequency f_{02} which are lower than the synchronous speed n_S and grid frequency f_{grid} (regulator position 1′), then, after connection to the grid, the generator's rotational speed will rise to the synchronous speed and the generator will force the prime mover to also increase its speed. When this happens, the generator will enter motor mode (point 2′ in Figure 6.23) and the prime mover will start to behave as an appliance. When motor mode is entered, the automatic protection system of a generator will deem the prime mover to be malfunctioning (e.g. with no steam being fed to a steam turbine) and will immediately shut the generator down. Thus, the connection of the generator to the grid will fail.

It is important to bear in mind that **a generator's frequency can be changed via the prime mover and a regulating mechanism only before the generator is connected to the grid. After connection to the grid, the same regulator will only change the active power supplied by the generator to the grid.**

This is because electrical grids contain a large number of generators, the combined power of which will be significantly higher than that of any single generator. While a single added generator can take a bit of the load off the generators that are already running, this will not significantly increase the grid frequency.

Additionally, changing the excitation current, and thereby the electromotive force E, of a single generator connected to the grid, will not significantly affect the terminal voltage of that generator. This is because the terminals of a generator are connected directly to the grid, which is supplied with voltage by a large number of parallel-operated generators. A single generator is unable to significantly change the voltage of all of the other generators.

When examining an individual generator, the electrical grid is considered to be a voltage source with an infinitely high power and zero internal resistance. The voltage and frequency of such a voltage source will always be constant (when examining the stability of the grid, however, such a simplification is untenable).

If the terminal voltage of a generator that is connected to a grid cannot change, then what *will* change in this generator if we change the excitation current? This is explained in the phasor diagram shown in Figure 6.24.

The active power of a generator that is connected to a grid depends solely on the position of the regulator of the prime mover. If the position of the regulator is not changed, then the active power remains constant. At a constant rotational speed and constant power, the torque, too, will be constant. It follows from equation (6.11) that if the torque, angular velocity, and voltage U are constant, the product $E\sin\theta$ must also be constant (the number of pole pairs p and the synchronous reactance X_S of the machine do not change).

If, when the excitation current is adjusted, the electromotive force E changes and $E\sin\theta$ must remain constant, then, consequently, the end point of the

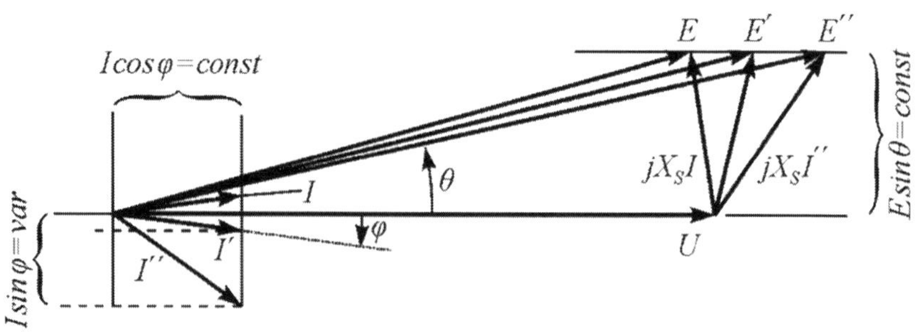

FIGURE 6.24 Effect on reactive power of changing the excitation current of a synchronous generator that is connected to a grid.

E–phasor in the phasor diagram must 'slide' along a straight line that is parallel to and spaced apart from the U–phasor by $E\sin\theta = \text{const.}$ (Figure 6.24). The phasor connecting the end points of the phasors E and U represents the voltage drop $X_S I$ that occurs across the generator's synchronous reactance X_S due to the generator's current I. When the electromotive force changes, both the direction and length of the phasor of this voltage drop will change, too. The direction of this voltage drop phasor is perpendicular to that of the generator current phasor I, which is why, when the electromotive force changes, the phase shift angle ϕ between the voltage U and current I will also change. Since active power is constant in this case, the product $I\cos\phi$, which corresponds to the active component of the current, must remain constant. At the same time, the figure shows that the product $I\sin\phi$, which is directly proportional to the reactive power, changes when the electromotive force E changes. In other words, increasing the excitation current of a generator that is connected to a grid with an infinitely high power causes the reactive power produced by the generator to increase.

Note, however, that voltage can be adjusted via the excitation current only if the generator is operating alone.

In conclusion, if a generator is operating in parallel with an infinitely powerful system, then

- the frequency and terminal voltage of the generator are determined by the system to which it is connected;
- the setpoint of the regulator of the generator's prime mover determines the active power supplied by the generator to the system;
- the generator's excitation current determines the reactive power supplied by the generator to the system.

This situation is more or less equivalent to what happens in the operation of a real generator that is connected to a very large power system.

7 DC Machines

Up until now we have been looking at electrical machines where rotating magnetic field poles are generated in a stationary stator. As we saw, these poles are made to move by either switching the current on and off in a set of constructional poles (in stepper motors) or by means of alternating current (in synchronous and induction machines).

Now let us take a look at direct-current motors, where moving poles are generated inside a rotating rotor and where the stator has stationary poles. This design is seen in conventional DC machines, which are, historically speaking, the oldest type of electrical machine. These machines contain a special component: the **commutator**. Through the commutator, direct current passes from the stationary part of the machine to the rotating rotor, and as it does so, the commutator converts the direct current into alternating current (in motor mode). The transmission and conversion of the current is achieved by means of mechanical contacts called **segments**, some of which are positioned on the rotating rotor. The remaining contacts, which are called **brushes**, are fixed. The brushes are usually made of graphite and are pressed against the rotating segments by springs. The segments are made of copper.

Commutators with moving contacts are not very reliable (they require regular maintenance). That is why so-called **brushless DC motors** have been introduced. In brushless DC motors, the commutator is replaced by semiconductor switches, which are controlled by signals from a **rotor position sensor**. The angle of rotation, speed, and acceleration of this type of motor can be controlled very precisely, which is why such motors are used in high-precision applications (e.g. in table-top metal cutting machines).

7.1 DESIGN AND OPERATING PRINCIPLE OF DC MACHINES

DC machines utilise a stationary magnetic field. The wires of the rotor winding are positioned inside this field. Let us first explore the forces acting on the rotating shell placed in the magnetic field (Figure 7.1).

Current is fed into the rotating shell through two slip rings, which rotate along with the shaft, and fixed brushes that are in contact with these rings. The rotating shell can rotate around the axis on the bearings. The rotating shell has two longer sides (a and b) with the length l. On side a of the rotating shell, the current is directed away from the slip rings, while on side b, the current is directed towards the slip rings. The sides a and b are separated by the distance D. The distance D is also the diameter of the cylinder formed by the rotation of the rotating shell. When the current passing through this rotating shell interacts with the magnetic field, a force is induced on all sides of the rotating shell. With regard to

DOI: 10.1201/9781003404088-7

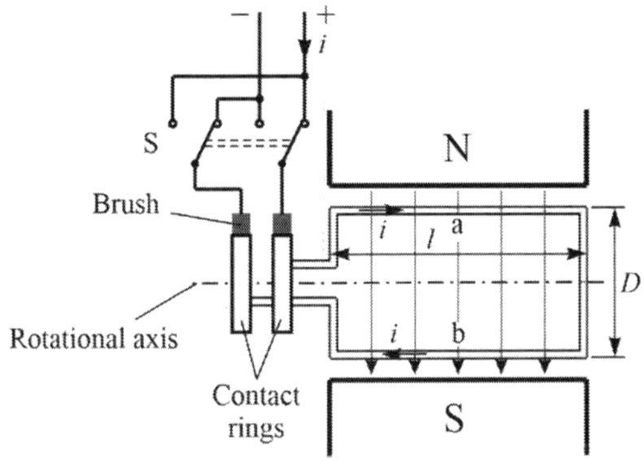

FIGURE 7.1 Side view of the rotating shell set on bearings and placed in the magnetic field.

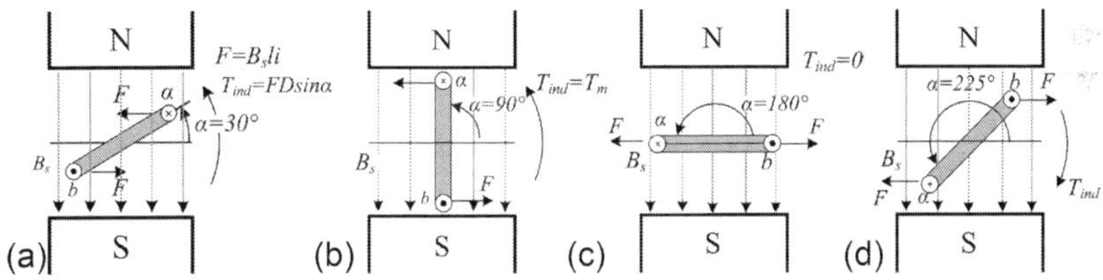

FIGURE 7.2 Change in the application points and direction of the torque-producing forces relative to the axis of rotation when the position of the rotating shell changes.

the rotation-inducing torque that occurs due to the induced forces, however, only the sides a and b are relevant. The total torque of the forces acting on the shorter sides can be demonstrated to be equal to zero in every position of the rotating shell. The production of the torque is best observed in a plane that is perpendicular to the axis of the rotating shell.

Figure 7.2 shows the forces F acting on the sides of the rotating shell at different angles of rotation α as viewed from the side of the slip rings (the slip rings themselves are not shown in the figure).

The magnitude of the force acting on one side of the rotating shell is

$$F = B_s li, \tag{7.1}$$

where B_s – magnetic flux density;

l – the length of the side of the rotating shell;

i – the current in the rotating shell.

If the current is constant and the angle of rotation α , the magnitude of the force F acting on the side of the rotating shell does not change; what does change, however, is the lever arm of the force pair F–F. **The induced torque** T_{ind} is dependent on the angle of rotation α as seen in the equation

$$T_{\text{ind}} = FD\sin\alpha \tag{7.2}$$

where D is the diameter of the rotating shell (as well as the distance between the longer sides). Combining equations (7.1) and (7.2) gives us

$$T_{\text{ind}} = B_s liD\sin\alpha = B_s iA\sin\alpha, \tag{7.3}$$

where $A = lD$ is the area of the rotating shell.

If the angle of rotation $\alpha = 0°$ when the current is switched on, the torque will be equal to zero and the rotating shell will not move. However, if, in the starting position, the angle of rotation$\alpha = 30°$, the rotating shell will begin moving and the angle of rotation will start to increase.

Maximum torque will occur when $\alpha = 90°$, after which the torque will begin to decrease (Figure 7.3a). When the angle of rotation $\alpha = 180°$, the torque will again be zero, because the forces F will be acting along the same line (Figure 7.2c). When the rotating shell passes through this position due to inertia, a torque that is opposite in direction to the initial torque will be produced, which will attempt to return the rotating shell to the position $\alpha = 180°$ (Figure 7.2d shows the position

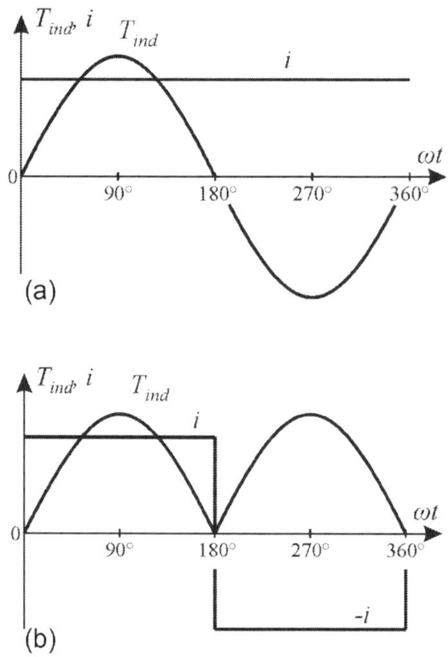

FIGURE 7.3 Relationship between the torque induced in a turn of wire in a homogeneous magnetic field and the angle of rotation: (a) the direction of the current in the turn of wire does not change; (b) the direction of the current in the turn of wire is changed after every half revolution.

$\alpha = 225°$). In the case of a one-way current, the rotating shell can only complete half a revolution, i.e. 180°.

One way to achieve constant rotation is by changing the direction of the current whenever the rotating shell reaches the positions 180° and 0°. In principle, the direction of the current could be changed by means of the changeover switch S shown in Figure 7.1. In practice, however, semiconductor switches (power transistors) and a rotor position sensor are used instead. Based on the signals received from the position sensor, the direction of the current is changed whenever the angle of rotation is such that the torque is zero (0° and 180° in our example). This is the principle of operation of brushless DC machines.

Note that in the situation depicted in Figure 7.2, in order to maintain the direction of the torque, the current in the side of the rotating shell that is closer to the south pole S (either a or b) must always be directed outwards from the plane of the figure. For this reason, it is necessary to change the direction of the current every time the rotating shell passes through the horizontal position (0°, 180°, etc.). When the direction of the current is changed, the force acting on the side of the rotating shell reverses, as does the torque (Figure 7.4). As a result of this periodic changing of the direction of the current, the direction of the torque developed by the current-carrying rotating shell will be identical at all times (Figure 7.3b).

The direction of the current can also be changed by means of a mechanical switching device, i.e. a commutator, mounted on the shaft of the rotating shell. This method is used in commutating DC machines, a simplified design for which is shown in Figure 7.5.

The poles 1 generate a magnetic flux, which passes through the two air gaps and the cylinder 2 positioned between them, which is made of electrical steel and is called **the armature**. The conductive rotating shell is constructed from winding wire inserted into the grooves, known as 'slots', on the outer surface of the

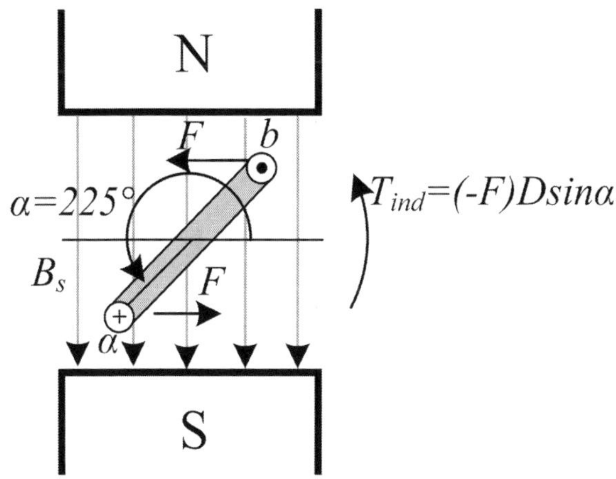

FIGURE 7.4 When the direction of the current is changed, the torque on the rotating shell reverses.

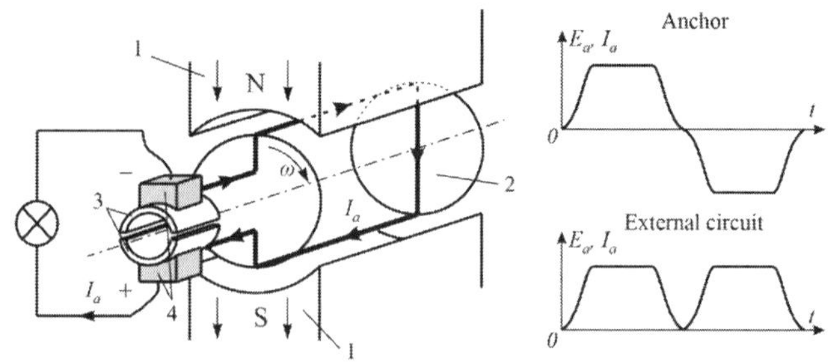

FIGURE 7.5 Simplest DC machine design: 1 – poles; 2 – rotor, i.e. armature; 3 – commutator; 4 – brushes.

cylinder. In the simplest DC machines, the commutator consists of two copper commutator segments, which are insulated from each other as well as from the shaft. These segments are connected to the ends of the armature winding. The rotating shell is in contact with two fixed brushes 4, through which the armature winding is connected to the external circuit. Such a DC machine can operate as both a **generator** and a **motor**. In Figure 7.5, the machine is operated in generator mode. In both generator and motor mode, **there is direct current in the external circuit** and **alternating current in the rotating shell** (which is called the armature coil) constructed of winding wire. The direction of the current in the winding changes when the rotation causes the brushes to lose contact with one segment and come into contact with another. The change in the direction of the current, which is called **commutation**, occurs when the plane of the winding coil is perpendicular to the direction of the axis of the pole (in the case shown in Figure 7.5, this happens when the winding coil is in the horizontal position). The figure also shows the curves for the instantaneous values of the electromotive force and the current in the armature and the external circuit. The shape of the curves is not perfectly sinusoidal because the shape of the ends of the machine's poles is not flat, as in Figure 7.2, but curved, and there is a cylindrical electrical steel armature between the poles. This results in an air gap where the direction of the magnetic flux density field is perpendicular to the path of the winding. In this section of the path, the current, electromotive force, and torque are more or less constant, which produces the horizontal part in the curves for the current and electromotive force.

If the machine is operated in generator mode, an external torque needs to be applied to make the armature of the machine rotate. The electromotive force on one side of the rotating winding shell when the shell is moving in an air gap is given by the following equation:

$$E_{\text{side}} = B_s l v, \tag{7.4}$$

where E_{side} – electromotive force in volts, V;

B_s – strength of the stator's magnetic flux density field in teslas, T;

l – length of the wire in metres, m;
v – velocity of the wire, m/s.

During rotation, the upper side of the turns of winding moves from left to right (Figure 7.5), while the lower side moves from right to left. Meanwhile, the magnetic field of the stator is directed downwards in both the upper air gap and the lower air gap. Due to this, the electromotive forces induced on the different sides of a turn of winding are opposite in direction (right-hand rule). If on the lower side of the turns of winding there is a positive potential on the commutator-side end, then on the upper side, the potential on the commutator-side end will be negative. Thus, the electromotive forces on the two sides of the rotating shell are in series and merge. The electromotive force induced in one turn is

$$E_{turn} = 2B_s lv, \tag{7.5}$$

When the machine is operated in generator mode, the commutator behaves as a rectifier. The armature winding is affected by an alternating electromotive force, but the polarity of the electromotive force transferred to the brushes is invariable.

Using linear velocity v to describe a rotating machine is not appropriate. In Section 1.3, we demonstrated that the relationship between linear velocity v and angular velocity ω is given by the equation

$$v = \omega r = \omega \frac{D}{2}, \tag{7.6}$$

where r is the radius of the trajectory of the moving point and D is the diameter of that trajectory (in this case, the diameter of the armature).

Based on equations (7.5) and (7.6), we can find the angular velocity of the electromotive force in a turn as follows:

$$E_{\text{turn}} = 2B_s l \frac{\omega}{2} D = B_s lD\omega. \tag{7.7}$$

The product $B_s lD$ is equal to the magnetic flux Φ passing through the turn, because $lD = A$ (A is the area of the turn) and $B_s A = \Phi$ (the magnetic flux is equal to the product of the flux density and the area). Therefore, the electromotive force in a turn as expressed through the magnetic flux is

$$E_{\text{turn}} = F\,\omega. \tag{7.8}$$

In real machines, the armature (Figure 7.6) has many turns that are connected in series, which is expressed mathematically by the so-called machine constant k_E. The electromotive force in a DC machine is thus given by

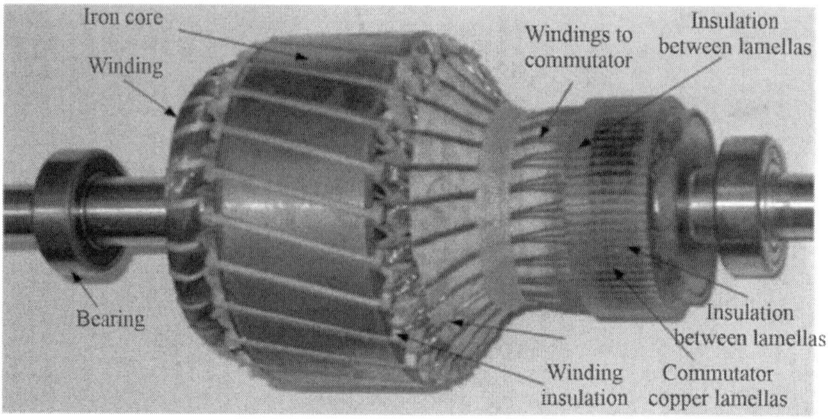

FIGURE 7.6 Armature of a DC machine.

$$E = k_E F \omega. \tag{7.9}$$

This tells us that the electromotive force and voltage can be increased in any of the following ways: **by using a greater number of turns, building a larger machine, or operating the machine at a higher speed.**

The torque produced for the DC machine under consideration in motor mode can be found on the basis of equation (7.3) above. In doing so, it is necessary to take into account that the direction of the magnetic flux density field in the air gaps is perpendicular to both the air gap and the path of the winding. Due to this, the forces induced in the air gaps are always tangential to the air gap and do not depend on the angle of rotation α of the shaft. The torque for one turn of wire is

$$T_{\text{turn}} = B_s A i = F i. \tag{7.10}$$

Since equation (7.10) contains the instantaneous value of the current, the value of the calculated torque, too, will be the instantaneous value. In real machines with many turns of wire, the current is a virtually uniform direct current, and thus the instantaneous value i is never considerably different from the effective value I. The torques of the many turns combine, which is expressed mathematically by the machine constant k_I. Hence

$$T = k_I F I. \tag{7.11}$$

This may be easier to remember if you note how this equation for the torque resembles that for the electromotive force: $E = k_E \Phi\ \omega$. It should also be remembered that **the torque is directly proportional to the current and the electromotive force is directly proportional to the angular velocity.**

In addition to the magnetic field of the stationary poles, which is also called the stator's magnetic field B_s, the winding or current-carrying shell on the rotating rotor, too, generates its own magnetic field B_r. Figure 7.7 shows the directions of these fields when the torque of the shell is at its maximum value.

To better understand how the torque is produced, imagine that instead of the current-carrying coil there is a permanent magnet that generates a field with the same direction as that of the coil. In the figure, this imaginary permanent magnet is indicated by the dashed line. In permanent magnets, as we know, like poles repel each other, while opposite poles attract. Figure 7.7c shows the pulling and pushing forces between the permanent magnets produced by the stator field and the imaginary permanent magnet representing the rotor field. We can see that

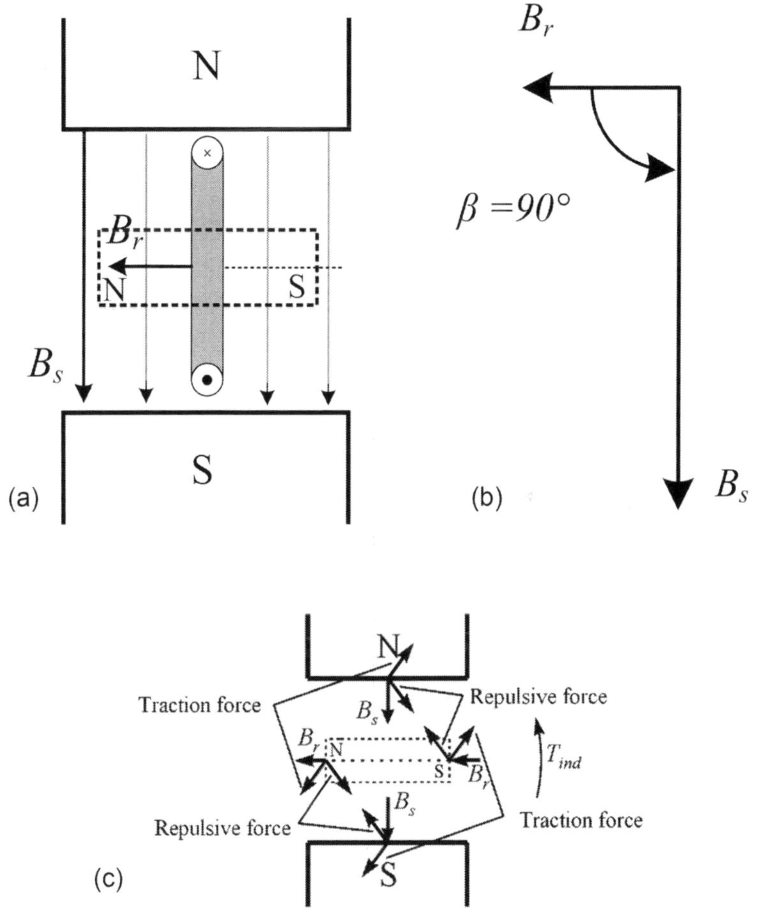

FIGURE 7.7 Relative positions of the stator's magnetic field B_s and the rotor's magnetic field B_r at maximum electromagnetic torque: (a) a turn of rotor winding and its magnetic field; (b) the vectors of the magnetic fields of the stator and rotor, and the angle between the vectors; (c) replacement of the magnetic field of the rotor winding with the magnetic field of an imaginary permanent magnet, and the pushing and pulling forces generated between the magnetic poles.

these forces are trying to move the rotor into a position where the north and south poles are facing each other. When this happens, the lines of force of the stator field B_s travel in the same direction as the lines of force of the field B_r in our imaginary permanent magnet. **The rotor tries to adopt a position where the direction of the rotor field matches that of the stator field.** Upon entering this position, the angle β between the rotor and stator fields (Figure 7.7b) becomes equal to zero, as does the torque. In order for the rotation to continue, the direction of the rotor field is changed by commutation of the rotor current in the DC machine, which produces a torque that causes another half revolution to occur.

The electromechanical torque T_{ind} can also be expressed through the magnetic flux densities for the stator and rotor B_s and B_r. Based on equation (7.3) and the fact that the rotor field

$$B_r = \frac{i\mu}{G}, \tag{7.12}$$

where μ is the magnetic permeability and G is a factor for taking into account the shape and size of the rotating shell, after some rearrangement we get

$$T_{ind} = kB_rB_s \sin\beta, \tag{7.13}$$

where $k = AG/\mu$ (A is the area of the rotating shell). The factor k describes the design of the machine.

Therefore, the torque in an electrical machine generally depends on four factors:

- the strength of the rotor's magnetic flux density field;
- the strength of the stator's magnetic flux density field;
- the sine of the angle between the direction of these two fields;
- the constant describing the design of the machine.

Equation (7.13) indicates that the moment of the electrical machine is greatest when the angle between the magnetic fields of the stator and the rotor is 90°. The stator's magnetic field travels from the pole shoe of the lower magnetic pole (Figure 7.8) to the rotor via the air gap and then leaves the rotor via the other air gap to enter the south pole on the upper side of the stator. The rotor winding and brushes need to be arranged in a way that ensures that the magnetic field generated by the rotor current is perpendicular to the magnetic field of the stator.

Figure 7.9 shows the appropriate arrangement of the rotor winding and brushes. To the right of the brushes, the current in the armature winding is directed outwards from the plane of the figure, while to the left, it is directed towards the plane of the figure (indicated by the cross symbol). The winding coils are inserted into the slots and connected to the segments in a way that ensures that the directions of the current described above are maintained in the windings despite the rotation of the rotor. To the right of the brushes, the currents are always in one

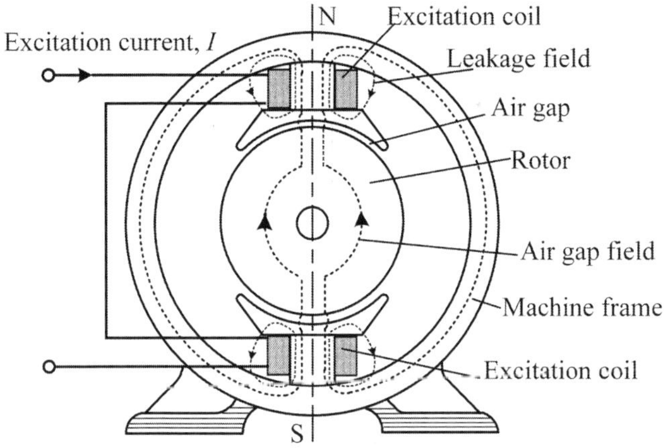

FIGURE 7.8 Introduction of excitation field to the rotor of a DC machine by means of constructional electromagnetic poles on the stator.

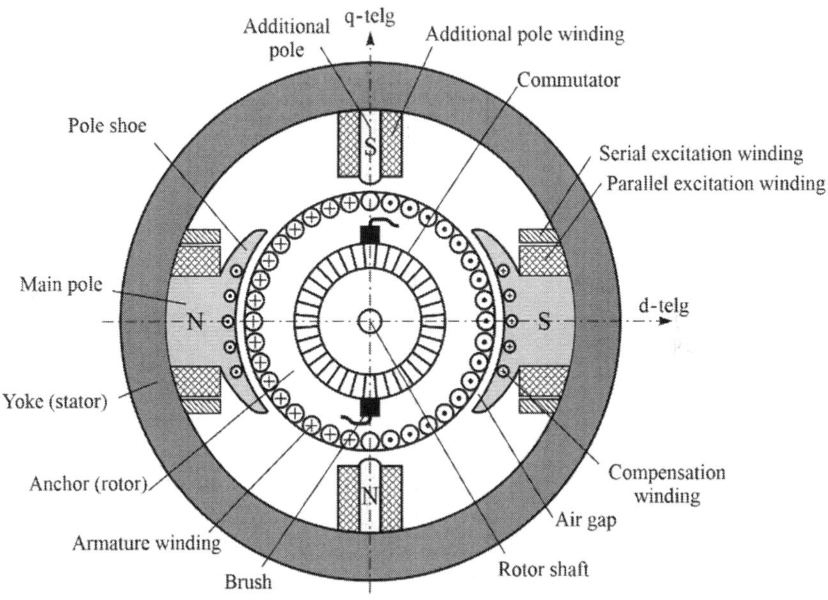

FIGURE 7.9 Cross-sectional view of a compound DC machine

direction, while to the left, they are in the opposite direction. During the rotation of the machine, the direction of the current in the winding coils changes when a segment connected to the end of the coil passes from one side of the brushes to the other side (the segment passes under a brush), but the overall distribution of the current shown in Figure 7.9 remains the same. With the current distributed in this way, the poles produced by the rotor current always lie on the q-axis. In the rotor shown in the figure, the direction of the currents in the rotor is such that the north pole produced by the rotor current lies at the top. For this reason, there is an additional pole, which is a south pole, on the upper side of the machine to compensate for the field of the rotor current. The magnetic field of the rotor is perpendicular to the field of the stator poles that lie on the d-axis. Therefore, the angle β in equation (7.13) is always 90° , and the torque is maximal.

Figure 7.10 shows one configuration for the winding coils and the arrangement of the poles and brushes relative to the winding coils. The rotor has 10 slots, and each slot houses two coil sides (this is a smaller rotor than the one in Figure 7.9). The number of coils is also 10, and a two-layer winding is utilised here. In this case, there are five slots per pole. The number of segments is also 10, and each segment is connected to two winding coil ends. One is the end wire of a winding coil (e.g. the end wire of the first coil, which proceeds from slot six to segment 2) and the other is the start wire of the next winding coil (that of the second winding coil, which proceeds from slot 2 and is also connected to segment 2). Through the segments, the first winding coil is connected in series to the second winding coil, the second winding coil is connected in series to the third, and so on up to

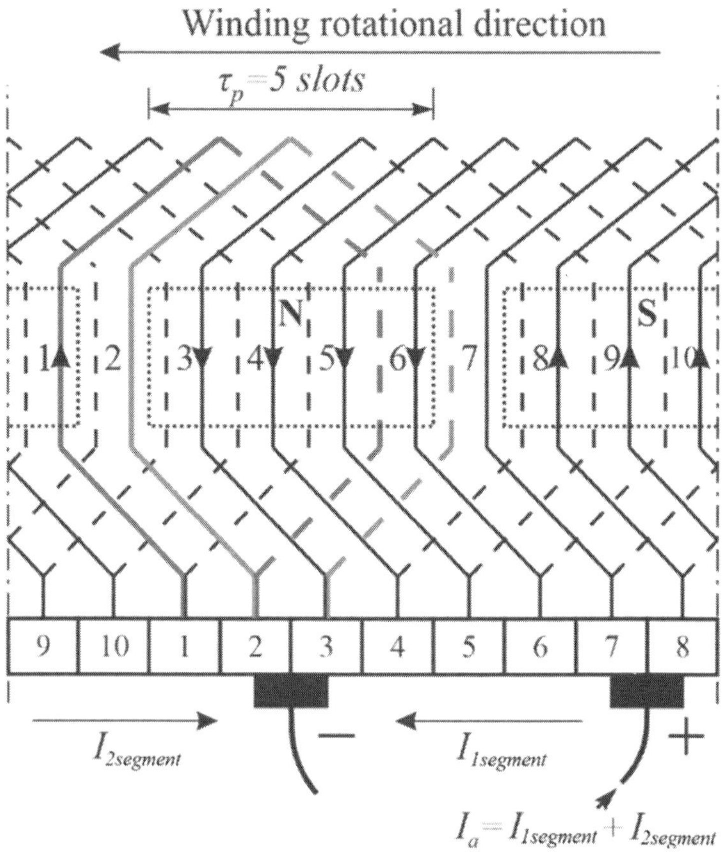

FIGURE 7.10 Lap winding in a two-pole DC machine.

the last coil, the end wire of which is connected to the start wire of the first coil. This completes the series circuit. This type of winding is called a lap winding and is common in low-power and medium-power machines. Higher-power machines also utilise other winding configurations.

The collector segments are connected to two brushes, which are set off by 180° in one-pole machines. From both brushes, the current flows both to the right and to the left, resulting in two parallel current branches of equal length (Figure 7.10). The brushes are arranged relative to the excitation poles in a way that ensures that, from each brush, the current passes via a collector segment to the coil sides positioned between two excitation poles, where there is no excitation field. The position of the brushes is fixed relative to that of the excitation poles on the stator, but can be adjusted slightly to find the position of minimum sparking during operation. The rotor rotates along with the slots, winding coils, and segments, but the current always flows directly from the brushes to the coils in only those slots which, at the given moment, are positioned between two stator poles, where the field is close to zero and no voltage is induced in the coils.

This design provides two important advantages. First, while two adjacent segments are connected to a brush (the brush short-circuits the coils between the segments), no voltage induced by the excitation field is received from the coils between the segments (at this point, the coils are positioned between two poles). As a result, there is no sparking (or significantly less sparking) when the current is transmitted from one segment to another, adjacent segment.

Secondly, during rotation, a situation is maintained where the current in the slots is directed one way to one side of the brushes and the opposite way to the other side. That is why, at any given time, the currents in the rotating rotor are distributed as shown in Figure 7.9, and the resulting torque is always maximal. When the rotor is rotating, its outer surface rotates in one direction, while the field of the rotor currents always rotates in the opposite direction at the same speed on that same surface. Thus, the field of the rotor currents is always stationary relative to the stator and is directed along the q-axis (Figure 7.9), regardless of whether the rotor is rotating or not.

In addition to the main poles on the d-axis, Figure 7.9 also shows a set of additional poles on the q-axis, plus a set of compensating windings set in the pole shoes of the main poles. These additional poles and compensating poles are used to reduce the distortion of the stator field by the rotor field, which is explored in Figure 7.11. This field distortion is called the armature reaction. In the figure, the outer surfaces of the stator and the rotor are depicted as straight lines. The figure also shows the windings of the stator and the rotor. The top graph shows the magnetomotive forces of the stator winding poles and rotor winding currents in the air gap along the circumference, while the bottom graph shows the sum of these magnetomotive forces (curve F_m), as well as the magnetic flux density distribution B resulting from these forces in the air gap along the circumference.

Here, we can see that where the magnetomotive forces separate, the flux density decreases in proportion to the decrease in the magnetomotive force; meanwhile, where the forces combine, the magnetic circuit becomes saturated and the

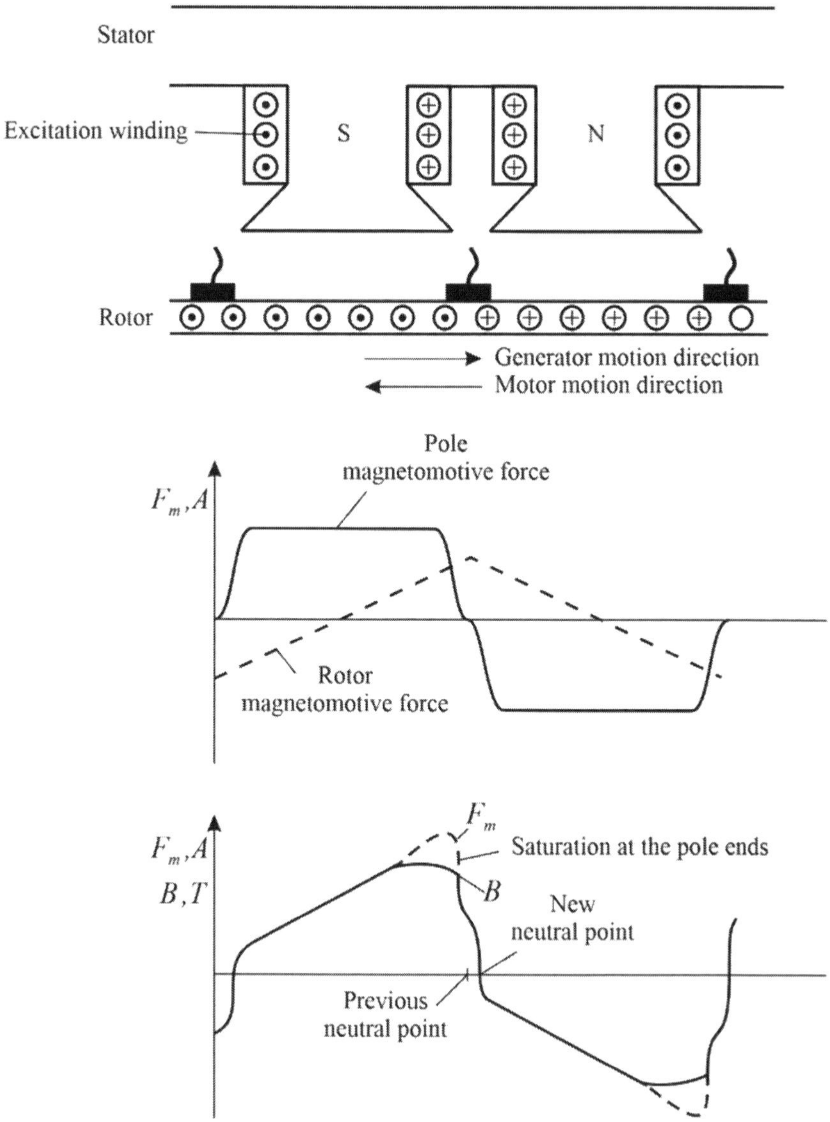

FIGURE 7.11 Distortion of excitation field by rotor winding current, and the magnetomotive forces responsible for this distorted field in a DC machine.

flux density increases less than the total magnetomotive force. This results in a decrease in the average total magnetic flux under the surface of the entire pole. A decrease in the flux creates problems for both generators and motors. In generators, it simply causes a decrease in the voltage under load. In motors, the effect is much more serious. When the flux in a motor decreases, the speed increases. As the motor's speed increases, the load, too, increases, leading to a further decrease in the flux, which can eventually result in a breakdown.

Figure 7.11 additionally shows that the zero point, or **neutral point**, of magnetic flux density has shifted due to the armature reaction. A shift in the neutral point leads to problems in commutation. The brushes are arranged in such a way that the coil sides connected to them via a segment lie in the neutral point, where the magnetic flux density is close to zero, as this prevents electromotive forces from arising in the winding coils that are connected directly to a brush and minimises sparking at the brushes. When the neutral point shifts, a coil that is connected to a brush will end up in the magnetic field. This causes sparking and more wear on the segments.

The harmful effects of the armature reaction can be prevented by means of the additional poles and compensating windings shown in Figure 7.9.

The additional poles are small poles positioned between the main poles of the machine, on the q-axis. The windings of the additional poles are connected to the rotor windings in series. When the rotor current and the magnetic field strength increase, so, too, do the current in the additional poles and the magnetic field of the additional poles, which is opposite in direction to the field of the rotor. Thus, the field of the additional poles balances out the rotor field, preventing the zero point of the stator field from shifting. This prevents increased sparking and lengthens the service life of the segments and brushes. The additional poles can be utilised in both motor and generator mode, because, when switching from one mode of operation to the other, the directions of the current will change in both the rotor and the additional poles. Note that the additional poles have no effect on the armature reaction beneath the surfaces of the main poles, which means that they also have no effect on the weakening of the flux of the machine.

In high-power machines, the armature reaction is also compensated for by means of compensating windings, which are arranged in slots in the pole surfaces to eliminate the field distortion and the resulting weakening of the flux of the main poles due to the armature reaction. The compensating windings are connected to the rotor windings in series, meaning that when the current changes in the rotor windings, so does the compensating effect of the compensating windings. The main drawback of using compensating windings is that they are expensive. Where compensating windings are used, the machine will also have additional poles, because compensating windings cannot prevent the shifting of the neutral point and the increased sparking this leads to.

7.2 OPERATING MODES OF A DC MACHINE

Figure 7.12 shows a simplified schematic for how a machine can be switched to motor mode, generator mode, or no load by adjusting the torque on the machine's shaft. A crank is attached to the shaft of the machine, by operating which rotation can be obstructed or accelerated. When obstructed, the machine enters motor mode, and when accelerated, the machine enters generator mode. If the shaft is left alone, the machine will operate at no load and will rotate at an angular

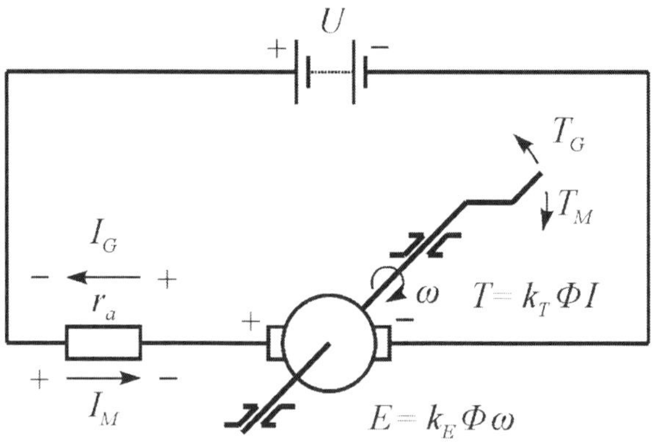

FIGURE 7.12 Torque and currents in a DC machine in motor mode and generator mode.

velocity where the resulting electromotive force $E = k_E\Phi\ \omega$ is equal to the supply voltage U. If the torque T is zero (ideal case), then, according to equation (7.11), the current I will also be zero. If, however, the torque differs from zero, then, according to (7.11), there will be a current

$$I = \frac{T}{k_T F}. \tag{7.14}$$

This current will produce a voltage drop equal to $I{\cdot}r_a$ across the resistance of the armature winding. In Figure 7.12, the resistance of the armature winding is depicted as a separate resistor connected in series to the armature. In motor mode, the speed will be less than the no-load speed, and so the electromotive force will also be lower than the voltage (the electromotive force changes with the speed). Due to this, a current will be generated, which will be directed away from the positive terminal of the power supply. In motor mode, the supply voltage U will balance out the sum of the armature voltage drop and electromotive force:

$$U = E + Ir_a. \tag{7.15}$$

In generator mode, the electromotive force will be greater than the supply voltage, and the current will flow in the opposite direction compared to motor mode. As a result, power will be transmitted from the generator to the power supply. If the power supply is a battery, the battery will be charged. The change in the direction of the current will also cause the direction of the voltage drop across the armature winding to change, which is why generator mode is described by the equation

$$U = E - Ir_a. \tag{7.16}$$

Problem 7.1

The shaft of a permanent magnet DC motor is loaded with a torque of $T = 20$ N m. A voltage of $U = 230$ V is applied to the terminals of the motor. The armature circuit resistance $r_a = 0.5\ \Omega$. The machine constant $k_E\Phi = 1$ V·s/rad and the constant $k_T\Phi = 1$ N m/A.

Find the rotational speed n and power P on the shaft of the motor (disregard mechanical losses and additional losses).

Solution

1. Motor current:

$$I = T/\left(k_T F\right) = 20/1 = 20\text{A}.$$

2. Voltage drop across the armature circuit:

$$\Delta U = I \cdot r_\text{a} = 20 \cdot 0{,}5 = 10\text{V}.$$

3. Electromotive force:

$$E = U - I \cdot r_\text{a} = 230 - 10 = 220\,\text{V}.$$

4. Angular velocity of rotation:

$$\omega = E/\left(k_E F\right) = 220/1 = 220\,\text{rad/s}.$$

5. Rotational speed:

$$n = \frac{30 \cdot \omega}{\grave{A}} = \frac{30 \cdot 220}{3{,}14} = 2102\,\text{p/min}.$$

6. Power on the shaft:

$$P_\text{mech} = \omega \cdot T = 220 \cdot 20 = 4400\,\text{W}.$$

7. Power consumed by the armature circuit:

$$P_{\text{el}} = U \cdot I = 230 \cdot 20 = 4600\,\text{W}.$$

Problem 7.2

A permanent magnet DC generator is made to rotate at a speed of 230 rad/s. A load $R_{\text{load}} = 11\ \Omega$ is applied to the terminals of the generator. The constants of the generator are as follows: $k_E\Phi = 1$ V·s/rad; $k_T\Phi = 1$ N m/A. The resistance of the generator's armature circuit is $r_{\text{a}} = 0.5\ \Omega$.

Find the power supplied by the generator to the resistive load, as well as the power received by the generator via the shaft.

Solution

1. Electromotive force:

$$E = \omega \cdot k_E F = 230 \cdot 1 = 230\,\text{V}.$$

2. Current:

$$I = E / \left(R_{\text{koormus}} + r_{\text{a}}\right) = 230 / \left(11 + 0{,}5\right) = 20\,\text{A}.$$

3. Torque:

$$T = I \cdot k_T F = 20 \cdot 1 = 20\,\text{Nm}.$$

4. Power of the prime mover:

$$P = T \cdot \omega = 20 \cdot 230 = 4600\,\text{W}.$$

5. Electrical power supplied to the load:

$$P = I^2 \cdot R_{\text{load}} = 20^2 \cdot 11 = 4400\,\text{W}.$$

7.3 ADJUSTING THE ROTATIONAL SPEED OF A DC MOTOR

Let us explore how the rotational speed can be adjusted in motors where the power supplied to the armature circuit and the excitation can be controlled separately. Such motors are called separately excited motors.

Plugging the electromotive force equation (7.9) and the current equation (7.14) into equation (7.15), we can get the equation for the angular velocity ω

$$\omega = \frac{U}{k_E F} - \frac{T r_a}{k_E k_T F^2}. \tag{7.17}$$

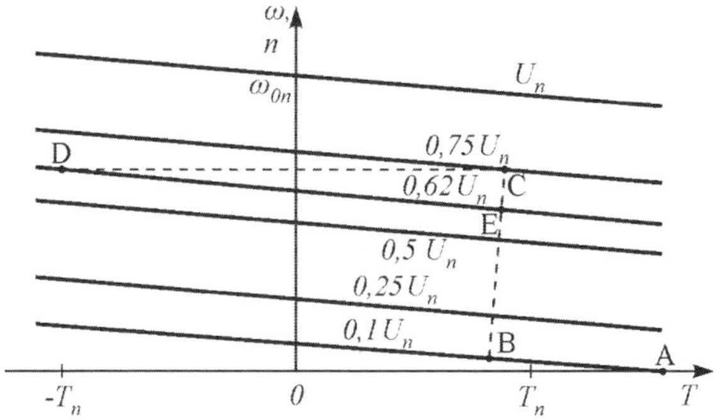

FIGURE 7.13 Mechanical characteristics of a DC machine when reducing the supply voltage.

Equation (7.17) tells us that the rotational speed of a DC motor can be adjusted in three ways: by changing the supply voltage U, by changing the magnetic flux Φ (which, due to saturation, can only be reduced), and by changing the resistance of the armature circuit (this can be done by connecting an additional resistance R_{add} to the armature in series). Figure 7.13 shows the adjustment of speed by changing the supply voltage of the armature circuit.

For start-up, a low voltage is selected for the motor (0.1 times the rated voltage in point A in Figure 7.13). At startup, the rotor is standing still, and the electromotive force of the machine is equal to zero. Thus, the intensity of the starting current is determined only by the resistance of the armature circuit: $I_{st} = U/r_a$. The resistance of the armature circuit is so low that if the rated voltage U_n were used for start up, the starting current would exceed the rated current by more than 20 times, which would destroy the motor. At the proposed low voltage, however, the current will remain within the acceptable range, and the machine will start to rotate slowly (point B). Thereafter, the voltage is slowly increased until the desired rotational speed is reached (point C). Changing the voltage rapidly would cause the current to increase, and care should be taken to ensure that the current does not get too high. For example, if we were to rapidly shift from a voltage of 0.75 U_n to 0.62 U_n (reducing the voltage), we would move abruptly from point C to point D, where braking would occur at the rated torque in generator mode. At this point, the current is equal to the rated current, but opposite in direction to the initial current. Due to the braking torque, the speed would drop, and we would end up in point E.

Figure 7.14 shows the adjustment of speed by reducing the excitation current. Equation (7.17) indicates that when the magnetic flux Φ of the poles decreases, the rotational speed increases. In Figure 7.14, we can see that as the pole flux drops, the speed becomes higher. Nevertheless, there is a limit to how much the speed can be increased. According to equation (7.11), as the magnetic flux decreases, so, too, does the torque developed by the motor. However, since the resistive torque of the appliance does not change, then in order to balance out this resistive torque, the motor current needs to increase. To prevent the motor from being overloaded

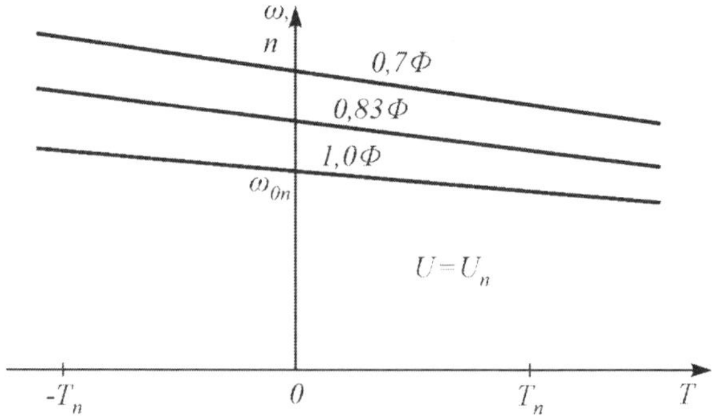

FIGURE 7.14 Mechanical characteristics of a DC machine when reducing the excitation flux of the poles.

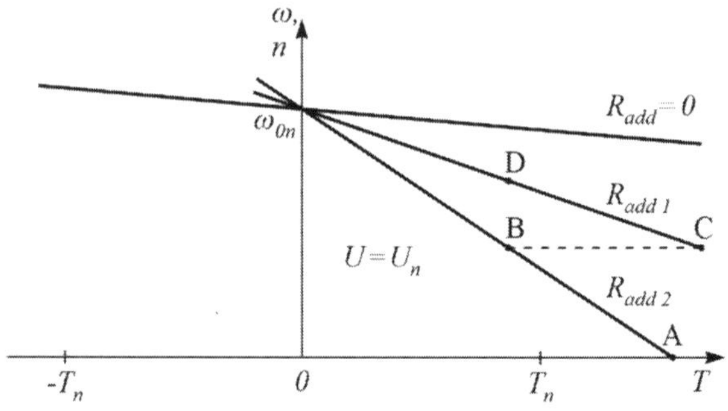

FIGURE 7.15 Mechanical characteristics of a DC machine when an additional resistor is connected in series to the armature.

with current, when the excitation field is reduced, the load torque needs to be lower than the rated torque. The more the speed is increased, the less the motor may be loaded.

An additional resistance does not cause the no-load speed to change, but when the machine is loaded with torque, the speed will begin to decrease more. The greater the additional resistance, the more the speed will drop (in Figure 7.15, $R_{\text{add2}} > R_{\text{add1}}$). An additional resistance can also be used when starting a motor. If, at start-up, a resistor R_{add2} is present in the rotor circuit, this resistor will limit the starting current, putting us at point A in Figure 7.15. As the speed begins to increase, we will reach point B, where the torque of the motor becomes equal to the load torque. To increase the speed further, we can reduce the resistance of the armature circuit by switching over to the resistance R_{add1}. This first puts us at point C, whereafter the speed increases and we get to point D.

Using an additional resistance for controlling the speed is not as efficient as changing the supply voltage. Additional resistances were used before the invention of semiconductor converters.

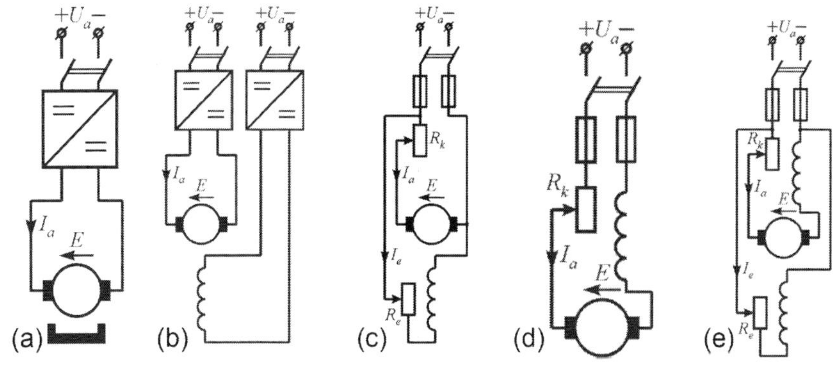

FIGURE 7.16 Excitation circuits for DC motors (a) permanent magnets, (b) independent field coil (separately excited DC motor), (c) shunt field coil (shunt wound DC motor), (d) series field coil (series wound DC motor), (c) compound field coil (compound wound DC motor)

There are also DC motors where the excitation current and armature current can only be adjusted together (Figure 7.16d), as well as motors where excitation is achieved by means of permanent magnets (Figure 7.16a). Figure 7.16 shows a variety of possible ways to generate an excitation field.

If excitation is achieved by means of permanent magnets mounted on the stator (Figure 7.16a), the magnitude of the excitation field cannot be changed and the motor speed can only be adjusted by changing the armature voltage.

In separately excited motors (Figure 7.16b), the armature circuit and excitation circuit have different voltage sources, in which the voltage can be adjusted separately. The effect of this adjustment was explored above.

Motors where the armature winding and excitation winding are connected in parallel and supplied from the same voltage source are called shunt motors (Figure 7.16c). Shunt motors can be started and controlled by means of variable resistors that are connected in series either with the armature only or with both the armature and the excitation winding. Shunt motors behave similarly to separately excited motors. The breaking of the excitation circuit due to a fault during the operation of the motor is dangerous for a shunt motor. If this happens, the armature current will increase abruptly, and the speed will begin to increase unfettered, as when the excitation current is lost, there will remain only a weak residual magnetic field. This can be prevented with an appropriate protective circuit.

Motors where the armature winding and excitation winding are connected in series are called series motors (Figure 7.16d). In series motors, the pole flux Φ is dependent on the current I. Ignoring saturation, $\Phi = k_\Phi I$, where $k\Phi$ is a factor for accounting for the relationship between the flux and the current. Rearranging equation (7.17) gives us the no-load speed of a series motor as

$$\omega = \frac{U}{k\sqrt{T}},$$

where k combines the factors k_{Φ}, k_E, and k_T. This equation also indicates that if the torque is reduced to zero, the speed should become infinitely high. For this reason, a series motor must always be loaded with torque. Otherwise, the speed may increase to a level where the motor is destroyed. As the torque increases, the speed drops by much more than in a shunt motor (the torque–speed characteristic curves have a steep slope, Figure 7.17b). Such characteristics are suitable for means of transport where a high torque is needed at low speeds. Speed is adjusted by changing the supply voltage. The characteristic curves corresponding to different supply voltages are shown in Figure 7.17a, b.

In Figure 7.16e, the motor has two excitation windings, one of which is connected in series to the armature, as in a series motor, while the other is connected in parallel. Such motors are called compound motors. The motor shown in Figure 7.9 is also a compound motor, where we can see two winding coils on the poles. In this motor, the excitation flux will be maintained even when the armature current is zero, and the speed will not become excessively high when the motor is unloaded. The slopes of the torque–speed characteristic curves of a compound motor are gentler than those of a series motor, but steeper than those of a shunt motor. Compound motors, too, are used for means of transport. Their speed is adjusted in the same way as that of separately excited motors: by changing the voltage of the armature circuit and reducing the current of the excitation winding that is connected in parallel to the armature.

7.4 AC SERVOMOTOR OR BRUSHLESS DC MOTOR

The design of AC servomotors is similar to that of synchronous machines. What sets them apart is that servomotors also contain a rotor position sensor. The stator winding is three-phase like that of conventional synchronous machines, but is powered by an inverter controlled by signals received from the rotor position sensor. In the process, the solid angle between the axis of the north pole of the rotor field and the axis of the north pole of the stator field is kept more or less constant. This angle is usually set at 60–90°. Position sensor-based control bestows similar properties on the machine as those of DC machines, where the commutator maintains a constant solid angle of 90° between the rotor and stator fields (thereby ensuring maximum torque). Due to this similarity, AC servomotors are also called brushless DC motors. In brushless DC motors, instead of mechanical commutation of the current, commutation is electronic – performed by an inverter controlled by the position sensor. The mechanism for producing the excitation field in a servomotor is located on the rotor, instead of the stator, and usually consists of permanent magnets. The commutated winding, conversely, is located on the stator, instead of the rotor, so that current would not need to be transmitted from the rotating part to the stationary part. This is known as an inverted design. To make the electronic commutator simpler, only three winding sections are used. In a conventional DC machine, such a small number of sections would result in an unacceptably high voltage between the commutator segments. In an electronic commutator, however, excessive voltage is prevented by means of an appropriate

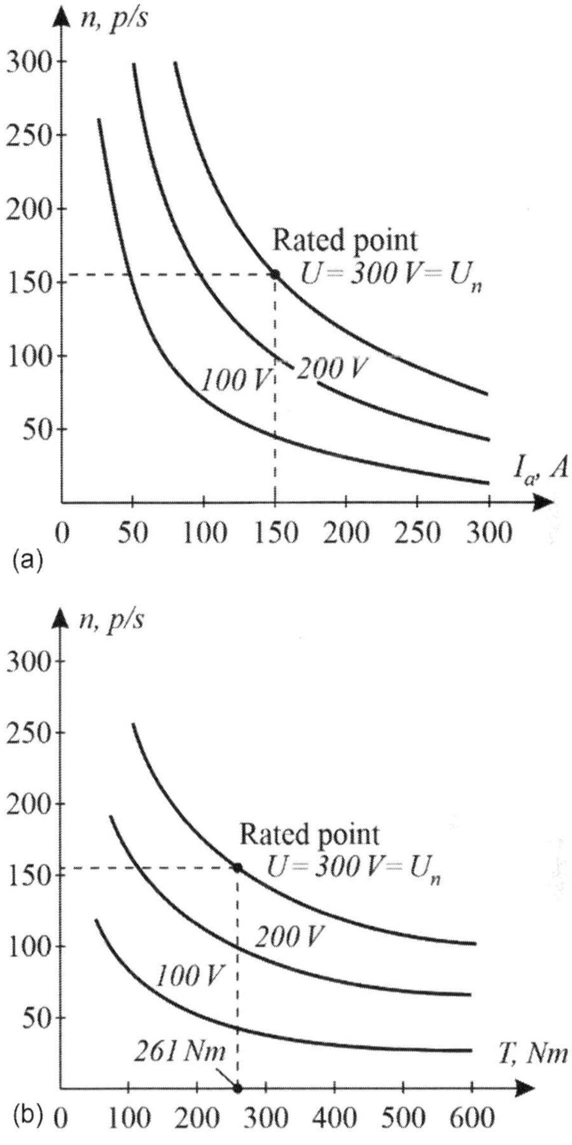

FIGURE 7.17 Speed–current characteristic curves (a) and torque–current characteristic curves (b) for a series motor.

circuit. The integrated position sensor makes it easy to use these machines in route control systems and tracking systems. And it is also why they are called AC servomotors.

Servomotors are supplied with adjustable voltage from a DC power supply via an inverter (Figure 7.18). In higher-power applications, a thyristor inverter with a DC choke L, which forms a current inverter, is used. In lower-power applications,

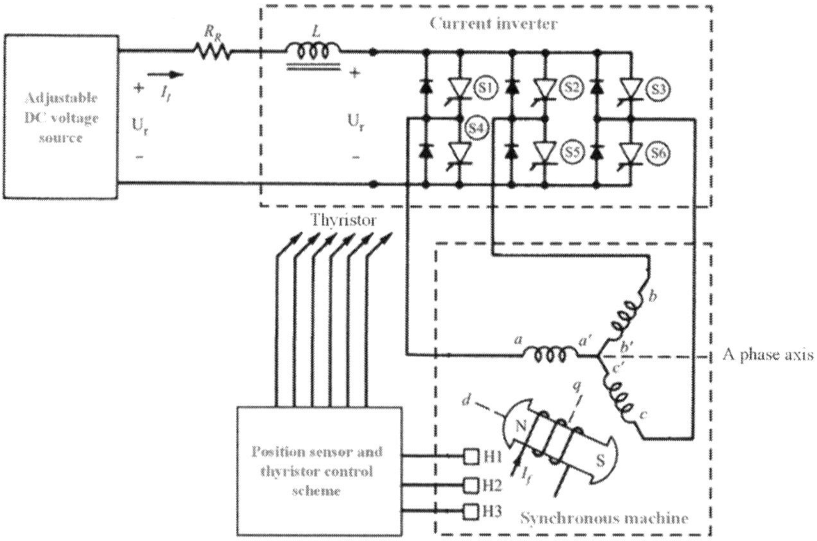

FIGURE 7.18 Circuit diagram of an AC servomotor and its power-supplying inverter (current inverter).

a voltage inverter with power transistors is used. When the DC voltage U_r supplied to the inverter is adjusted, the motor's speed changes in a similar manner as when the armature voltage is adjusted in a DC motor. The inverter thyristors S1–S6 are connected in parallel to counter-current diodes. When a thyristor is closed, the motor winding current cannot disappear instantly, because this current is linked to the energy of the stator's magnetic field. A counter-current diode enables the winding current to continue flowing until the energy of the magnetic field has been returned fully to the DC side of the inverter.

Figure 7.19 shows the shapes of the current curves for all three motor phases. Continuous lines mark the ideal current curves: current is generated and lost instantaneously. Dashed lines mark the actual current curves, where the increase and decrease of the current follow an exponential trajectory. The horizontal part of the current curves is caused by the DC choke L in the circuit. The intensity of the current passing through the choke L is determined both by the voltage U_r of the DC voltage source and by the balancing electromotive force of the synchronous motor and the voltage drop across the inverter valves. When using a thyristor inverter, it is expedient to over-excite the synchronous machine so that the phase of the current would lead that of the voltage. This is advantageous when closing the thyristors. For this purpose, the angle between the fields of the stator and rotor is selected to be slightly less than 90° (angle β in Figure 7.20).

Small servomotors utilise voltage inverters in which there is no choke L. The rotor in most cases is fitted with permanent magnets. In order to avoid overloading the inverter and motor windings with current, voltage pulse-width modulation, for which power transistors are utilised in place of thyristors (S1–S6 in Figure 7.18),

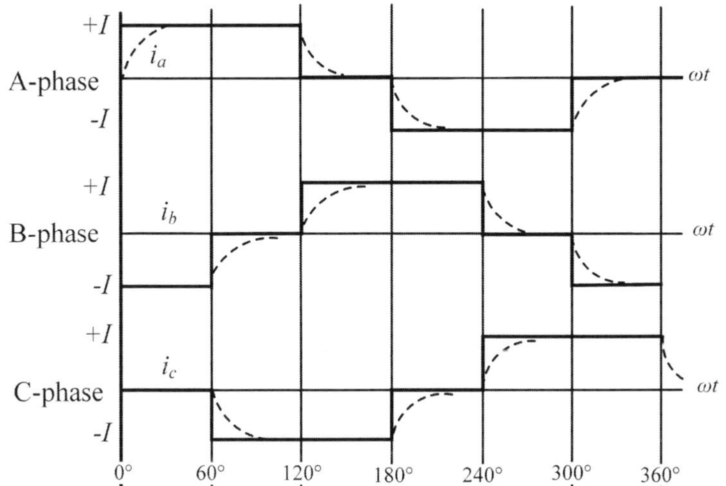

Sensor code with 120° offset sensors	1 0 1	1 0 0	1 1 0	0 1 0	0 1 1	0 0 1	H1 H2 H3
Sensor code with 60° offset sensors	0 0 0	0 0 1	0 1 1	1 1 1	1 1 0		H1 H2 H3
Open thyristors	S1, S5	S1, S6	S2, S6	S2, S4	S3, S4	S3, S5	

FIGURE 7.19 Phase current instantaneous value curves for a servomotor, and the thyristor numbers and sensor codes used to obtain the phase currents.

is used either in the power supply rectifier or inverter. In normal operation, however, pulse modulation is used to adjust speed. In very small machines, the same integrated circuit can contain both an inverter and a control circuit.

Inverter control is achieved by means of a rotor position sensor. This may be an LED sensor or a Hall effect sensor. Two essentially equivalent LED arrangements are used: either three LEDs at every 120 electrical degrees (H1, H2, and H3 in Figure 7.20) or three LEDs at every 60°. The parallel code of the signals of these LEDs (the LED signals come in parallel and separately) and the numbers of the thyristors that are switched on in the case of each code combination are shown in Figure 7.19.

Figure 7.20 shows one possible design for an LED sensor. The opaque code wheel, which rotates along with the shaft, is shaped in such a way that it does not prevent light from reaching the photosensor from the LEDs across an arc of 180°, but does prevent it across the remaining 180°. When the sensor is illuminated, the logic level is 1, and when unilluminated, it is 0.

The figure shows a point in time $\omega t = 60°$ when the sensor code 101 is changing to 100. Before this point in time, control was exerted by the thyristors S1 and S5, and current entered from the beginning of phase A and left from the beginning of

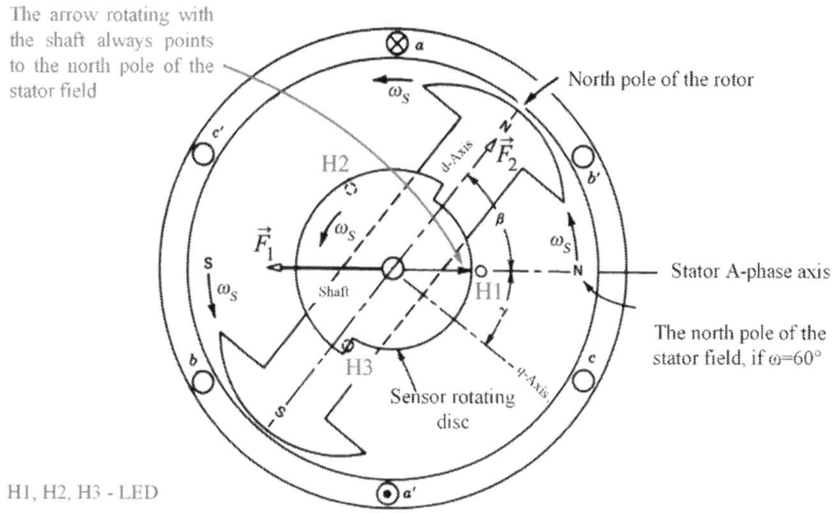

FIGURE 7.20 Position of the rotor and code wheel of a servomotor when the stator current is commutated from phase B to phase C.

phase B. In phase C, there was no current. After this point in time, the thyristor S5 was switched off and the thyristor S6 was switched on. This resulted in a change in the path of the phase B current. The current stops flowing through the thyristor S5 and instead begins to flow through the counter-current diode connected in parallel to the thyristor S2, from where it proceeds to the positive terminal of the power supply. The intensity of this current starts to decrease, while the phase C current starts to increase. This transfer, or commutation, of current from phase B to phase C changes the direction of the stator field by 120°. The stator's magnetic field turns in the same direction and by the same angle as the code wheel and the rotor, which is fixed rigidly to the latter, along with its magnetic field. As a result, the angle between the magnetic fields of the stator and rotor is kept such that a torque is constantly induced (this angle is close to 90°).

In a conventional DC motor, this angle is maintained by a mechanical commutator consisting of collector segments and brushes. Here, the same task is performed by a thyristor bridge, which may be termed an electronic commutator. The coding wheel is usually stamped with an arrow that always points towards the north pole of the stator's magnetic field. The stator field rotates synchronously with the rotor field, as it does in synchronous machines. Unlike in synchronous machines, however, the rotational speed and the frequency of the stator current are not constant in this machine.

Appendix 1 Simple Calculation for Small Transformers

A.1 BASE DATA

1. Apparent power S of the transformer in volt-amperes.
2. Rated primary voltage U_{1n} in volts.
3. Secondary voltage at no load U_{20} in volts.
4. Operating frequency f in hertz.

A.2 CHOICE OF MAGNETIC CIRCUIT

The choice of magnetic circuit depends firstly on the operating frequency. For a frequency of 50 or 60 Hz, the magnetic circuit is usually made from either a set of punched E-shaped or U-shaped 0.35-mm thick sheets of transformer steel or two U-shaped magnetic circuit parts wound from strips and placed side by side. Converters typically use frequencies ranging from 20 kHz to 1 MHz. The lower end of this range corresponds to higher-power applications. At power ratings of 1–10 kW, the frequency is usually 30–40 kHz. Converters utilise ferrite cores. There are several types of ferrites available. At frequencies of 20–40 kHz, the maximum magnetic flux density is around 0.2 T (or 0.2 Wb/m^2, which equals 2,000 gauss). At higher frequencies, the maximum magnetic flux density is lower.

Before performing calculations for a transformer, it is advisable to make an initial decision regarding the magnetic circuit. Magnetic circuit manufacturers typically display a wide range of magnetic circuits of different sizes in their product catalogues. There, they will also indicate the power loss per unit of volume or weight in the core (W/cm^3 or W/g), depending on the chosen operating frequency and maximum magnetic flux density. Transformer calculations will determine the suitable size of the core for the desired transformer.

A.3 DESCRIBING THE GEOMETRY OF A MAGNETIC CIRCUIT

Transformer magnetic circuits have two key parameters. These are

- the cross-sectional area of the leg of the magnetic circuit A_{Fe};
- the window area A_{win}.

These can be determined on the basis of the leg width a and the factors K_1, K_2, K_3, and K_{Fe} (Figure A.1).

In the case of a steel sheet or strip-wound core, the cross-sectional area of the magnetic circuit will be smaller than that of the leg, as some of the space will be taken up by insulation. This can be accounted for with the iron space factor K_{Fe}:

$$A_{Fe} = K_{Fe} \cdot A_{sa}. \tag{A.1}$$

The factors K_1, K_2, and K_3 can be found based on the data presented in the manufacturers' product catalogues on the leg width a, leg thickness b, window width w_{win}, and window height h_{win} ($K_1 = b/a$; $K_2 = w_{win}/a$; $K_3 = h_{win}/a$).

In the case of steel sheeting or strips, the iron space factor may be presumed to be $K_{Fe} = 0.9$, while in the case of a ferrite core, it may be presumed to be $K_{Fe} = 1.0$.

The cross-sectional area of the leg is determined by the maximum magnetic flux Φ_m, because the maximum magnetic flux density B_m is limited either by saturation (at 50 Hz) or by losses (at higher frequencies).

$$F_m = A_{Fe} \cdot B_m. \tag{A.2}$$

Transformer calculations are based on the electromotive force equation

$$E_1 = U_1 = 4,44 f w_1 F_m. \tag{A.3}$$

The primary voltage U_1 and frequency f are pre-defined for transformer calculations. The product $w_1\Phi_m$ needs to be selected:

$$w_1 F_m = \frac{U_1}{4,44 f}. \tag{A.4}$$

Here we can see clearly that the higher the selected maximum magnetic flux Φ_m (which is determined by the cross-sectional area of the magnetic circuit A_{Fe}), the lower the number of turns w_1 in the primary winding must be. If the number of turns is very small, part of the window will be empty. This is not ideal, as the core will then be needlessly large. Conversely, if the number of turns is too large, the winding will not fit in the window, and the core will need to be made larger. This makes the transformer design process one of trial and error. However, there is a better way. It is possible to derive a mathematical equation where the size of the magnetic circuit can be determined from the leg width a, if the factors K_1, K_2, K_3, and K_{Fe} as well as the winding current densities j and the window space factor for copper K_{ws} are known. The window space factor for copper is defined as follows:

$$K_{ws} = \frac{A_{Cu}}{A_{win}}, \tag{A.5}$$

where A_{Cu} is the combined cross-sectional area of all winding turns in the window.

The cross-sectional area A_{Cu} can be found by multiplying the cross-sectional area of one wire in the primary winding by the number of turns in the primary winding and adding to the result the product of the cross-sectional area of the wire in and the number of turns in the secondary winding. In reality, the aforementioned products are essentially equal for the primary and secondary windings, and A_{Cu} can be calculated from the equation

$$A_{Cu} = \frac{2w_1 I_1}{j}, \tag{A.6}$$

where I_1 – rated current of the primary winding, $I_1 = S/U_1$;
j – current density in the winding.

In the case of a copper winding, the recommended current density is 2–4 A/mm^2 (or $2 \cdot 10^6$–$4 \cdot 10^6$ A/m^2). 4 A/mm^2 is suitable for very small transformers (of only a few watts), while larger transformers will require a lower current density, as they have lass cooling area per unit of power.

The recommended value for the window space factor K_{ws} is 0.3. This is because experience shows that the window of a magnetic circuit can only be filled with winding copper by about 30%. The remaining 70% is taken up by the coil base, insulation, and various air gaps.

A.4 GENERAL EQUATION FOR THE WIDTH OF THE LEG OF A MAGNETIC CIRCUIT

The general equation for the leg width a can be derived on the basis of the transformer electromotive force equation (A.3). In equation (A.3), the number of turns w_1 is an unknown value. This number of turns w_1 features in equation (A.6), but to use this equation, we first need to find the cross-sectional area for copper A_{Cu}. Based on equation (A.5) and Figure A.1,

$$A_{Cu} = K_{ws} \cdot A_{win} = K_{at} \cdot K_2 \cdot K_3 \cdot a^2. \tag{A.7}$$

By setting the right sides of equations (A.6) and (A.7) as equal and taking into account that $I_1 = S/U_1$, we get

$$\frac{2w_1 S}{jU_1} = K_{ws} \cdot K_2 \cdot K_3 \cdot a^2,$$

which then gives us

$$w_1 = \frac{jU_1 \cdot K_{ws} \cdot K_2 \cdot K_3 \cdot a^2}{2S}. \tag{A.8}$$

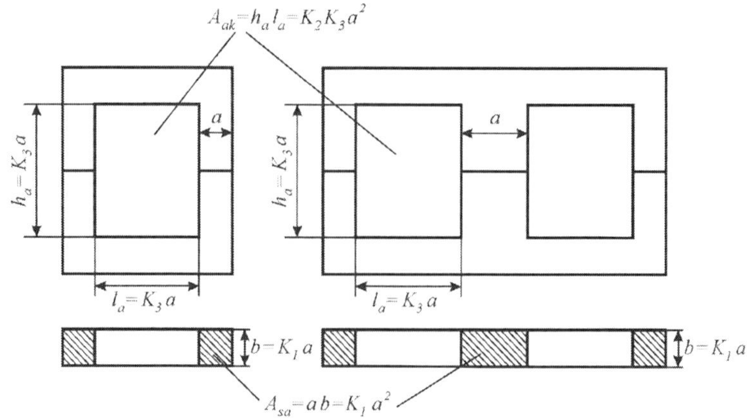

FIGURE A.1 Determining the cross-sectional area of the window A_{win} and of the leg A_{leg} based on the leg width a and the factors K_1, K_2, and K_3.

The second unknown value in the transformer electromotive force equation is the maximum magnetic flux Φ_{m}. Based on equations (A.1) and (A.2) and Figure A.1, we get

$$F_m = A_{Fe} \cdot B_m = K_{Fe} \cdot A_{leg} \cdot B_m = K_{Fe} \cdot K_1 a^2 \cdot B_m. \tag{A.9}$$

Substituting the expressions given in (A.8) and (A.9) for w_1 and Φ_{m} in the base equation (A.3) gives us

$$U_1 = 4{,}44 f \cdot \frac{jU_1 \cdot K_{ws} \cdot K_2 \cdot K_3 \cdot a^2}{2S} \cdot K_{Fe} \cdot K_1 a^2 \cdot B_m. \tag{A.10}$$

Then, rearranging equation (A.10) gives us the equation for the leg width a (in metres):

$$a = \sqrt[4]{\frac{S}{2{,}22 f \cdot j \cdot K_{ws} \cdot K_1 \cdot K_2 \cdot K_3 \cdot K_{Fe} \cdot B_m}}, \tag{A.11}$$

where S – apparent power of the transformer, VA;
- f – frequency, Hz;
- j – current density in the windings, A/m^2, ($j = 2{\cdot}10^6$–$4{\cdot}10^6$ A/m^2);
- K_{ws} – window space factor for copper ($K_{\text{ws}} \approx 0.3$);
- K_1 – ratio of leg thickness b to width a ($K_1 = b/a$);
- K_2 – ratio of window width w_{win} to leg width a ($K_2 = w_{\text{win}}/a$);
- K_3 – ratio of window height h_{win} to leg width a ($K_3 = h_{\text{win}}/a$);
- K_{Fe} – iron space factor (for steel sheeting and strips: $K_{\text{Fe}} \approx 0.9$; for ferrite: $K_{\text{Fe}} = 1$);

B_m – maximum magnetic flux density in the core, Wb/m^2 (for electrical steel: 1.0–1.6 Wb/m^2; for ferrite, depending on the make and frequency: 0.1–0.3 Wb/m^2).

Equation (A.11) shows that the dimensions of a transformer increase when the apparent power S increases and decrease when the frequency f increases. For example, a 10-fold increase in frequency while the dimensions remain the same will increase the power of a transformer 10-fold, if the peak magnetic flux density B_m does not change. As the frequency increases, however, iron losses go up significantly, and to limit this, it is usually necessary to reduce the maximum magnetic flux density B_m. Nonetheless, increasing the frequency is the most radical way to bring down the weight and size of a transformer.

A.5 FINDING THE REMAINING PARAMETERS

Once the calculated leg width a has been established, it will be necessary to select the nearest actual leg width and, with it, the magnetic circuit listed in the manufacturer's catalogue. Once this has been done, we can find the number of turns in the primary winding based on equations (A.3) and (A.9):

$$w_1 = \frac{U_1}{4{,}44 f \cdot K_{Fe} \cdot K_1 a^2 \cdot B_m}, \tag{A.12}$$

where a is the leg width selected from the catalogue.

The number of turns in the secondary winding can be calculated based on the secondary voltage:

$$w_2 = w_1 \cdot \frac{U_2}{U_1}. \tag{A.13}$$

The cross-sectional area of the copper in the windings can be found based on the current and current density in the corresponding winding. The calculated cross-sectional area of the primary winding A_{w1} (m^2):

$$A_{w1} = \frac{I_1}{j} = \frac{S}{U_1 j}. \tag{A.14}$$

Similarly, the calculated cross-sectional area of the secondary winding A_{w2} (m^2):

$$A_{w2} = \frac{S}{U_2 j}. \tag{A.15}$$

After finding the calculated cross-sectional areas of the windings, it will be necessary to select the nearest cross-sectional area from the product catalogues of winding wire manufacturers.

When selecting winding wire, it should be borne in mind that while for 50(/60) Hz transformers it is suitable to use a single wire with the appropriate cross-sectional area, this is not an option for higher frequencies due to the skin effect. At higher frequencies, it will be necessary to use several electrically insulated wires with a smaller cross-sectional area that are woven together and set in an additional common insulating shell. This type of winding wire is called Litz wire (German: *Litzendraht*). Alternatively, a thin copper sheet may be used instead.

Transformer core manufacturers usually also list the weight or volume, or both, of the core in their product catalogues. Additional values that may be listed there include the specific power losses (in either W/cm^3 or W/g) at different frequencies f, magnetic flux densities B_m, and core operating temperatures. Based on these details, it is also possible to find the power loss in the core by multiplying the specific power loss by the weight or volume, respectively. For this calculation, the operating temperature will be an estimated value, which means that the calculation results will be merely approximate.

A.6 CALCULATION EXAMPLE

1. **Base data**

1. Rated power 3,300 VA.
2. Rated primary voltage 210 V.
3. Secondary voltage at no load 3,300 V.
4. Operating frequency 20 kHz.

2. **Choice of magnetic circuit type and manufacturer**

Since the secondary voltage is high, we will choose a UU-shaped magnetic circuit (two U-shaped magnetic circuit parts placed side by side). Then the primary and secondary windings can be placed on separate legs, which will make it easier to insulate the high voltage. For the core, we will choose a type P ferrite from the company Magnetics, for which the recommended maximum magnetic flux density at 25 kHz is 200 mT (0.2 Wb/m^2.

To find the factors K_1, K_2, and K_3, we will first consider a medium-sized core 0–49316UC, where the leg width a = 28 mm; leg thickness b = 16 mm; window height $h_{win} = 2 \times 48$ mm, and window width w_{win} = 36 mm. For these dimensions, we get:

$$K_1 = b / a = 16 / 28 = 0,571;$$

$$K_2 = l_a / a = 36 / 28 = 1,286;$$

$$K_3 = h_a / a = 96/28 = 3{,}429.$$

3. **Finding the leg width**

For equation (A.11), we will make the following choices:

We will choose a current density of $j = 2{\cdot}10^6$ A/m^2, since the power is quite high.

We will choose a window space factor of $K_{ws} = 0.25$ in order to leave some extra space between the windings.

For a ferrite, $K_{Fe} = 1.0$.

We will choose a maximum magnetic flux density of $B_m = 0{,}2$ Wb/m^2.

Now we can find the leg width:

$$a = \sqrt[4]{\frac{3300}{2{,}22 \cdot 20 \cdot 10^3 \cdot 2 \cdot 10^6 \cdot 0{,}25 \cdot 0{,}571 \cdot 1{,}286 \cdot 3{,}429 \cdot 1.0 \cdot 0{,}2}}$$

$$= 0{,}0233\text{m} = 23{,}3\text{mm}.$$

The nearest offered leg widths are 20 mm and 28 mm. We will choose 28 mm (0.028 m), which is the leg width for the core with the product identifier 0-49316UC. This means that the core we initially chose at random was, in fact, the closest to what we needed.

4. **Number of turns in and cross-sectional areas of windings**

Number of turns in primary winding

$$w_1 = \frac{U_1}{4{,}44 f \cdot K_{Fe} \cdot K_1 a^2 \cdot B_m} = \frac{210}{4{,}44 \cdot 20 \cdot 10^3 \cdot 1{,}0 \cdot 0{,}571 \cdot 0{,}028^2}$$

$$= 26{,}4 \text{ turns}.$$

We will choose $w_1 = 27$ turns.

Number of turns in secondary winding

$$w_2 = w_1 \cdot \frac{U_2}{U_1} = 27 \cdot \frac{3300}{210} = 424{,}2 \text{ turns}.$$

We will choose $w_2 = 424$ turns.

Cross-sectional area of copper in primary winding

$$A_{w1} = \frac{S}{U_1 j} = \frac{3300}{210 \cdot 2 \cdot 10^6} = 7{,}88 \cdot 10^{-6}\, m^2 = 7{,}88\, mm^2.$$

Cross-sectional area of copper in secondary winding

$$A_{w2} = \frac{S}{U_2 j} = \frac{3300}{3300 \cdot 2 \cdot 10^6} = 5 \cdot 10^{-7}\, m^2 = 0,5\, mm^2.$$

5. **Double-checking the calculations**

To double-check our calculations, let us find the window space factor for copper based on the chosen core and the calculated cross-sectional areas of the windings.

The total cross-sectional area for copper is:

$$A_{Cu} = w_1 \cdot A_{w1} + w_2 \cdot A_{w2} = 27 \cdot 7,88 + 424 \cdot 0.5 = 425 mm^2 = 4,25 cm^2.$$

The cross-sectional area of the window is:

$$A_{win} = h_a \cdot l_a = 96 \cdot 36 = 3456\, mm^2 = 34,56\, cm^2.$$

The window space factor is:

$$K_{ws} = \frac{A_{Cu}}{A_{win}} = \frac{425}{3456} = 0,123.$$

Therefore, the window space factor is about twice as low as initially specified. The difference is mainly due to the fact that we increased the leg width from 23.3 millimetres to 28 millimetres, i.e.

$$\frac{28}{23,3} = 1,2 \text{ times}.$$

For the window space factor, however, this difference in leg width applies in the power of four:

$$1,2^4 = 2,08.$$

If we wanted to use materials more optimally, we could try differently shaped magnetic circuits from the same manufacturer, or magnetic circuits from a different manufacturer altogether.

Bibliography

Chapman, S. J. (2005). *Electric Machinery Fundamentals.* 4th ed. New York: McGraw-Hill.

Jänes, H., Kaasik, P., Puusepp, E., Voldek, A. (1961). *Elektrimasinad.* Tallinn: Eesti Riiklik Kirjastus.

Lehtla, T. (2007). *Elektriajamid.* Tallinn: Tallinna Tehnikaülikooli Kirjastus.

McPherson, G., Laramore, R. D. (1990). *An Introduction to Electrical Machines and Transformers.* 2nd ed. New York: John Wiley & Sons.

Mohan, N., Undeland, T. M., Robbins, W. P. (1995). *Power Electronics: Converters, Applications, and Design.* 2nd ed. New York: John Wiley & Sons.

Voldek, A. (1972). *Elektrimasinad, I. Alalisvoolumasinad ja transformaatorid.* Tallinn: Valgus.

Voldek, A. (1973). *Elektrimasinad, II. Vahelduvvoolumasinad.* Tallinn: Valgus.

Võrk, R., Mägi, V. (1989). *Elektrotehnika.* 3rd, revised ed. Tallinn: Valgus.

Index

Printed in the United States
by Baker & Taylor Publisher Services